MW00846204

Introduction to Quantitative Statistical Analyses

Second Edition

By William P. Wallace
and Jill A. Yamashita

University of Nevada, Reno
California State University, Monterey Bay

cognella™
San Diego, CA

Bassim Hamadeh, CEO and Publisher
Michael Simpson, Vice President of Acquisitions and Sales
Jamie Giganti, Senior Managing Editor
Miguel Macias, Graphic Designer
Kristina Stolte, Senior Field Acquisitions Editor
Natalie Lakosil, Licensing Manager

First published in the United States of America in 2016 by Cognella, Inc.

Printed in the United States of America

ISBN: 978-1-63487-369-7 (pbk) / 978-1-63487-370-3 (br)

cognella™

www.cognella.com 800.200.3908

Dedication

With love and appreciation to our
immediate and extended families

Contents

Preface to the Second Edition

The first edition of this text was written to replace a favorite textbook both authors had used for several years. The late Geoffrey Keppel was the lead author for that book ("Introduction to Design and Analysis: A Student's Handbook" by Keppel, Saufley, and Tokunaga), which was last revised in 1992. Hence, we decided to create our own book in the Keppel tradition, but with a number of variations we had developed over the course of teaching statistics for many years.

Initially a brief and inexpensive pamphlet was prepared, resulting in a 200-page booklet with an amateurish appearance and many typing errors. Too much class time that first semester was spent correcting "typos;" however, most students did not complain (probably because the pamphlet was available at cost approximately $90 less than a new copy of the Keppel book). Of course, more time was needed to repair the damage and do a better job on the pamphlet, and in contacting a professional publisher about just putting a nice cover on what we produced, an editor suggested we develop it into a full-scale book, assuring us they could do so while keeping the price of the book below most competitors on the market. So we did a trial run, and still found errors that needed to be corrected. The first trial run was followed by a "revised" first edition. Happily, we only found a few errors we needed to correct with this version. In addition, students in our classes provided valuable feedback, much of which got us thinking about changes (e.g., additions, deletions, expansions, clarifications) that should be considered in the future.

The invitation to prepare a second edition has given us the opportunity to address these changes; and we are hopeful they will result in an improved textbook in terms of clarity, readability, accuracy, and representativeness. Our goals continue to include preparing serious students who are required or choose to take higher-level statistics courses and helping all students understand and appreciate statistical thinking and analysis.

Our approach and emphasis in teaching a beginning course in statistics have not changed. We continue to stress the combination of a "hands-on," step-wise approach to carrying out various statistical analyses with the use of a hand-held calculator; and doing more complex statistical analyses using computer software (with illustrations using SPSS in this text). For the first edition, we were aware that we took some short-cuts with descriptive statistics and in leading into the Analysis of Variance (the inferential statistical procedure that gets the most attention). We have tried to develop these sections more slowly in this edition. Some sections have been added, deleted, rewritten, or moved (e.g., the

Cochran Q test was moved to an appendix so it would be easy for instructors to skip if they choose). Also, at chapter ends we have expanded some "Model Problem/Practice Problem" sections. Many students who have done well in our classes have used the book both as a text and as a "workbook." Solicited and unsolicited comments and evaluations from our students have indicated that they found the methodical, step-wise organization in working through problems helpful, and some students even asked for more practice problems.

A Note to Students

The authors have taught beginning statistics courses for many years; and for most of those courses somewhat unconventional textbooks by Geoffrey Keppel and his associates were adopted [*Introduction to Design and Analysis: A Student's Handbook*, by Keppel & Saufley, 1980 (1st Edition), and by Keppel, Saufley, & Tokunaga, 1992 (2nd Edition)]. These textbooks were unconventional in the sense that they focused on statistical procedures most widely used in research settings. Consequently some of the early "baby-steps" common to many introductory statistics textbooks were sacrificed in order to introduce more fully the **Analysis of Variance**; which is currently the most frequently used statistical test procedure for "evaluating" data (results) from **experimental** research in psychology. Since prior courses were developed for use with the two editions of Keppel's books (adopting his approach, emphases, and notation), as well as his advanced textbook (Keppel & Wickens, 2004), it should come as no surprise that our introduction to statistical analyses closely follows his approach.

The task of writing this book was initiated for two principle reasons. Perhaps the main reason resulted from the fact that the most recent 2nd edition by Keppel, Saufley, and Tokunaga has a 1992 publication date. While it is true that statistics textbooks do not appear to get "out of date" as rapidly as substantive area textbooks, the present authors felt somewhat uneasy about assigning a book for a statistics course that had advanced beyond the ripe old age of 10 years old. A secondary motivation for developing this new statistics book resulted from the discovery of how much the aging 1992 textbook by Keppel, et al. cost students (over $100 at the University of Nevada bookstore).

Example problems are provided throughout the book (within the main text, at the end of all but the first chapter, and in the four practice tests). Step-by-step solutions accompany problems in the text, as well as the one or two model problems that begin the "Practice Problems" sections at the end of each chapter. Solutions have been deliberately omitted for the practice problems that follow the model problems. It is our recommendation that you attempt as many of these practice problems as you and/or your instructor deem appropriate. For several practice problems you may be able to confirm your answers by entering the data in an SPSS spreadsheet and instructing the computer to perform the required analysis. As far as the four Practice Tests are concerned, most students will find that the tests are too long for what could be reasonably completed within an hour-long class session. The tests are presented primarily to illustrate exam-type problems, rather than to duplicate an actual exam. A more reasonable time to expect for completing any of the given practice exams may be more in the neighborhood of 75–100 minutes. Although solutions to problems on the practice exams have not been detailed, answers are included **in Appendix B** at end of the book.

This note to students will close with a bit of a "pep" talk. We believe that both "reading" and "doing" are essential components for success in an introductory statistics course. Thus, the emphasis is on working through problems by hand (with the aid of a simple hand calculator that has a "statistics" function, and one that **you are familiar with and can use efficiently**). In "real laboratory life" (possibly an oxymoron) computers are used to replace computations done on a hand-held calculator, and computer applications (using a relatively common statistical software package—SPSS) are illustrated for more advanced problems in this textbook. It is recommended that you review the model problems and attempt solving practice problems that appear at the end of most chapters. Further, we suggest that where it is appropriate, you solve the practice problems first by using your hand-held calculator, and then check your work by using SPSS (or a suitable computer software alternative with which you are familiar).

It should be obvious that we agree with the arguments and approach advanced by Keppel and various others, *viz.*, that working through simple versions of various statistical tests by hand (the "doing" part mentioned earlier) facilitates understanding and the facility with which research interests direct the process of data analysis. The practice problems at chapter ends are provided to afford the opportunity for **doing** statistics after you read and understand the required operations. Perhaps it is human nature, but it is not uncommon to have an impression that one understands what he or she has read, only to discover later that problems arise with direct attempts to put that understanding into practice. The practice problems are intended to give students the opportunity to confirm that they truly understand the material presented. If you experience difficulty, it is far better to discover it while working through a practice problem than to make that discovery in class while trying to solve a problem on an exam!

It is also important not to fall behind in your course because current material builds upon skills and understanding mastered earlier. The subject matter may be particularly difficult for those whose style is to "crack the book" for the first time the night before an exam. As a final cautionary note, we wish to emphasize that it is important not to become complacent; especially if the material in the early chapters appears simple and obvious. For the vast majority of students we have worked with, concepts and problems become more complex and difficult as the text progresses. Thus, an excellent score on the first exam in the class should not be taken as a sign that one can relax and easily sail through the rest of the course (i.e., early success should not be construed to mean that your statistics course will be "an easy A").

Chapter 1

Experimental Research

What we have to learn to do, we learn by doing.
—Aristotle

I. A "Sow's Ear"

Statistical methods introduced in this textbook involve fundamental procedures and techniques central to many research activities. They enable researchers to summarize and communicate information acquired as a result of their research; in many situations, they play an important role in decisions about the reliability of research results (i.e., do observed differences in performance among different groups of individuals reflect more than merely the operation of chance or random factors?). As such, statistics may be fairly regarded as a research **tool**, much like lathes, staple guns, and impact drivers are tools used in carpentry that may require some instruction. Although learning proper uses of statistics may be more complex than learning proper uses of carpenter's tools, neither carpentry tools nor statistical tools may be a particularly exciting topic for a semester-long college course. However, the value of statistics for advanced study in most research fields cannot be denied. Pursuing advanced study in a research discipline without knowing statistics would be like trying to compete in today's marketplace without being computer literate. Perhaps a fair testimony to this analogy may be found by simply checking catalogs from most universities and colleges. You will certainly find a number of different statistics and research methods courses offered at both graduate and undergraduate levels.

The field of statistics represents a legitimate field of study within mathematics, as well as its tool-use applications within many scientific and research disciplines. Hopefully, your course experience will be more than just a not-so-pleasant-tasting medicine that has to be taken. However, "statistics are statistics," and an old adage may apply here: "You cannot make a silk purse out of a sow's ear." No attempt will be made to disguise a first course in statistics and make it appear to be what it is not or insult your intelligence by using a demeaning title for the text (titles such as "Statistics for Dummies," "Statistics for Those Who Hate Math," or "Statistics for Those Who Don't Like Anything Very Much"). So be prepared: the topic may be dry, with many artificial and contrived references to research results called upon for the purpose of illustrating various statistical indices developed to communicate information; and various statistical test procedures to decide objectively (according to agreed-upon ground rules) whether or not observed research results should be attributed to the operation of factors other than just randomness or chance.

II. Organization

The primary goal of this introductory chapter is to provide the necessary context for introducing statistics as a tool for describing and evaluating psychological research. Books the authors have used in teaching statistics in the past (e.g., Aron and Aron 1997; Keppel and Saufley 1980; Keppel, Saufley, and Tokunaga 1992; Keppel and Wickens 2004) have no doubt influenced the organizational format, coding schemes, analytical approach, and emphasis present in this book. A belief shared with Keppel is that working through calculations by hand is important for understanding various statistical techniques. Importantly, it is essential to complement your understanding by using labor-saving computer techniques not only to confirm class exercises done by hand, but also to replace hand calculations when dealing with large data sets. Since a major focus in this textbook has to do with experimental research, we will begin with a brief exposition on the language and anatomy of the experiment. In addition, it is essential to appreciate the logic involved when one draws causal implications from the results of experimental research.

III. What Is an Experiment?

If a large part of our effort entails describing and evaluating **experimental** research, it is important to identify what constitutes an experiment. Experiments involve observations (data) that are recorded under a **minimum of two** different conditions, or from at least two groups of individuals who received different experiences or treatments. Most often these differences are contrived or artificial, created by the experimenter for the purpose of recording the observations at a given time and place within a rigorously controlled environment or laboratory. The simple requirement (at least in theory, if not in practice) is that the two conditions or groups are **equivalent** in all respects except one. The reasoning from this contrived situation (we call an experiment) is that if the observed behaviors of research participants (responses) differ between the two treatment conditions, then the behavioral effects must be attributed to these differences in corresponding experiences or treatments. We will see later how additional conditions or groups can be introduced into the experiment without violating this fundamental principle. In the language of experimental research, the recorded observations or data constitute the *dependent variable*, and the difference in treatments introduced by the experimenter defines the *independent variable*.

The basic logic of experimental-control group research was described by John Stuart Mill in his treatise on "A System of Logic" (1888). Mill noted that if only one common element existed between two conditions, the occurrence of a specific phenomenon for each condition implied that the common element caused or was a necessary part of the cause for the phenomenon. Similarly, if all elements are common between two conditions with the exception of one element, the absence of the specific phenomenon implied that this element caused or was a necessary part of the cause for the phenomenon. Combined, these two canons anticipated the experimental-control group logic characteristic of modern experimentation. In an experiment, an independent variable may involve providing a specific treatment condition for an experimental group and withholding the specific treatment condition from the control group. A treatment that causes a given behavior to occur would result in the presence of that behavior for the experimental group and the absence of that behavior for the control group.

The language, logic, and virtues of an experiment may best be understood by analyzing actual experiments. The basic dictum of the experiment is that observations are made under different conditions that are equivalent in all respects except one. Thus, if there are observed differences in the behavior of interest among participants across different treatment conditions in an experiment, they may be attributed to the one factor that was different across conditions. When the experimental method is suited to given research questions, it provides a very powerful technique for arriving at clear causal statements.

While this "all-things-equal-but-one" principle appears simple in theory, it is not always that easy to achieve in practice. The following illustration of a classic experiment by Loftus and Palmer (1974) may provide a convenient starting point for illustrating the care and attention paid to detail in experimental design. In addition, the experiment should prove helpful for introducing some terminology and issues relevant for a detailed illustration of an actual experimental design. Following this illustration, we will consider some potential obstacles that may be encountered when designing experiments.

A final note about dependent variables is in order before moving on. Assessing behavior may be done using one of four kinds of measurement scales: Nominal, Ordinal, Interval, or Ratio. **Nominal** scales involve data that are categorical (e.g., gender, political affiliation), and basically do little more than **name** classification categories. **Ordinal** scales encompass numbers that define ordered arrangements of categories. Rankings consumers give to different products and ranking of job candidates by hiring committees represent two examples of ordinal scales. The major limitation of an ordinal scale is that separation of ranks is not represented meaningfully by numbers (i.e., we know the top rank is above the second rank, the second rank is above the third rank, etc.; however, differences between ranks are not represented on a standard numerical scale). **Interval** scales are involved for data with separations between successive categories or values assumed to be equal (e.g., temperature, and also Likert scales, a type of scale that entails a small number of response alternatives to select from as appropriate for each statement in a set, such as seven "equal" steps from an extreme negative response of "strongly disagree" to an extreme positive response of "strongly agree"). However, an interval scale lacks a true zero point. For example, consider a temperature of zero degrees. A value of zero has a different meaning for Fahrenheit than for Centigrade, and in neither case does it represent **absence** of temperature. **Ratio** scales consist of numerical scores having a true zero point. That is, zero presumably represents absence of something (e.g., zero seconds to react to a stimulus; zero distance separating two points). Much of statistical work proceeds with the assumption that data being analyzed approximate conditions meeting requirements for interval or ratio scales; data assumed to be on a continuous, rather than discrete dimension.

IV. Component Parts of an Experiment

Although somewhat dated, a simple experiment by Loftus and Palmer (1974, Experiment 2), should serve our purpose because: (1) the experimental design is easy to follow; (2) it addresses a research question that arguably is quite relevant; and (3) the research report is "reader-friendly," that is, it is presented adequately in common language without resorting to sophisticated statistical analyses, technical jargon, or complex theoretical models. Loftus and Palmer were interested in whether or not subtle changes in the phrasing of a question about an observed event would influence "memory" for the event; certainly an important question in the legal field where interrogating witnesses and suspects is a central activity. Translating this interest into an experimental question involved many decisions about specific independent variables, dependent variables, participants, tasks, and procedures; basically in making sure that groups were equivalent in all respects with the exception of one factor (i.e., differences defined by the independent variable). The following excerpts from Experiment 2 reported by Loftus and Palmer illustrate these considerations (page 587).

> One hundred and fifty students participated in this experiment, in groups of various sizes. A film depicting a multiple car accident was shown, followed by a questionnaire. The film lasted less than 1 min; the accident in the film lasted 4 sec.

This initial description identifies the experimental participants (150 students), the manner in which they participated in the experiment (groups of various sizes watched a film), and some details about the stimulus material (a brief film showing a multiple-car accident).

> At the end of the film, the subjects received a questionnaire. … The critical question was the one that interrogated the subject about the speed of the vehicles. Fifty subjects were asked, "About how fast were the cars going when they smashed into each other?" Fifty subjects were asked, "About how fast were the cars going when they hit each other?" Fifty subjects were not interrogated about vehicular speed.

Following the initial description, the experimenters proceeded to identify the independent variable. We are informed that the 150 students were separated into three groups of 50 students each, and presumably the only systematic difference among the three groups involved a single question from a common questionnaire: a requested speed estimate for only two of the groups, with either the verb "smashed" or "hit" describing the collision shown in the film. Intuition and common sense also come into play here, as the underlying assumption is that the verb "smashed" implies a more violent collision than the verb "hit." The third group served as a control: a question about vehicle speed was absent from the questionnaire for students in this group. One dependent variable is also identified: the various speed estimate responses from students who were asked the "smashed" and "hit" versions of the critical question.

> One week later, the subjects returned, and without viewing the film again, they answered a series of questions about the accident. The critical question here was, "Did you see any broken glass?" which the subjects answered by checking "yes" or "no." This question was embedded in a list of 10 questions, and it appeared in a random position in the list.

A second dependent variable is identified in the preceding paragraph, namely, the yes–no responses all students gave to a question about the presence of broken glass. Again, common sense comes into play as it would be expected that the more violent the collision, the greater the likelihood that broken glass would be present at the scene. Additional procedural details are provided to inform the reader that the second questionnaire occurred one week after viewing the film, and the position of the critical question in the 10-item questionnaire was determined "at random." Presumably, subjects were also assigned to the three groups at random, which is a standard procedure for forming "equivalent" groups. Randomization ensures that every individual has an equal opportunity to be assigned to each condition of the experiment, with chance alone determining the final assignment (e.g., like drawing the number 1, 2, or 3 out of a hat to determine whether an individual is assigned to the "smashed" group, the "hit" group, or the group that did not have a speed-estimate question). Thus, for differences among groups in response to the speed estimate and broken glass questions, only variations for the critical question (the independent variable) and "chance" (referred to as "random error") could account for any observed performance differences among groups. It is important to note at this point that the "all-things-equal-but-one" principle is always violated by the fact that "chance" or random error represents a second source contributing to differences that may be observed among experimental conditions. We will see later that statistical procedures, in effect, enable one to assess the random-error contribution in a probabilistic sense, and evaluate whether or not it alone could reasonably account for all observed performance differences.

The authors did not report actual speed of the vehicles in this film, although they did refer to a prior study in which a film depicted vehicles traveling at 12 miles per hour (mph). With regard to speed estimates (only data for the "smashed" and "hit" groups were available since the control group did not have a speed-estimate question), the average speed estimate for students in the "smashed" group was 10.46 mph, and the average speed estimate for students in the "hit" group was 8.00 mph. At this point we have a simple illustration of an important descriptive use of statistics. Although there were 50 speed estimates for subjects in the "smashed" group and 50 more speed estimates for subjects in the "hit" group, only two numbers (not all 100) were reported: a "representative" estimate in the form of a mean or average speed estimate for each group.

We can see from these representative scores that speed estimates tended to be a little higher when "smashed" was used in framing the question, as opposed to when "hit" was used. The authors go on to note that the two means "are significantly different." This latter phrase illustrates the second function of statistics. Clearly, a difference between means of 10.46 vs. 8.00 is pretty small. Unfortunately, we cannot know for certain whether this is a "real" difference—one that would reliably appear with this independent variable for similar experiments—or just a chance fluctuation. Statistical tests are used to help decide the issue in a probabilistic way. For now, you merely need to know that if the probability is very small that chance (random error) alone could account for group differences as large as those reported, we must look elsewhere to find the cause for the difference. For the well-designed experiment, the only remaining option for "looking elsewhere" for the source of the effect is the independent variable.

The second dependent variable assessed whether participants remembered seeing broken glass at the scene of the accident. Although broken glass was not present in the film clip, 16 subjects in the "smashed" group answered "yes" to the broken glass question, compared to 7 subjects in the "hit" group and 6 subjects in the control group who answered "yes" to this question. Based on an appropriate statistical test, the authors reported that the differences among groups were "significant." It should be noted that the term "significant" is used statistically in reference to decisions about the reliability of differences among experimental groups; it is not a comment about the importance of differences as a common language use of the term might imply.

V. Confounding Variables

Problems in experimental design arise whenever equivalence across groups (or more generally, conditions) on all factors except the independent variable cannot be assured. We must live with the fact that random error (chance) is a potential source for disrupting equivalence. This source will be dealt with statistically. Systematic treatment differences exist among groups in accordance with the specific independent variable(s) investigated, and an independent variable ought to represent the only other potential source for disrupting equivalence for the behavior of interest (the dependent variable). Other potential independent variables that are not adequately controlled pose a threat to the integrity of the experiment. As such, they represent *confounding variables*, and if they vary systematically in an experiment rather than randomly, then any observed effect on behavior could just as readily be attributed to the confounding factor(s) as to the independent variable(s) (or to a combination of independent and confounding variables). So, you might say, let's just move on and agree that we will not permit confounding variables to intrude in designing experiments. However, for at least two reasons, confounding variables need to be addressed in more detail. First, keeping them at bay is not always as easy as it sounds. A confounding variable can "sneak" its way into an experiment, even for experiments conducted by experienced and highly qualified researchers. Second, there is a wide array of research problems for which it is known that the independent variables of interest are confounded; however, the research proceeds, usually with demonstrations that attempt to discredit on intuitive or empirical grounds the attribution of effects to the confounding variable(s). Designs with deliberate confounding variables are referred to as *quasi-experimental designs*.

Psychologists have long been interested in subject variables (e.g., comparing performance among different age groups, among different patient groups). Unfortunately, comparisons of subject variables that are not directly "manipulated" by the experimenter are problematic. If subjects cannot be assigned to groups at random, groups are most likely not equivalent on other relevant factors (e.g., can you think of any variables besides age that would differentiate 20-year-olds from 70-year-olds?). So what does the researcher do—abandon his or her interests and move on to other research problems? Of course, there are clear advantages for experiments that can build equivalent groups with random assignment; however, many quasi-experimental designs can be informative and should be pursued.

The basic failing with quasi-experiments is that there is competition among hypotheses (the independent variable and one or more confounding variables) with regard to what is responsible for the destruction of group equivalence. Consequently, a general strategy with quasi-experimentation involves attacking the tenability of the rival confounding-variable hypotheses (Cook and Campbell 1979). Often, intuition and theory can make life difficult for rival hypotheses. For example, one might speculate that many elderly individuals perform more poorly on simple memory tests because they are overly sensitive about saying the wrong things, thus they do not respond (counted as errors of omission) unless they are quite certain they are correct. Given this hypothesis, one can design a complex experiment involving two independent variables so that an "interaction" would be expected (we will have more to say about interaction between independent variables in later chapters). The idea is that performance of young and elderly adults on a memory task would be compared in a standard situation (one in which young adults were known to perform better) and in a "free-responding" situation where guessing and responding with the first thing that comes to mind are encouraged. According to the research hypothesis, differences in performance between young and elderly adults should be attenuated in the latter condition. The presence of a complex interaction can place an extra burden on a rival hypothesis because that hypothesis would have to account for a difference under one set of conditions, but not under a different set of conditions.

VI. When Statistics Come into Play

Statistical analyses and techniques have two basic functions in research: *descriptive* and *inferential*. The descriptive function involves describing and summarizing large bodies of data (recorded observations, such as SAT scores for high school seniors nationwide). Even small-scale experiments generate more data than can be readily communicated simply by reporting every single observed score. Thus, researchers organize and summarize their data using graphs and charts, and report numerical indices providing information about important characteristics such as where the center point is, how dispersed or spread out scores are, the degree to which pairs of scores are related, etc. Almost all researchers who are in the business of collecting data rely on descriptive statistics when reporting results of their research.

The inferential function of statistics involves reliability assessment—that is, for many researchers, various statistical techniques are employed when it comes to making objective decisions about reliability of research results, especially for experimental research. Remember, a good experiment is set up so that only one potential variable (called the independent variable) is systematically different across groups. To keep all other variables equivalent, at some point we normally rely on random assignment. This reliance means that these other variables would be equal in the "long run"; however, there will be chance variations for observations in any "short-run" experimental setting. So when differences in performance measures are observed among groups, the experimenter must decide whether differences can be reasonably attributed entirely to chance or random error. If the decision is that it is not likely that random error could account for all of the observed differences, then by default, the independent variable is credited with accounting for a share of these differences. The branch of statistics employed for deciding this issue is commonly referred to as **inferential** statistics—statistical tools that provide the basis for drawing inferences about research results. Although it may be convenient to think of inferential statistics in terms of a shortcut to replication (repeating experiments to demonstrate that observed results can be produced a second, third, etc. time), technically, it may not be possible to determine replication probability from the results of inferential statistical tests (Miller 2009).

Descriptive statistics are a part of most research efforts (at least empirical research involving records of what is observed). However, there is some variation in the extent to which researchers rely on inferential statistics. On the descriptive side we will consider some simple techniques for organizing and condensing large data sets into more meaningful displays; we will introduce computational procedures for determining center points (means, medians), for assessing the degree of dispersion or spread amongst a set of scores (variances, standard deviations), and later, for measuring the degree to which pairs or sets of scores are related (correlation coefficients). While the descriptive side of statistics is arguably the more important of the two functions noted here, the majority of the chapters in this text focus on the inferential function. The justification for this uneven distribution is based on complexity, as the inferential extension involves a more complex rationale based on probability theory, and a more complex set of computational operations. The inferential technique we will spend the most time on is the **analysis of variance (ANOVA)**, which appears to be the most frequently used inferential technique applied in the statistical analysis of experiments in psychology. The test statistic in the analysis of variance is abbreviated with an upper case F. If you look for an "F" somewhere in the spelling of "analysis of variance" or "ANOVA," you will not find it (except in the small preposition "of"). Sir Ronald A. Fisher is credited with introducing analysis of variance, and the test statistic with this procedure was coded as "F" in his honor. As the name implies, analysis of variance is all about calculating variances. The F test statistic is simply a *ratio* of two variance estimates (one variance estimate divided by another variance estimate). Thus, it should be obvious that understanding both the concept and computational operations for the descriptive index of **variance** is pivotal to later requirements of performing ANOVAs on selected data sets.

VII. Summary

Although this text is designed for a beginning undergraduate course in statistics, we began our treatment of this topic with a brief description of experimental research methodology. The justification for this introduction is based on the fact that the experiment provides the context for statistical applications that will occupy most of the space in this text. Developing carefully reasoned experiments with sound methodology that will enable you to answer questions of interest is of primary importance. Various statistical techniques that communicate information effectively and assist in evaluating the reliability of results are important tools when used appropriately and reasonably; however, they cannot produce meaningful research hypotheses, generate productive theories, or salvage poorly designed and executed experiments.

An experiment requires a minimum of two groups or conditions that ideally are equivalent in all ways except for one systematic difference introduced by the researcher. Some other factors may be kept in check by deliberately eliminating or equating them (e.g., controlling gender by using only female participants). However, at some point we rely on randomization (e.g., randomly assigning subjects to groups) in the effort to establish equivalence among conditions on "other" factors (potential sources of influence that we might suspect are important and unknown sources). Randomization establishes equivalence in the long run (for extremely large samples); however, the procedure does not guarantee equivalence for real experiments with a finite set of recorded observations. We are left with the possibility that some observed differences in performance reflect "luck" or "chance." Statistical applications can help determine the likelihood that observed differences can be reasonably explained by chance alone. Some common research language was introduced in the course of this discussion identifying the factor that the experimenter introduces to differentiate conditions as the independent variable or treatment factor, and the recorded observations (scores) as the dependent variable.

Unwanted variables that are not excluded from the experiment through use of careful design and control procedures are called confounding variables. Confounding variables may be troublesome because they coexist and may co-vary with the independent variable, and hence may present plausible reasons for the differences among experimental conditions that are observed. In some cases they can be tolerated and experiments can proceed. Experiments that include confounding variables are referred to as **quasi**-experiments; extra caution must be exercised in design and interpretation of data from quasi-experiments.

Given that this text is designed for a statistics course, it seemed only fair to close the introductory chapter by identifying the two major statistical uses in psychology. Statistics come into play in the experimental research enterprise primarily to describe research results in a meaningful and understandable way. That is, in response to a question of "What did you find in your research?" summary answers may be given that do not require detailed descriptions of each individual response. A second role for statistical tools involves drawing inferences from observed results primarily by providing rules for deciding if it is likely that the experimenter's independent variable contributed to observed differences over and above what could be reasonably expected by chance factors alone.

VIII. Appendix I.A: Coding Symbols and Notation

Probably all statistics textbooks involve some mathematical notation and symbols. Math notation and symbols are necessary for at least two reasons. First, they provide a shorthand abbreviation, conserving both time and space in communicating information clearly and concisely. Second, they provide instruction about what to do with a set of numbers. Specific examples that we will frequently encounter appear below:

A. **X** and/or **Y** are codes for response measures or scores that vary or take on different values for different experimental participants. Subscripts are used to identify different individuals within the set; e.g., Y_1 is the score for the 1st subject, Y_2 is the score for the 2nd subject, etc., with Y_i used for the general case.

B. **n** (lower case) is a code for the number of individuals in a group.

C. $\bar{X}$ and/or $\bar{Y}$ (upper case X or Y with a bar above) represent an average or mean score—one may also encounter an upper case M as a code or abbreviation for the mean.

D. s^2 represents a sample variance; σ^2 represents a population variance. A sample is a subset of a population.

E. **s** represents a sample standard deviation and σ represents a population standard deviation. The same letter (or Greek letter) code for variance and standard deviation is deliberate, as the standard deviation is the square root of the variance: $s = \sqrt{s^2}$; $\sigma = \sqrt{\sigma^2}$.

F. There are several "operators" that provide instructions regarding the computational steps required in working with numbers in a data set. Specific operators that provide instructions we routinely follow without even thinking about them are + (for addition), – (for subtraction), × (for multiplication), ÷ (for division; also, X ÷ N may be written as X/N), and √ (for taking the square root). Perhaps a less familiar one that we encounter frequently in statistics is ΣY_i, which instructs us to add (as an operator, Σ means take the sum of) all the Y_i scores. Subscripts may also be included beneath and above the summation sign (Σ) if it is necessary to limit the summing operation to a subset of Y_i scores (e.g., a 1 below and 5 above the summation sign instructs us to sum scores for subjects 1–5 only).

G. If you see ΣY_i^2, what does this instruct you to do? Importantly, is a ΣY_i^2 instruction any different from a $(\Sigma Y_i)^2$ instruction? The answer to this last question is **YES**. If you have the following five Y_i scores (n = 5): 3, 4, 7, 2, and 5, what does ΣY_i^2 instruct you to do, and what does $(\Sigma Y_i)^2$ instruct you to do?

$$\Sigma Y_i^2 = 3^2 + 4^2 + 7^2 + 2^2 + 5^2 = 9 + 16 + 49 + 4 + 25 = \mathbf{103}$$
$$(\Sigma Y_i)^2 = (3 + 4 + 7 + 2 + 5)^2 = 21^2 = \mathbf{441}$$

H. A verbal instruction for calculating a mean might be "The mean equals the sum of all the Y_i scores divided by the number of Y_i scores that were summed." Alternatively, we could express this instruction as $\bar{\mathbf{Y}} = \Sigma \mathbf{Y}_i \div \mathbf{n}$.

I. Two equivalent instructional formulas will be introduced later for determining the variance for a set of scores:

$$\mathbf{s^2 = \Sigma(Y_i - \bar{Y})^2 \div (n - 1)}$$

$$\mathbf{s^2 = ([\Sigma Y_i^2] - [\{\Sigma Y_i\}^2 \div n]) \div (n - 1)}$$

Can you follow both formula instructions and compute s^2 for the set of five numbers used earlier (3, 4, 7, 2, 5)? In the first case, the steps are to: (1) calculate the mean, $\bar{\mathbf{Y}}$; (2) subtract the mean from each Y_i score so that for each Y_i score you have a new $Y_i - \bar{\mathbf{Y}}$ deviation score; (3) square each of the new deviation scores; (4) add the squared deviation scores; and (5) divide the sum of squared deviation scores by n – 1. For the second formula, the steps are to: (1) square each Y_i score; (2) add the squared Y_i scores; (3) add the Y_i scores; (4) square the sum of the Y_i scores; (5) divide the squared sum of the Y_i scores by n; (6) subtract the number obtained in #5 from the number obtained in #2; and (7) divide the number obtained in #6 by n – 1. Note: **The solution in both cases will be a positive number!** Do you know why this must be true? Can you justify your answer both conceptually and arithmetically? While more steps were required for a verbal description of the computations for the second formula, it turns out that this formula is quicker and easier to conduct on a hand calculator. However, whichever one you use, your answer should be $s^2 = 3.7$.

J. **N!** (Read as "N factorial"): An operational instruction that will be considered briefly when we lay the probability foundation for using statistics in inferential work. The N! instruction calls for the following series of multiplications: $(N) \times (N - 1) \times (N - 2) \times (N - 3) \times \ldots \times (1)$. For example, if N = 5, then for 5! we have $5 \times 4 \times 3 \times 2 \times 1 = 120$.

K. Some algebraic definitions to note are that $0! = 1$; for any constant C, $C^0 = 1$; when summed over "n" observations, the sum of a constant equals n times that constant ($\Sigma C = nC$), and the sum of a constant times a variable equals that constant times the sum of the variable ($\Sigma CY_i = C\Sigma Y_i$).

L. Remember when you square fractions or multiply two fractions, you must perform the multiplication or squaring operation on both numerator and denominator: e.g., $(2/5) \times (2/3) = 4/15$.

Chapter 2

Descriptive Statistics I

The average human has one breast and one testicle.
—McHale

I. Displaying Data in Tables and Graphs
II. Locating the Center for a Set of Scores
   a. The Median (Mdn)
   b. The Mean (coded as $\bar{Y}$, or $\bar{X}$, or sometimes as M)
III. Summary
IV. Practice Problems

I. Displaying Data in Tables and Graphs

Two branches of statistics were identified in the first chapter: (1) Descriptive Statistics and (2) Inferential Statistics. The goal of descriptive statistics is to summarize a large body of information in a meaningful way. Experimental research generally involves relatively small samples of cases (e.g., subjects) to observe; however, even with relatively small samples it may be difficult to understand what recorded observations revealed merely by looking at all of the raw data (individual responses).

Organizing and condensing information into meaningful tables, graphs, and descriptive indices provides a necessary aid in communicating, in summary form, what an experiment revealed—and this is the important role played by descriptive statistics. To illustrate we will begin by considering the set of 60 numbers presented in Table 2.1. To keep things in perspective, remember that the total data set described in the simple experiment by Loftus and Palmer (1974) included nearly twice as many speed estimates and nearly three times as many responses to the broken glass question.

Table 2.1: Fictitious Data

7	13	21	9	10	16
11	12	8	12	15	25
15	9	15	6	17	14
3	9	19	14	13	14
9	5	17	20	10	15
9	8	19	17	5	24
10	6	11	9	30	13
4	2	7	15	22	22
12	5	16	8	9	25
18	4	7	9	17	7

Assume the numbers in Table 2.1 are simply listed in chronological order according to the order in which observations were recorded, thus the first thing we can do is to impose further order on this array of data. A first step in organization is accomplished in Table 2.2 by rearranging the same set of 60 scores in order of magnitude, going from lowest to highest. Hopefully you will agree that this represents a little improvement in terms of getting a picture of the results; however, the reader still must make sense out of a rather large array of data: i.e., the full set of 60 original scores.

Table 2.2: Fictitious Data with Scores Arranged in Order of Magnitude

2	7	10	14	17
3	8	10	14	18
4	8	10	15	19
4	8	11	15	19
5	9	11	15	20
5	9	12	15	21
5	9	12	15	22
6	9	12	16	22
6	9	13	16	24
7	9	13	17	25
7	9	13	17	25
7	9	14	17	30

It is rather cumbersome to deal with large numbers of scores; consequently, another summarizing move we can make is to group like scores together. Thus, instead of writing a score of "9" eight times, we will simply list the score once, with the number of subjects who had that score listed next to it (the frequency of occurrence). At this point we have not sacrificed any information with this change in display; however, you can see that we have managed to condense a table of 60 numbers into a display involving a listing of 24 scores, with each score paired with the number of individuals who had that score (i.e., each obtained score is paired with a frequency count of how often it occurred). Table 2.3 presents the data in what is referred to as a *frequency distribution*. Alternatively, if scores were listed in sequence from 0 to 30, which would include a zero frequency of occurrence for some possible outcomes (e.g., a score of 1 was not recorded for any of the participants), then the frequency distribution table would have 31 paired scores, ranging from a score of 0 with 0 frequency up to a score of 30 with a frequency of 1.

Table 2.3: Frequency Distribution

Scores	Frequency (f_i)	Scores	Frequency (f_i)
2	1	16	2
3	1	17	4
4	2	18	1
5	3	19	2
6	2	20	1
7	4	21	1
8	3	22	2
9	8	24	1
10	3	25	2
11	2	30	1
12	3		
13	3		
14	3		
15	5		

So far, no information has been sacrificed. For each of the tables you can determine the number of scores (n = 60, which is arrived at by simply counting the numbers in Tables 2.1 and 2.2, or by adding the numbers in the "Frequency" or f_i column in Table 2.3). Similarly, the sum of the 60 scores is the same regardless of which tabled data you work from: $\Sigma Y_i = 753$ from Tables 2.1 and 2.2; and $\Sigma f_i Y_i = 753$ from Table 2.3. In the latter case, each score is multiplied by the number of times it occurred (f_i): $\Sigma f_i Y_i = (1)(2) + (1)(3) + (2)(4) + \ldots + (1)(30) = 753$.

As we begin summarizing information further, we also begin to lose some detail. A further summary step depicted in Table 2.4 is to present scores in a **Grouped Frequency Distribution**. For this table, score "categories" have been created, and the number for each category is simply a frequency count of how many scores occurred within that category. For example, one category may be 8–11, a category that includes scores of 8, 9, 10, and 11 (technically, the category boundaries are 7.5–11.5; referred to as the *real category limits or boundaries*). Since there are three scores of 8, eight scores of 9, three scores of 10, and two scores of 11, this category contains a total of 16 scores. Table 2.4 is more compact than the preceding tables; however, reducing the data set to a more manageable size comes at the expense of some of the details. Looking only at Table 2.4, a reader can see that there are 16 scores that were 8s, 9s, 10s, or 11s; however, this table alone does not include information regarding how these 16 scores were distributed within the category bounded by a low score of 8 and a high score of 11.

There are some important features that you will need to address when constructing a grouped frequency distribution. First, you have to decide on the number of intervals that you want to use—most of the time researchers settle on between 5 and 15 intervals (probably closer to 15 if you have a large data set and closer to 5 if you have a small data set). Second, the width of each category interval must be the same for all categories; and a good rule of thumb for approximating category width is to subtract the lowest score (2 in this example set) from the highest score (30 in this example set). If we decided to have about 7 or 8 categories, then we would simply divide the difference between the highest and lowest scores by the number of desired categories, and if reasonable, round off to the nearest whole number). For the data in Table 2.2 the difference between the highest and lowest scores is 28, and 28 ÷ 7 (number of target categories) = 4; thus we will use a category width of 4. Of course, you cannot have overlapping categories (e.g., 5–8, 8–11, 11–15) because some scores (e.g., 8 and 11) would then be ambiguous regarding a category assignment. Also, rather than always using the lowest score obtained as the lowest interval boundary, we will decide where to start by taking the lowest multiple of the category width that yields a starting category that includes at least one score. Given these considerations, we could set the lowest category at either "0–3" or "1–4," and then build upwards until we reach the highest category beyond which there are no higher scores. The grouped frequency distribution in Table 2.4 was constructed as described using "0–3" as the lowest starting category. Given this description for constructing a grouped frequency distribution, you can see that there may be several acceptable ways to group the data.

Table 2.4: Grouped Frequency Distribution

Category Number	Category Interval	Frequency
VIII	28–31	1
VII	24–27	3
VI	20-23	4
V	16–19	9
IV	12–15	14
III	8–11	16
II	4–7	11
I	0–3	2

Picture displays are often more user-friendly than tables, so it is quite common for researchers to present summary information graphically instead of using table displays. The information in Table 2.4 is displayed twice: first, in a bar graph (Figure 2.1), a bar is constructed over each respective category

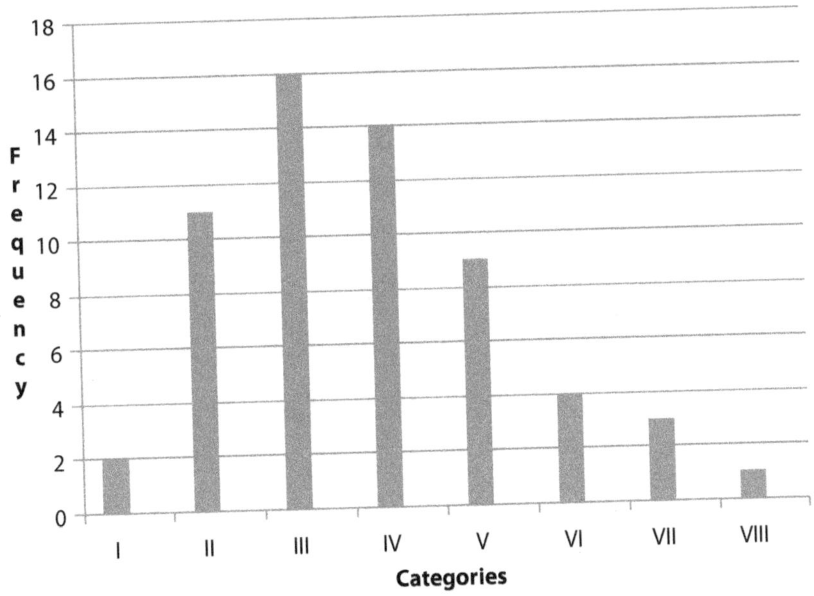

Figure 2.1: Bar Graph Presentation of Example Data

interval, with the height of each bar corresponding to the frequency of scores in the category; and second, in a line graph (Figure 2.2), the height of the points in a line graph correspond to frequency of scores, with each point located above the *middle* of each category interval. Technically, a line graph with lines connecting the points should only be used if the numbers on the baseline (the X axis) represent points on a continuous dimension, as connecting the points implies continuity. Since space is precious in research journals, when a table and graph present the same information, either one or the other of the two formats (not both) would be selected for summarizing the research results.

It is important to note that certain ground rules need to be followed when constructing graphs so that the information presented is accurate and not misleading. For example, the X (horizontal) and Y (vertical) axes must always be labeled; numbers on X and Y axes should increase uniformly, and if there is a break in the uniform increase in numbers along either axis, it should be clearly denoted (e.g., a small open space in the vertical axis marked off by two short horizontal lines); and steps along the Y axis should reflect "reasonable" variations in response measures so that small differences in scores among various groups do not appear to be "large," and large differences in scores among various groups do not appear to be "small." For example, differences between annual salaries of men and women faculty members would appear to be relatively small visually in a graph if each inch along the vertical axis represented an increment of $10,000, and relatively large if each inch represented an increment of only $1,000.

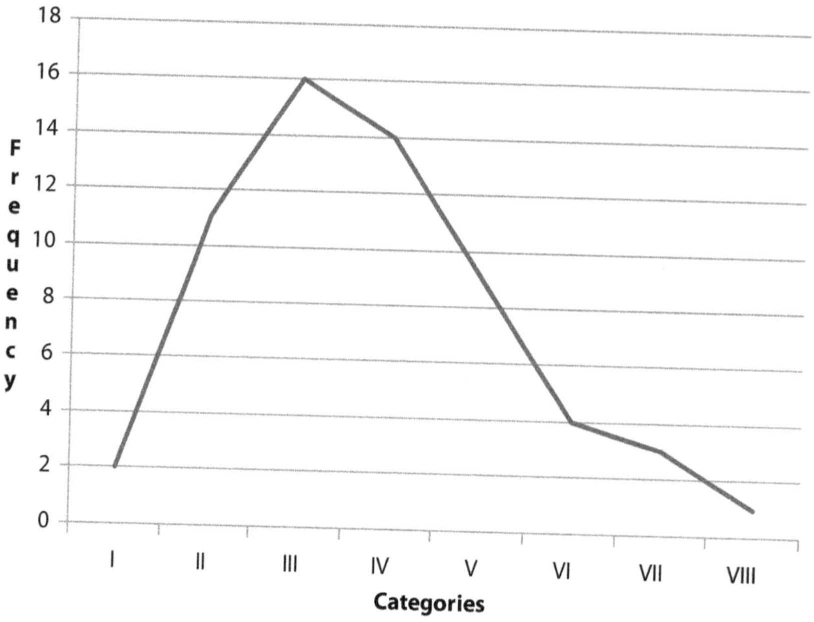

Figure 2.2: Line Graph Presentation of Example Data

II. Locating the Center of a Set of Scores

Probably the ultimate achievement in simplifying and refining all of the scores in a data set would be to reduce the raw data to a single number that provided a measure or index of an important feature of the data. Two such indices that are especially useful for both descriptive and inferential statistics provide information about (1) **central tendency** and (2) **dispersion or variability**. In the remainder of this chapter we will introduce two measures of central tendency: the *median* (center point) and the *mean* (arithmetic average). Measures of dispersion will be described in the next chapter. Before proceeding with these two more frequently used indices of central tendency, we should at least identify a third measure, namely, the *mode*. The mode is the score in a distribution of scores that occurred most often. It is computed by simply counting the number of times each score occurred, and the one with the highest count "wins" the title of *mode*. In most cases, this measure does not make use of very much information in a data set; that is, it is determined by only one of the observed scores (the one that occurred most often). You can quickly scan the data in Table 2.3 and see that the score of "9" had the highest frequency count (8), thus the mode for this data set is 9. Occasionally we encounter a large data set that has two peaks (high points in terms of frequency counts), and even though the frequency counts may not be exactly the same, the data are referred to as "bi-modal," or as having two modes.

a. The Median (Mdn)

In experimental research the mean is the measure of central tendency most often reported; however, there are situations for which the median is preferred. We will conclude this section with a contrived example to illustrate a type of situation where it would make more sense to use a median rather than the mean to represent a meaningful center for a set of scores. For now we will begin with a definition of the median and description of procedures for finding it. The median is a point within a set of scores that divides the top half of the distribution from the bottom half of the distribution. That is, the median is the middle point for a set of scores such that half of the scores are above and half are below this point. If you have an odd number of scores in a distribution, then the median is the middle score once the scores have been arranged in order of magnitude (e.g., from lowest to highest, as was done in the Table 2.2 that represented a re-organization of the original data set).

What does the median equal for the following set of 11 numbers: 8, 5, 3, 9, 2, 1, 1, 4, 10, 7, and 9? The first step in answering this question is to rearrange the numbers in order of magnitude: 1, 1, 2, 3, 4, **5**, 7, 8, 9, 9, and 10. The median is the 6th highest score, since there will be 5 scores below and 5 scores above this number [a formula for determining the middle of the distribution is (n + 1) ÷ 2, where n is the number of scores in the data set]. In this case, n = 11, and (11 + 1) ÷ 2 = 6, so the median will be the 6th highest score. The number 5 is the 6th highest number, therefore, Mdn = 5 (the number typed in bold for the ordered display of the 11 scores).

If the data set in Table 2.2 had included only 59 scores (omitting the last entry of 30), the middle score would be the 30th highest score [(59 + 1) ÷ 2 = 30], since there are 29 scores lower than this number and 29 scores higher than this number. One of the scores of 12 is in this 30th position, so we would be content to conclude that the Mdn = 12. We could spend more time on the concept and measurement of medians by introducing somewhat esoteric procedures designed to produce a more precise estimate (e.g., an estimate between 11.5 and 12.49), but we will pass on this exercise since we have more important issues to worry about.

Unfortunately, not all data sets are conveniently arranged to have an odd number of scores. Indeed, the original data set in Table 2.2 included 60 scores; hence the median is a point with 30 scores above it and 30 scores below it. A minor complication arises because we do not have an actual score from the data set in this position. To illustrate, let's begin with the small set of 11 numbers considered earlier and add one more score (13) to the set: 8, 5, 3, 9, 2, 1, 1, 4, 10, 7, 9, and *13*. The first step in determining the median is to rearrange the numbers in order of magnitude: 1, 1, 2, 3, 4, **5**, **7**, 8, 9, 9, 10, and 13. We are looking for a number that has 6 scores below it and 6 scores above it [recall the formula for determining the middle point is (n + 1) ÷ 2. In this case, n = 12, and (12 + 1) ÷ 2 = 6.5. We have a 6th highest score (**5**) and a 7th highest score (**7**), but we do not have a score in a 6.5 position. So we will estimate the median by taking a point midway between the 6th (**5**) and 7th (**7**) highest scores. In this example, **Mdn = (5 + 7) ÷ 2 = 6**. This formula looks like the formula presented earlier for determining the **middle position**; however, you should note that when determining the median for an even number of scores, once the middle position number has been identified (e.g., position 6.5), actual score values of 5 and 7 (not positions) are used in the formula [(5 + 7) ÷ 2 = 6]. Now, returning to the original data in Table 2.2, the middle position is determined by (60 + 1) ÷ 2 = 30.5. This means the median is the point halfway between the 30th highest score and the 31st highest score. For this data set, both the 30th and 31st highest scores are **12**. Again, we will not worry about procedures for fine tuning a median estimate, and we will just go with: **Mdn = (12 + 12) ÷ 2 = 12.**

b. The Mean (coded as $\bar{Y}$, or $\bar{X}$, or sometimes as M)

Like the median, the mean for a set of scores is a single number designed to locate the center of the distribution. It is the arithmetic average, calculated by adding all scores and then dividing by the number of scores. In shorthand formula notation, the instruction for calculating a mean is $\bar{\mathbf{Y}} = \Sigma \mathbf{Y}_i \div \mathbf{n}$. It is a balancing point insofar as if you subtract the mean from each individual score, the sum of the negative scores will equal the sum of the positive scores. The mean for the 12-number set used in the example for the median is (1 + 1 + 2 + 3 + 4 + 5 + 7 + 8 + 9 + 9 + 10 + 13) ÷ 12 = 72 ÷ 12 = 6. Subtracting the mean (**$\bar{Y}$ = 6**) from each individual score gives values of -5, -5, -4, -3, -2, -1, +1, +2, +3, +3, +4, and +7. The negative numbers (5, 5, 4, 3, 2, and 1) sum to 20 and the positive numbers (1, 2, 3, 3, 4, and 7) also sum to 20.

You may have noted that the mean and the median were both equal to 6 for the set of 12 scores used in this example. It is often the case that these two measures of central tendency will be close in value. We saw earlier that for the set of 60 scores in Table 2.2, the **median ≈ 12** (≈ means approximately equal). What does the mean equal for the data in Table 2.2? This calculation can be tedious since we have to add 60 scores and then divide the resulting sum by 60. But be forewarned—more tedious hand calculations are on the horizon. For $\Sigma Y_i \div n$, we have (2 + 3 + … + 30) ÷ 60 = 753 ÷ 60; thus the mean equals **12.55**.

Although the mean is more widely used than the median, as it has an essential role in the analysis of experimental data, there are times when the median may be more appropriate. For example, a small number of extreme scores (referred to as outliers) could have considerable impact on the location of the center of a distribution when indexed by the mean; however, they would have a relatively small impact if indexed by the median. Consider a fictitious data set based on responses of junior high school boys to a question about the number of different girls they have "dated" (this is the silly illustration alluded to at the beginning of this chapter). To conserve space, these fictitious data are presented in a frequency distribution in Table 2.5. This table reveals responses from the 150 junior high school boys, and inspection of the table indicates that the bulk of the scores are 0, 1, 2, or 3. However, 10 boys wrote down 100, perhaps raising suspicions that they were less than serious in responding. What would be the better representative score here because the mean and median are quite different? The median is the middle score occupying the position between the 75th and 76th highest scores. Since both are scores of "1," without a more esoteric fine-tuning of the median, we can report, **Mdn ≈ 1.0**.

Table 2.5: Fictitious Data on Number of Girls Junior High School Boys Have "Dated"

Responses	Frequency	Responses	Frequency
0	30	4	1
1	50	5	2
2	40	10	2
3	15	100	10

The mean takes the actual **value** of all scores into consideration. The formula used here is an adaptation of $\Sigma Y_i \div n$ for a grouped frequency distribution, as each score is simply multiplied by the number of times it occurred (e.g., 50 boys gave the response of "1," so instead of writing "1" 50 times in the summation formula, we simply multiply 1 by 50).

$$\Sigma Y_i = ([0][30] + [1][50] + [2][40] + [3][15] + [4][1] + [5][2] + [10][2] + [100][10])$$
$$\Sigma Y_i = (0 + 50 + 80 + 45 + 4 + 10 + 20 + 1000) = 1{,}209$$

Dividing ΣY_i by the n = 150 gives us the mean, which is equal to 1,209 ÷ 150 = **8.06**

Why do these two indices of central tendency produce such discrepant estimates of the location of the center of the distribution (Mdn = 1.0; $\bar{Y}$ = 8.06)? The answer is that the impact of an extreme score of 100 is minor in calculating the median; in effect, one score of 100 simply cancels out one score of 0 as you work your way to the middle point of the distribution. In calculating a mean, the actual value of 100 comes into play, and the 10 scores of 100 contribute a sum of 1,000 in determining the mean (the sum of the remaining 140 scores is only 209, and re-calculating the mean omitting these 10 extreme scores of 100 equals 209 ÷ 140 = 1.49). If one is inclined to comment on dating behavior of junior high school boys from data like these, then the choice of a representative index could be pretty important. Perhaps a second significant point comes across with this contrived illustration. Statistics are nothing more than a tool. They can do little to salvage a poor experiment, and they can mislead when applied inappropriately.

III. Summary

In this chapter we began the task of describing results of an experiment. Presenting data in a frequency distribution creates a more orderly presentation, and at the same time conveys essential information in a smaller space and more readable format compared to reporting every individual score. Organizing data into categories or groups of similar scores reduces a complex data array even further; as long as a reasonable balance is maintained in category width and number of categories, the gist of the set of scores can be presented without too great a cost in loss of detail. With regard to "reasonable balance" in number of categories, just keep in mind that using categories only one unit in width (e.g., 1–2) will result in a grouped frequency distribution that is no different from a frequency distribution, and using a category that is so large that it includes all the scores in a distribution would not be particularly helpful.

While condensing data by grouping operations for table or graphical displays is useful for summarizing experimental results, reducing a large number set to a single meaningful score would be the ultimate in simplifying. Two such indices derived from data are widely used. One of these indices provides information about central tendency; that is, the location of a center for the array of scores. Another useful single number index provides information about dispersion or spread among the scores. In this chapter we addressed the first of these two important single number indices, identifying two different approaches for locating a center. The median is the middle score in a set of scores when scores are arranged in order from lowest to highest (or highest to lowest if you prefer). Determining a median mathematically is not complex. Once you determine where the middle is for a set of n scores, you merely have to count scores to get to the middle one (if n is an even number, then there is the additional requirement of computing a point midway between the scores on either side of this middle point).

The mean is the arithmetic average, and represents a center based on magnitude of individual scores. It is arrived at by adding all the scores and dividing by n: $\bar{Y} = \Sigma Y_i \div n$. The actual value of a score is relevant for determining the mean, whereas the actual value of a score is relevant for determining the median only in a very limited way: for example, in determining the median for the fictitious data from junior high-school boys, the highest score would have the same impact whether it was 100, 1,000, or 100,000.

2.1: Model Problem. To illustrate the primary tasks introduced in this chapter, we will do three things for the set of numbers below:

Number Set:

2	12	7	13	5	5	4	7	3	10
1	1	13	5	4	3	11	3	7	8
9	8	2	14	7	5	5	11	5	8
0	8	5	8	10	14	5	5	2	7

A. The first problem involves rearranging the 40 numbers in the table into a grouped frequency distribution. In order to do this we have to decide on the approximate number of intervals (categories) we want and the width of the intervals. Of course, we have to capture all of the scores in the set, so it is important to note that the lowest score is 0 and the highest score is 14. There are 15 whole numbers from 0 to 14, so an interval width of 2 would require about 7 categories and an interval width of 3 would require about 5 categories (the set of whole numbers separating the lowest and highest scores divided by interval width ≈ number of categories). For this illustration, we will use an interval width of 3 (you may want to repeat this exercise using an interval width of 2 for comparison purposes). All that remains is determining a starting point, building the categories, and tallying scores in each category (to avoid negative numbers, we will start with a 0–2 category). You should note that the sum of the numbers in the "Frequency" column is 40, so it essential to count all 40 scores and to make sure that no score is counted more than once. A grouped frequency distribution for these data appears in Table 2.6.

Table 2.6: Grouped Frequency Distribution of a Set of 40 Scores

Categories	Frequency Tallies	Frequency
0–2	||||||	6
3–5	||||||||||||||	14
6–8	||||||||||	10
9–11	|||||	5
12–14	|||||	5

B. What is the median for this set of scores?

Step 1. Organize the data arranging the scores from lowest to highest

Number Set:

0	2	3	5	5	**7**	7	8	10	13
1	2	4	5	5	7	8	8	11	13
1	3	4	5	5	7	8	9	11	14
2	3	5	5	**5**	7	8	10	12	14

Step 2. Determine the position of the middle score: $(n + 1) \div 2 = (40 + 1) \div 2 = 20.5$
Step 3. Find the 20th and 21st highest scores in the set (scores of 5 and 7 in bold and underlined)
Step 4. For an even number of scores, the median is the point midway between the two scores on each side of the middle position (position 20.5 in this case). **Mdn = $(5 + 7) \div 2 = 6$**

C. Calculate the mean: $\bar{Y} = \Sigma Y_i \div n$

Step 1. Sum the 40 scores: $2 + 1 + 9 + \ldots + 7 =$ **262**
Step 2. Divide this sum by $n = 40$: **$\bar{Y} = 262 \div 40 = 6.55$**

2.2: Practice Problem 1. For this problem you are on your own. There are 50 scores in Table 2.7 ($n = 50$).

Table 2.7: Data Set of 50 Scores

44	37	33	50	37	24	32	15	33	42
38	33	25	38	14	15	13	33	38	44
17	39	41	14	41	36	16	18	20	24
19	38	39	31	33	46	26	43	45	49
10	49	12	44	20	33	37	10	36	21

A. Display these data in a grouped frequency distribution using 9–12 as the lowest or starting interval (remember to keep interval widths the same throughout).

B. What does Mdn =?

C. What does $\bar{Y}$ =?

2.3: Practice Problem 2. How do the means and medians compare for the fictitious midterm and final exams for an imaginary statistics class (perhaps the most popular type of statistics class)?

Midterm Exam Scores	Final Exam Scores
75	80
49	71
65	90
90	99
82	76
25	68
95	95
98	88
33	65
65	71
40	70
85	80
82	90
77	61
59	84
75	80
66	60
91	70
86	72
73	
91	
100	

2.4: Practice Problem 3. For the following 16 scores, calculate both the mean and the median. Normally these two measures of central tendency are relatively close in value; however, this does not appear to be the case for this number set. Discuss why this is so, and indicate which measure you think provides a more "representative" index for a center around which scores tend to cluster.

The 16 Scores: 4, 6, 16, 5, 8, 3, 2, 2, 75, 10, 15, 12, 4, 90, 3, 16

2.5: Practice Problem 4. The data in Table 2.8 are presented in a "Grouped Frequency Distribution." Since information about each individual score is not presented, you cannot calculate the exact mean for this set of scores. However, you can still come up with a pretty good estimate of what the mean should be. Even though procedures for calculating a mean from a grouped frequency distribution were not described in the text, if you think about this you ought to be able to provide an estimate of $\bar{Y}$ for these data. Describe how you arrived at your estimate of $\bar{Y}$.

Table 2.8: Grouped Frequency Distribution

Scores	Frequency
86–94	2
77–85	5
68–76	9
59–67	14
50–58	12
41–49	16
32–40	10
23–31	4
14–22	5
5–13	3

Chapter 3

Descriptive Statistics II

A small error at the beginning can lead to great ones at the end.
—Aquinas

I. Measuring Dispersion: The Degree to Which Scores in a Data Set Are Spread Out

In reducing data to single numerical indices, a goal is to capture two important features of a data set. In the last chapter the first of these was introduced, *viz.*, a single number that provided information about the center for a given distribution of scores. This central tendency index was an actual **reference point** within the set of scores: a representative center point. The second feature of a data set we wish to describe is the dispersion or variability represented in the set of scores. It is most likely that all scores are not identical (constant), and it is of particular interest to assess the degree to which scores differ from one another or are spread out (measuring distances that separate scores from a given reference point). Thus, the concern is with developing an index that provides information about distances (common terms relevant to this concept are *spread, dispersion, deviation*, and *variability*; with specific indices identified by *range, variance, standard deviation*, and *standard error*). The concept of a distance measure is a little more abstract than we encountered with the concept of central tendency. The numerical index reported for measures of dispersion may be unlike any of the scores in the data set. For example, the data may consist of a set of scores on a standard intelligence test from 200 college students. Most all of the scores will be 3-digit numbers; however, a dispersion index such as a standard deviation may be a number quite different from any score in the data set (e.g., a standard deviation of 10).

To convert individual scores to distance scores, a question that must be addressed involves a reference point; that is, in order to measure distance, we must be able to answer the question "Distance from what?" Obviously, an answer to the question "How far away is San Francisco?" differs depending on whether the reference point is New York, Chicago, or Los Angeles. The two indices of dispersion we will be most concerned with (the *variance* and *standard deviation*) use the mean of a set of scores as the reference point, as a mean represents a central location in a distribution of scores, and because it has useful properties for comparing data sets. A third measure that should at least be identified is the *range*; a measure of distance between the highest and lowest scores in a data set, computed by subtracting the lowest score from the highest score. It has limited usefulness as it is based on information about only two scores from an entire set of scores.

III. The Sample Variance (s^2)

a. Logical Development and Definitional Formula

An objective for an index of central tendency was a number that represented a center point in a distribution of scores. We have a similar objective for an index of dispersion regarding a representative index; however, in this case the objective is a number that is representative of the distances that separate individual scores from a specific reference point. The best reference point for measuring variance is the mean. This logic leads to a reasonably direct route: What better way to get a representative index of variability than to take an "average" of distances of individual scores from the mean—that is, $[\Sigma(Y_i - \bar{Y})] \div n$.

Unfortunately, an average of the distances-from-the-mean score is not useful because it will always be 0. This should be apparent if you think about it for a minute, as we showed earlier when discussing the balancing property of the mean that the sum of the scores below the mean (negative signed scores) was exactly the same as the sum of the scores above the mean (positive signed scores).

The proof that an average deviation index is always 0 involves just a smidgen of algebra:

a. The formula given above for calculating an average of distance-from-the-mean scores is $(\Sigma[Y_i - \bar{Y}]) \div n$. This formula may be rewritten as $(\Sigma Y_i \div n) - (\Sigma\bar{Y} \div n)$.
b. The first term in the expression, $\Sigma Y_i \div n$, is the formula for the mean, $\bar{Y}$.
c. The second term, $\Sigma\bar{Y} \div n$, can be rewritten as $(n)\bar{Y} \div n$ since $\bar{Y}$ is a constant and the sum of a constant is equal to n times that constant; and $(n)\bar{Y} \div n = \bar{Y}$, as the n in the numerator and n in the denominator cancel out.

d. Given that $\Sigma Y_i \div n = \bar{Y}$ (step 2) and $\Sigma\bar{Y} \div n = (n)\bar{Y} \div n = \bar{Y}$ (step 3), therefore, $(\Sigma[Y_i - \bar{Y}]) \div n = \bar{Y} - \bar{Y} = 0$

How can we retain the general concept of a variance index that reflects a representative distance measure from the mean and still have a useful index of dispersion (i.e., one that will not always be 0)? While not the only solution to this problem, the one that is used involves a squaring operation, since a negative number squared is a positive number. Recall that the problem with summing distance scores was that negative and positive difference scores were equal. Thus, squaring each deviation score (individual score minus the mean) before taking an average gets rid of the negative numbers, as a negative number squared equals a positive number. Basically, the variance index is computed by taking an average of the **squared** distance-from-the-mean scores: Variance $(s^2) = (\Sigma[Y_i - \bar{Y}]^2) \div n$

In Appendix I. A of Chapter 1, two formulas for the variance were given. The first of these is referred to as a definitional formula involving actual distance scores. It bears a close resemblance to the formula above; however, on closer inspection you will notice a minor discrepancy in the denominator. The formula for the variance we will use has (n – 1) in the denominator instead of n. The reason for this is that in the long run, the average squared deviation score taken from a sample (a subset of a population) slightly underestimates the variance in the population from which the sample was drawn; dividing by n – 1 instead of n corrects this bias. Mercifully, the algebraic proof of this claim is not included in this text. The second formula below is referred to as the computational formula, so named because with a hand calculator, the computational task is faster and easier with this formula. The n – 1 bias correction appears only in the final denominator of the computational formula. When calculating a variance, either of the two formulas below is acceptable:

Definitional Formula: $s^2 = (\Sigma[Y_i - \bar{Y}]^2) \div (n - 1)$

Computational Formula: $s^2 = ([\Sigma Y_i^2] - \{[\Sigma Y_i]^2/n\}) \div (n - 1)$

b. An Application

While you cannot be expected to do an "eyeball" test to determine a value for the variance, an understanding of the concept of the variance should enable you to do an eyeball comparison between two small data sets to arrive at a relative conclusion regarding which data set reflects a greater degree of variability. Nothing sophisticated is meant here by the phrase "eyeball test"; it simply means a relative judgment about variability is arrived at by visual inspection. There are two sets of scores in Table 3.1 that follows. Each set has six scores (n = 6); and Set A and Set B have identical means of 9.0. Deviation scores arrived at by subtracting the mean from each individual score $(Y_i - \bar{Y})$ appear next to each set, since deviation scores are necessary for calculating the variance using the definitional formula. Based on your eyeball test, do you think Set A or Set B will have the larger variance?

Table 3.1: Two Fictitious Data Sets to Illustrate Variance Calculations

Set A		Set B	
Scores	$(Y_i - \bar{Y})$	Scores	$(Y_i - \bar{Y})$
4	-5	1	-8
7	-2	3	-6
8	-1	6	-3
10	1	7	-2
11	2	15	6
14	5	22	13
ΣY_i **54**	$\Sigma(Y_i - \bar{Y})$ **0**	ΣY_i **54**	$\Sigma(Y_i - \bar{Y})$ **0**

For those readers who are skeptics, please note that the sum of the $Y_i - \bar{Y}$ columns (the average deviation when divided by n) is 0 for both Set A and Set B. For both sets, we will calculate the variance twice: once using the definitional formula and once using the computational formula in order to confirm that our eyeball judgment was correct (i.e., that we can see from inspection that the scores in Set B are more "spread out" than the scores in Set A).

Definitional Formula	Computational Formula
$s^2 = (\Sigma[Y_i - \bar{Y}]^2) \div (n - 1)$	$s^2 = ([\Sigma Y_i^2] - [\{\Sigma Y_i\}^2 \div n]) \div (n - 1)$
Set A: $n - 1 = 6 - 1 = 5$ $\Sigma(Y_i - \bar{Y})^2 = (-5)^2+(-2)^2+(-1)^2+(1)^2+ (2)^2+(5)^2 = 60$ $\mathbf{s_A^2 = 60 \div 5 = 12.0}$	Set A: $n - 1 = 6 - 1 = 5$ $\Sigma Y_i^2 = 4^2 + 7^2 + \ldots + 14^2 = 546$ $(\Sigma Y_i)^2 \div n = 54^2 \div 6 = 2{,}916 \div 6 = 486$ $\mathbf{s_A^2 = (546 - 486) \div 5 = 60 \div 5 = 12.0}$
Set B: $n - 1 = 6 - 1 = 5$ $\Sigma(Y_i - \bar{Y})^2 = (-8)^2+(-6)^2+(-3)^2+(-2)^2+(6)^2+(13)^2 = 318$ $\mathbf{s_B^2 = 318 \div 5 = 63.6}$	Set B: $n - 1 = 6 - 1 = 5$ $\Sigma Y_i^2 = 1^2 + 3^2 + \ldots + 22^2 = 804$ $(\Sigma Y_i)^2 \div n = 54^2 \div 6 = 2{,}916 \div 6 = 486$ $\mathbf{s_B^2 = (804 - 486) \div 5 = 63.6}$

c. Computational Components in Calculating s^2

The computational formula for the variance, $\mathbf{s^2 = ([\Sigma Y_i^2] - \{[\Sigma Y_i]^2/n\}) \div (n - 1)}$ has three component parts: two in the numerator and one in the denominator. For the numerator we need to calculate ΣY_i^2 and $(\Sigma Y_i)^2/n$; for the final denominator, we need to determine the value of $(n - 1)$. The numerator component is commonly referred to as the *sum of squares* (SS), and the final denominator component is referred to as the *degrees of freedom* (df). You should also note that the SS has two separate computational parts, ΣY_i^2 and $(\Sigma Y_i)^2 \div n$. The sum of squares and degrees of freedom are part of the language of the Analysis of Variance (the common acronym is ANOVA), a major statistical procedure in the hypothesis-testing tradition that we will consider in detail when we move from descriptive statistics to inferential statistics. It is necessary in doing ANOVAs to calculate two or more variances, and when doing these on a hand-held calculator, it is helpful to carry out the computations in an organized, step-by-step manner. The computational steps are illustrated here for the Set B sample data:

Step 1: $\Sigma Y_i^2 = 1^2 + 3^2 + \ldots + 22^2 = 804$

Step 2: $(\Sigma Y_i)^2 \div n = 54^2 \div 6 = 2{,}916 \div 6 = 486$

Step 3: The Sum of Squares: $SS_B = \Sigma Y_i^2 - ([\Sigma Y_i]^2 \div n) = 804 - 486 = 318$

Step 4: The Degrees of Freedom: $df_B = n - 1 = 5$

Step 5: $s^2 = SS_B \div df_B = 318 \div 5 = 63.6$ (**NOTE**: in the analysis of variance, variance estimates are referred to as *Mean Squares* [coded as MS]; thus, $s^2 = MS = SS \div df$)

III. The Sample Standard Deviation (s)

While the variance is often the dispersion measure of choice for inferential test statistics (e.g., the ANOVA), the standard deviation (or a related measure called the standard error of the mean) is preferred for **describing** the degree of dispersion in a data set. Recall that the central ingredients in our measure of variability are deviation scores $[\Sigma(Y_i - \bar{Y})]$. We saw earlier that it was necessary to square each deviation score to avoid the mathematical problem that $\Sigma(Y_i - \bar{Y}) = 0$ for all data sets. The squaring operation led to the development of the variance as a measure of dispersion. Undoing this operation after the variance computations have been completed by taking the positive square root of the variance provides the only additional work required for calculating a standard deviation. Importantly, the standard deviation is in the same units of measurement as the original observations recorded in the experiment, and it provides a numerical index of the typical or representative (e.g., standard) deviation from the mean.

Means and standard deviations (or *standard errors*—a close cousin of the standard deviation) are normally reported together in research publications. So does this mean we have to gear up for understanding a new concept and get ready for a new set of computational procedures in order to crunch numbers down to a new standard deviation index that measures dispersion? If you have read this far, you deserve some good news. In this case, the good news is that if you understand the variance, you will have no difficulty understanding the standard deviation; if you can calculate a variance, you will have no difficulty calculating a standard deviation.

a. Conceptual and Computational Notes

On the computational side, all you have to do to calculate a standard deviation is to calculate a variance and then take the square root of that number. Lower case "s" is used as a symbol for the standard deviation for a sample of scores, so if you use s^2 or MS as a symbol for the variance, then the standard deviation, $s = \sqrt{s^2} = \sqrt{MS}$.

Variability was measured with the mean as a reference point (distances of individual scores from the group mean) and the desired information was a representative distance score. To achieve this goal, each distance score (an individual score minus the mean) had to be squared, with a "sort of" average arrived at by summing the squared deviation scores and dividing the sum by n – 1. So if we "unsquare" this index (more conventionally referred to as taking the square root), we get back to the same units of measurement we started with. If each individual score is measured in seconds or errors, the standard deviation index provides a distance measure in seconds or errors, whereas a variance index is necessarily a representation in squared units. Certainly, it makes more sense to **describe** variability in a data set with numbers that represent the same unit of measurement as the recorded observations. That is, if performance in a visual tracking task is recorded in terms of the number of seconds a participant is able to keep a pointer on a moving target, then the descriptive advantage of reporting the standard deviation is that it is in the same units of measurement (seconds) as the original responses.

b. Standard Error of the Mean

At the beginning of this section another measure of dispersion was mentioned: the standard error (or standard error of the mean). The standard error (coded as s_E) is related to the standard deviation, and since it is often reported in research publications instead of the standard deviation, a brief descriptive comment is in order. Computationally, the standard error poses no problem. If one can calculate a variance, then it is a simple matter to calculate a standard deviation (an added step of taking the square root of the variance), and if one can calculate a standard deviation, then it is a simple matter to compute an estimate of the standard error of the mean (an added step of dividing the standard deviation by the square root of the sample size: $s_E = s \div \sqrt{n}$).

Moving conceptually from the variance to the standard deviation does not seem to present a particularly troublesome stumbling block. However, moving conceptually from the standard deviation to the standard error may require a bit more of a strain on abstract thinking. First, think of the standard error as simply an estimate of a standard deviation for a different type of data set that would be too tedious and too boring to collect. Second, the imagined data set would necessarily involve collecting a very large number of samples for each condition in an experiment, rather than collecting just one set of scores for each condition. Expanding on this imaginary part, if you had repeated an experiment 50 times, you would have been able to compute 50 means: one for each sample. Now we can talk about the different type of data set, one that had 50 scores; however, each of these 50 scores would actually be a sample mean rather than a score from an individual subject. We are getting close as there is one last imaginary operation to perform on these imaginary data of 50 sample means. The last step involves an imaginary calculation of a standard deviation on the distribution of sample means. Most students find imaginary calculations easier to do than actual calculations, so hopefully this was the easy part. The standard error calculation takes data from a single sample and creates a new index that provides an estimate of what the standard deviation would be for a distribution of sample means (called a *sampling distribution*). Thus, like the variance and standard deviation, the standard error of the mean provides an index of dispersion; however, in this case it provides an estimate of variability among hypothetical sample means.

The numbers in Table 3.2 represent 10 different random samples (Samples I–X) of the nine digits 0–9. This number set is designed to enable us to move from the imaginary to the real. This is obviously a limited set of digits in order to keep the tasks manageable. A mean, a standard deviation, and a standard error are reported for each sample. From these data, we will take the 10 sample means and compute a standard deviation (remember, we call a standard deviation for a sampling distribution of group means a standard error). We can then compare the 10 individual sample estimates of the standard error ($s_i \div \sqrt{n}$) with one actually computed from the 10 sample means. Each of the 10 samples provides an estimate of s_E; thus, each s_E computed for each of the 10 samples should be close in value, but, of course, not identical to s_E calculated by taking the standard deviation of the population of means from the 10 groups. A standard error of the mean is estimated for each of the 10 samples that appear in Table 3.2, and the s_E values for each sample range from .86 to 1.42, with an average of 1.12. The direct computation of the standard error of the mean (s_E: the standard deviation of the 10 sample means) appears below the table, and for this data set it is close to the top of the 10 sample estimates ($s_E = 1.39$).

Table 3.2: Fictitious Data and Descriptive Statistics for 10 Samples (n = 9)

Samples	I	II	III	IV	V	VI	VII	VIII	IX	X
	12	11	7	6	15	4	9	4	3	2
	4	9	17	13	5	7	6	12	8	2
	9	5	6	8	15	7	9	4	3	6
	15	3	12	7	9	3	5	11	10	5
	5	1	10	7	11	5	10	6	8	9
	1	8	4	7	14	11	12	12	6	8
	7	6	15	4	7	6	12	7	6	10
	7	10	9	9	6	12	7	15	4	4
	10	8	8	10	6	5	5	9	9	7
$\bar{Y}$	7.78	6.78	9.78	7.89	9.78	6.67	8.33	8.89	6.33	5.89
s	4.27	3.31	4.24	2.57	4.09	3.04	2.74	3.89	2.60	2.89
s_E	**1.42**	**1.10**	**1.41**	**.86**	**1.36**	**1.01**	**.91**	**1.30**	**.87**	**.96**

Using the **10 mean scores**, computational steps for the standard deviation are:

1. $\Sigma Y_i^2 = 7.78^2 + 6.78^2 + \ldots + 5.89^2 = 627.72$

2. $(\Sigma Y_i)^2 \div n = 78.12^2 \div 10 = 6{,}102.73 \div 10 = 610.27$

3. $SS = (\Sigma Y_i^2) - ([\Sigma Y_i]^2 \div n) = 627.72 - 610.27 = 17.45$

4. Although n = 9 for each of the 10 samples, there are 10 sample means, thus $n_M = 10$ for determining the variance and standard deviation for this set of 10 mean scores.

5. s_E^2 (the variance calculated from the 10 sample means) $= SS \div (n - 1) = 17.45 \div 9 = 1.94$

6. The standard error is the standard deviation for the distribution of group means; thus, $s_E = \sqrt{s_E^2} = \sqrt{1.94} = \mathbf{1.39}$

IV. Measuring Different Variances: A Preview for the Analysis of Variance

Can more than one variance (with a mean as the reference point) be calculated for a given set of numbers? At the risk of appearing wishy-washy, we will go out on a limb and answer this question with "yes and no." Following the computational instruction given by the formula for the variance, and applying it to all the scores in the data set, there will be a single correct value for s^2. However, if the data are organized into different subsets (groups), then of course, one could do variance calculations restricted to the different groups; that is, s^2 could be computed for all of the data taken collectively and a different s^2 could be calculated for each subgroup taken individually. Further complicating matters is the fact that it is possible to use a mean score from each group and compute an s^2 treating each group mean as a separate score (based on deviations of group means from an overall mean computed from all scores). To illustrate, we will use a fictitious data set pretending that the numbers represent scores from 20 3rd grade children on a 10-word vocabulary test. The number correct for each student is presented in Table 3.3.

Table 3.3: Number Correct on 10-Word Vocabulary Test

Students	Correct	Students	Correct
1	10	11	7
2	7	12	3
3	10	13	5
4	10	14	8
5	8	15	5
6	9	16	4
7	8	17	6

8	10	18	9
9	6	19	2
10	9	20	6

We could proceed to calculate a variance using either the definitional formula or the computational formula. Using the computational formula for the variance we only need to know each individual Y_i score (these are given in Table 3.3) and the sum of all 20 scores ($\Sigma Y_i = 142$). The computational formula is $s^2 = ([\Sigma Y_i^2] - \{[\Sigma Y_i]^2/n\}) \div (n - 1)$; thus, we have:

$$s^2 = ([\Sigma 10^2 + 7^2 + \ldots + 6^2] - \{[142]^2/20\}) \div (20 - 1)$$

$$s^2 = (1{,}120 - 1{,}008.2) \div 19 = \mathbf{5.88}$$

Now suppose these data came from an experiment designed to evaluate whether or not it was helpful for students to have a study guide available before the test. Assume that such a study guide was available for students 1–10, but not for students 11–20. The data in Table 3.3 are re-written in Table 3.4 taking this group distinction into account.

Table 3.4: Number Correct on Vocabulary Test for Students With and Without a Study Guide

Study Guide Group		No Study Guide Group	
10	9	7	4
7	8	3	6
10	10	5	9
10	6	8	2
8	9	5	6

Given the meaningful organization of the data in Table 3.4 we can ask several questions about variability (e.g., why isn't every score the same?). Earlier we assessed variability with an index that was relevant to variability among all 20 scores in reference to a mean based on all 20 scores (a mean of 7.10; a variance of 5.88 for these 20 scores). Even though 10 students were given an identical study guide prior to the test, there was variability among the test scores those 10 students received. The mean for this group is **8.70** ($\Sigma Y_i \div n = 87 \div 10$), and the variance for the 10 scores around this "Study Guide" mean is

$$s^2 = ([\Sigma 10^2 + 7^2 + \ldots + 9^2] - \{[\Sigma 87]^2/10\}) \div (10 - 1) = (775 - 756.9) \div 9 = \mathbf{2.01}$$

Similarly, the mean for the "No Study Guide" group is **5.50**, and the variance around the "No Study Guide" mean is

$$s^2 = ([\Sigma 7^2 + 3^2 + \ldots + 6^2] - \{[\Sigma 55]^2/10\}) \div (10 - 1) = (345 - 302.5) \div 9 = \mathbf{4.72}$$

We are not quite through with the idea that multiple variance estimates can be calculated for this data set. We calculated an overall mean of 7.10 based on all 20 scores, as well as two other means: 8.70 for the Study Guide group and 5.50 for the No Study Guide group. Obviously, the group means differ from one another and from the overall mean based on all 20 scores. Even though there are only two group means, a variance capturing the deviations of these two group means can be calculated (with the overall mean based on all 20 scores serving as the reference point).

An F statistic for the analysis of variance test procedure that we will soon focus on is nothing more than a ratio of two variance estimates; however, for now we will merely content ourselves with the fact that different variances can be calculated for a given data set in relation to different conditions in an experiment. For example, in this problem, one variance can be calculated based on an average of the different vocabulary test scores among the students in the study-guide group and among the students in the no-study-guide group in reference to their respective group means. In addition, a second variance can be calculated based on different mean vocabulary test scores for the study guide group and the no-study guide group in reference to the mean based on all scores in the data set regardless of group designation. We will delay introducing analysis of variance procedures until after we have considered the logic of the analysis of variance and inferential statistics, and after a brief consideration illustrating the use of probability concepts in the context of theoretical comparison (*reference*) distributions.

V. Summary

Developing an index designed to reflect dispersion or distances among a set of numbers is probably not as straightforward as developing an index designed to locate a number that represents a center or middle point. A compelling approach would be to create a representative distance index similar to what was done for the first descriptive measure that was concerned with a representative center. However, a distance index requires a specification of a reference point; after all, to consider distances, we need to have a common point from which to measure distance. For both intuitive and applied reasons, the mean is the reference point of choice.

The simplest approach might be to convert each score in a data set to a deviation-from-the-mean score by simply subtracting the mean from each score and then average these deviation scores. Unfortunately, there is a problem with this direct approach, as it was shown that an average deviation score will be 0 for any set of numbers. A way around this problem, while still preserving the concepts of deviation scores and averaging, involved squaring each deviation score before taking the average. This operation got us very close to the index of dispersion known as the sample variance (s^2). Except for a minor adjustment in the final denominator, the definitional formula for s^2 instructs us to take an "average" of the squared deviation scores:

$$s^2 = \Sigma(Y_i - \bar{Y})^2 \div (n - 1)$$

In order for this formula for the sample variance s^2 to serve as an estimate of a population parameter σ^2, the averaging concept is compromised a little as the final division is by n – 1 instead of by n. Without justification, it was just noted that this slight reduction in the denominator represented an adjustment to correct for a bias because in the long run, the sample variance with "n" in the denominator slightly underestimates the variance in the population from which the sample was drawn. When computing variances on a hand-held calculator, an equivalent version of this formula is usually more expedient:

$$s^2 = ([\Sigma Y_i^2] - \{[\Sigma Y_i]^2/n\}) \div (n - 1)$$

A second index used more commonly than the variance as an index of dispersion for **descriptive purposes** is the standard deviation (s). It is computed quite simply once the variance has been calculated as $s = \sqrt{s^2}$. Thus, the variance and standard deviation share the same logical development and justification. For summarizing the results of an experiment, one feature that contributes to the standard deviation being preferred to the variance as a descriptive index reported in combination with a mean score is that the standard deviation is in the same units of measurement (not in squared units) as the mean and the individual scores.

The chapter closed with a preview of the type of variance computations that are relevant for the analysis of variance—an inferential statistical procedure commonly used for drawing inferences from experimental data. The main point emphasized here was that while a variance may be computed based on all scores in a distribution, different meaningful organizations of the data justify computations of additional variances (e.g., limited to specific groups or subsets of data, and we will see later, what might be considered as a higher level variance based on distances of group means from a grand mean based on all scores).

3.1: Model Problem. Calculate (a) the variance and (b) the standard deviation for the number set given in the Model Problem for Chapter 2 (pages 29–30).

a. The computational formula for the variance is $s^2 = ([\Sigma Y_i^2] - \{[\Sigma Y_i]^2/n\}) \div (n - 1)$. The necessary calculations may be completed in the following 5 steps:
Step 1. The 1st term in the numerator, $\Sigma Y_i^2 = 0^2 + 1^2 + \ldots + 14^2 = \mathbf{2{,}260}$

Step 2. The 2nd term in the numerator, $(\Sigma Y_i)^2 \div n = (0 + 1 + \ldots + 14)^2 \div 40 = 262^2 \div 40$
$(\Sigma Y_i)^2 \div n = 68{,}644 \div 40 = \mathbf{1{,}716.10}$

Step 3. The numerator is referred to as the sum of squares (SS), and SS = 2,260 – 1,716.10 = **543.90**

Step 4. The denominator is the degrees of freedom (n – 1), and df = 40 – 1 = **39**

Step 5. Substituting the computed values in the formula for the variance, we have

$$\mathbf{s^2 = 543.90 \div 39 = 13.95}$$

b. The standard deviation is the square root of the variance: $s = \sqrt{s^2}$

$$\mathbf{s = \sqrt{13.95} = 3.73}$$

3.2: Practice Problem 1. Calculate (a) the variance and (b) the standard deviation for the data in Practice Problem 1, Chapter 2 (page 30).

3.3: Practice Problem 2. Assume the data from Model Problem 2.1 (n = 40) represent scores obtained from two groups of 20 subjects. The scores organized into two groups are presented in the Table that follows (as you can see, we have merely taken the 1st 20 scores from the number set in Problem 2.1 and labeled them as Group I scores, with the remaining 20 scores labeled as Group II scores).

Data from Practice Problem 2.1 Organized in a Two-Group Format

Group I Scores		Group II Scores	
2	2	5	11
1	5	3	5
9	13	5	3
0	5	14	7
12	14	4	5
1	8	11	2

8	5	5	10
8	4	5	8
7	7	7	8
13	10	3	7

A. For Group I, what does s^2 =?

B. For Group II, what does s^2 =?

3.4: Practice Problem 3. Calculate s^2, s, and s_E for the following set of numbers: 12, 4, 7, 7, 1, 0, 6, 11, 14, 3, 5, 12, 2, 8, 9, and 7.

3.5: Practice Problem 4. Forty Y_i scores are given in the table below.

3	9	5	7	12	17	9	21
7	5	5	7	10	15	10	17
5	7	6	8	16	16	11	15
6	8	6	9	15	19	8	17
4	3	9	5	11	14	6	20

A. What does ΣY_i^2 = ?

B. What does $(\Sigma Y_i)^2 \div n$ = ?

C. What does $(\Sigma Y_i^2) - [(\Sigma Y_i)^2 \div n]$ = ? [This solution is the numerator component of a variance, and is referred to as the "sum of squares" (abbreviated SS).]

D. What does $([\Sigma Y_i^2] - \{[\Sigma Y_i]^2/n\}) \div (n - 1)$ = ? [This solution is the variance (s^2), and with the statistical test procedure known as the analysis of variance, it is referred to as a "mean square" (abbreviated MS).]

3.6: Practice Problem 5. Determine the values of s^2, s, and s_E for the 24 numbers below:

9, 4, 8, 4, 6, 7, 8, 12, 6, 7, 6, 9, 15, 6, 5, 7, 5, 7, 4, 5, 10, 9, 13, 8

3.7: Practice Problem 6. Below are three columns of numbers. Given this organization, you can calculate five different meaningful variances (actually, a sixth variance can be calculated from this data set, however we won't get to that until a later chapter). Coding the columns as C_1, C_2, and C_3, respectively, and coding the total set of 18 scores as T, you can determine s_{C1}^2, s_{C2}^2, s_{C3}^2, and s_T^2. In addition, you can calculate an average variance of the three column variances $[(s_{C1}^2 + s_{C2}^2 + s_{C3}^2) \div 3]$. The average of the variance within each group will play a prominent role in statistical analyses that will be introduced later.

It is coded as $\mathbf{s_{WG}}^{\mathbf{2}}$ and referred to as **within-group variance**. For each of these five variances, replace the question marks following the equal signs with the corresponding s^2 values.

Column 1	Column 2	Column 3
4	7	6
9	3	10
12	8	7
7	5	12
6	4	15
8	2	16

$s_{C1}^2 = ?$

$s_{C2}^2 = ?$

$s_{C3}^2 = ?$

$s_T^2 = ?$

$s_{WG}^2 = ?$

Chapter 4

Testing an Experimental Hypothesis

The theory of probabilities is at bottom nothing but common sense reduced to calculus.
—Laplace

I. A Review of the Logical Objective

In any experimental setting there are generally factors or variables known to influence performance, and there are most certainly unknown factors that influence performance. Ideally, we have measuring techniques sensitive enough to detect differences in responses; the experimental design ensures that both known and unknown factors that are not of interest are adequately controlled. For example, if we know or suspect that wide variations in age are relevant for performance on a given task, and if we are not interested in the age variable, then age can be controlled by equating or balancing it within certain reasonable limits for each group in the experiment (e.g., using subjects who are over 18 and under 25 years old). Remember, except for chance or random error, the goal is to start with experimental groups or conditions that are equivalent on all factors (known and unknown), so that with the exception of chance or random error, the independent variable is the only non-equivalence factor that could potentially influence the observed responses.

II. Other Variables

a. Controlling Known and Unknown Variables

Factors that are not controlled loom as potential confounding variables. In many cases, attempting to equate groups on all factors known or suspected to influence performance can be a rather daunting task. And even if this could be accomplished, we would still need to deal with the class of unknown factors that influence performance. Fortunately, the control procedure of *randomization* comes to the rescue, given the understanding that "in the long run" (very large numbers of observations) distributing other factors (known and unknown) randomly will result in groups that are initially equivalent.

While randomization gets us out of one mess, it may create a problem of its own. Equivalence through randomization only really works in the long run, and, unfortunately, experimental research involves relatively small samples of cases. As a result, random assignments result in chance variations (referred to as *random error*), so it appears that the well-intentioned experiments we strive to achieve are at best "confounded" by random error. Thus, differences observed among experimental groups may be influenced by both the independent variable(s) of interest **and** by random error. The experimenter is faced with a perplexing problem: Did observed differences in scores among groups occur because of the different conditions defined by the independent variable, or were they the result of chance? Before discussing how this problem gets resolved, we will consider an illustration of random assignment procedures and chance variations.

b. An Illustration of Random Assignments

Table 4.1 that follows lists rows and columns of random digits (0–9). This exercise involves creating lists of random numbers to use in selecting scores from three sets of 10 numbers and the collective set of all 30 numbers. Although random numbers may be generated by computer and on many hand calculators, we will use an old-fashioned procedure for purposes of illustration. The first set of 10 numbers (Set A) in **Table 4.2** merely lists the numbers 1 through 10. The 2nd set (Set B) adds 2 to each number in Set A. The 3rd set (Set C) multiplies each number in Set A by 2. Thus, we have known variables differentiating these three sets of numbers, and if numbers are drawn from the three sets at random, then there is also the possibility for chance influences to be operating. If we were to draw two random samples from Set A, the two sample means would be identical in the long run, but would likely differ with a limited sample size (e.g., 20 numbers drawn at random for each sample). Random error would also account for differences if we drew the 1st sample from Set A and the 2nd sample from Set B or from Set C. However, in addition to random error, there are known factors operating that in the long run would produce higher means for Set B compared to Set A, and still higher means for Set C. The following exercise is designed to illustrate these "random" and "known" factor differences, and you may repeat it drawing different

Table 4.1: An Excerpt from a Table of Random Digits

6366555741	9073282004	2853463709	6742890906	1483819643	1908913714
7309785591	00***14***388085	2948745221	2796239277	4604928036	7507771341
3589434416	0466380632	2724138840	9944140040	7489218745	9197401539
2021347420	4097621897	0192427444	987010481**2**	1615065914	0212054866
0581248230	8068373244	5794724826	7118813010	8847023447	4970080660

random numbers to see if your approximation works better or worse than the random draws used for this illustration.

The numbers used for Set A are simply the numbers 1–10. The numbers in Sets B and C are then derived from the numbers in Set A, either by adding 2 to each Set A digit to create Set B or by multiplying each Set A digit by 2 to create Set C. The table of random numbers (Table 4.1) may be used for **randomly drawing scores** from each of the three sets. Table 4.2 lists the scores for each set. The numbers we will draw at random will correspond to *ordinal positions* (1st, 2nd, etc. position of scores in a set). For Set A, scores and positions are identical; however, this correspondence does not hold for Sets B and C. A number drawn at random from Table 4.1 identifies an **ordinal position** for a score that will be selected: drawing **5** from Table 4.1 would result in a score of 5 from Set A; a score of 7 from Set B; and a score of 10 from Set C.

Table 4.2: "Scores" in Set A, Set B, and Set C

Set A	Set B (A_i + 2)	Set C ($A_i \times 2$)
1	3	2
2	4	4
3	5	6
4	6	8
5	7	10
6	8	12
7	9	14
8	10	16
9	11	18
10	12	20

For this illustration we will first draw four samples of 20 scores at random from Table 4.1 (0 in Table 4.1 represents ordinal position 10). Numbers drawn from Table 4.1 are only used to identify a "score" from Table 4.2. That is, if we are drawing a random sample of scores from Set C and we drew a 9 from Table 4.1, then the score occupying the 9th position in Set C (18) would be selected. Before starting, we need to decide how to proceed through the numbers listed in Table 4.1. First, find a starting location by closing your eyes and pointing to a number in the table (this may have to be done several times if your pointer misses a number in the table, or misses the book, or perhaps even misses the table on which the book lies). We followed this procedure and landed on the number **2** in row 4, column 40. We have now selected our first number; however, we still need to identify ordinal positions for 19 more numbers to complete our first sample of 20 scores. Rather than pointing blindly every time we need a new number, we decided to proceed by going down the columns, taking the numbers as they appear in sequence. Since we need two samples from Set A and one each from Sets B and C, we need a total of four sets of 20 random numbers. Table 4.3 lists the random numbers, **scores corresponding to ordinal positions for these random numbers**, and a mean score for each of our four samples.

Table 4.3: Four Random Samples of Scores

	Group A-1		Group A-2		Group B		Group C	
	Random #	Score	Random #	Score	Random #	Score	Random #	Score
	2	2	5	5	9	11	2	4
	10	10	7	7	4	6	9	18
	1	1	8	8	4	6	10	20
	4	4	9	9	3	5	10	20
	7	7	2	2	4	6	9	18
	1	1	10	10	1	3	1	2
	8	8	10	10	4	6	7	14
	4	4	1	1	3	5	8	16
	6	6	2	2	6	8	7	14
	4	4	1	1	5	7	7	14
	6	6	6	6	4	6	2	4
	8	8	2	2	7	9	10	20
	8	8	9	9	1	3	9	18
	10	10	8	8	7	9	7	14
	8	8	8	8	9	11	10	20
	1	1	5	5	10	12	10	20
	4	4	3	3	4	6	1	2
	3	3	6	6	9	11	7	14
	4	4	10	10	5	7	10	20
	9	9	7	7	1	3	5	10
$\bar{Y}_i$		**5.40**		**5.95**		**7.00**		**14.10**

Table 4.4: Two Random Samples of 20 Scores from the 30 Scores from Table 4.2

Sample 1		Sample 2	
Random #	Score	Random #	Score
14	6	27	14
6	6	23	6
8	8	24	8
28	16	16	8
8	8	14	6
8	8	13	5
2	2	12	4
7	7	12	4
20	12	7	7
20	12	22	4
6	6	28	16
18	10	14	6
2	2	18	16
4	4	9	9
25	10	21	2
2	2	10	10
28	16	14	6
29	18	26	12
27	14	7	7
1	1	27	14
$\bar{Y}$	**8.40**		**8.20**

This example of randomly selecting scores from four samples resulted in different mean values for the two samples from Set A (5.40 and 5.95), which must be the result of "random error." Means for the random samples from Set B and Set C are both higher than the two samples from Set A, which should be the case since 2 units were added to each score in Set A to create Set B and each score in Set A was doubled to create Set C. Even though known factors ("2 plus" and "2 times") account for differences in means as expected, the mean for the Set B is not two units larger than either of the two Set A sample means, nor is the mean from the Set C twice as large as the means for the two Set A samples. It appears that both random error and the known factors contributed to differences observed among the Set A, Set B, and Set C sample means.

Now let's take this one step further. What should we observe if we draw two random samples of 20 scores if this time we draw the 20 scores from the entire collection of 30 scores? Even though known factors (2+ and 2x) influence subsets of scores, sampling scores randomly from the entire set of 30 scores should distribute both known effects and random error equally across different samples. Of course, we would expect that random error will result in different means for these two small samples. Since we will be selecting 2-digit numbers (for positions 1–30 from Table 4.1), we will again begin the process by pointing to Table 4.1 with eyes closed to get a new starting point. For the total collection of 30 scores, we will assign ordinal position 1–10 to the 10 scores in Set A; positions 11–20 to the 10 scores in Set B; and positions 21–30 to the 10 scores in Set C. We will restart the sequence for 31 through 60 and again for 61 through 90 (e.g., 31 = 1, 72 = 12). Numbers 91–00 in Table 4.1 are excluded.

Pointing at the numbers in Table 4.1 with eyes closed, the selected starting point was the number 14 in the 13th and 14th columns in the 2nd row. An ordinal position of 14 corresponds to a score of 6 (from Table 4.2, the 4th score in Set B is actually in the 14th position, as positions 1–10 correspond to the 10 scores in Set A, positions 11–20 correspond to the 10 scores in Set B, and positions 21–30 correspond to the 10 scores in Set C). The number immediately below 14 in the random numbers table is **66.** Since we are restarting the sequence at 31 and at 61, **66** is equivalent to the 6th position and a score of 6 from Set A. The number in Table 4.1 that follows **66** is **97** (in the 91–00 range that is excluded). So we will drop down to the next number, which is **68.** An ordinal position of 68 is the same as an ordinal position **8**, and a corresponding score of 8 from Set A. At this point we are at the bottom of Table 4.1, so we will continue by moving to the top of the table and the next two columns to the right, which is the number **28.** This ordinal position identifies the 8th score for Set C, which is 16. We will continue through Table 4.1 in this manner to obtain 36 more random numbers (ordinal positions) in order to complete the two random samples of n = 20. Sample 1 and Sample 2 that resulted from this series of random draws are shown in Table 4.4. As expected, the means for these two samples (**8.40** and **8.20**) are quite close in value.

III. Within and Between Group Variances

Using numbers from different tables, with some representing scores and others representing ordinal positions is understandably confusing, so a brief review may be in order. The means for two random samples taken from Set A were 5.40 and 5.95. Since the two samples came from the same population of 10 scores, all we can say is the difference between the two sample means is the result of chance or random error. The mean for the random sample from Set B (7.00) and the mean for the random sample from Set C (14.10) are noticeably larger than the means for the Set A samples. These variations come as no surprise since the Set B population was created by adding the constant 2 to each score in the Set A population, and the Set C population was created by doubling each score in the Set A population. That is, known factors (the addition and multiplication operations) contributed to the differences between the sample means for Set B and Set C, compared to the sample means for Set A. Of course, chance factors based on the random selection of numbers from Sets B and C also accounted for some of the observed differences.

At this point we may draw a distinction between variability for a set of scores within the same group and variability between representative scores (means) from different groups. The former is referred to as within-group or within-subject variability, and the latter is referred to as between-group or between-subject variability. This distinction is very important for the task of separating variability from a selected known factor (an independent variable) from variability from other factors that may come from both known and unknown sources (random error). What can be exploited for these two sources of variability (within- and between-subject) is the fact that random error influences both, whereas only between-group variability is influenced by a known factor of interest (the independent variable, which in this simple illustration was "plus 2" for Sample B and "times 2" for Sample C). Remember, a basic objective in designing experiments is that groups are equivalent in all ways except for the independent variable. It appears that the best that can be done in practice is to have groups that are equivalent in all ways except for the independent variable and random error. The statistical task involves "neutralizing" random error.

Given that between-group variance (variability between group means) includes effects of both the independent variable and random error, and within-group variance (variability among individuals within a given group) includes only effects of random error, then dividing an estimate of between-group variability by an estimate of within-group variability should be informative. If the independent variable does not affect scores, then this ratio should be at or near a value of 1.0 [if we represent the effect of the independent variable as α and random error as ε, then the between-subject-to-within-subject ratio of variance estimates reduces to $(\alpha + \varepsilon) \div \varepsilon$; and if $\alpha = 0$, then $(0 + \varepsilon) \div \varepsilon = 1.0$]. If all this seems a bit confusing, it may be worthwhile to review this section before moving on. Ratios between variance estimates that fit this model define the F test in the analysis of variance that will be the focus of future chapters.

IV. The Null Hypothesis (H_0) and the Concept of a Reference Distribution

A traditional approach in inferential statistics involves testing the *null hypothesis* (abbreviated H_0). The null hypothesis is the hypothesis that the groups or conditions in an experiment **do not differ**, or more specifically, that the true difference among groups is 0, with any small observed differences being the result of chance fluctuations or random error. The null hypothesis is not really the hypothesis that researchers are interested in; although H_0 testing is heavily relied upon in current research, it has come under its fair share of criticism for a number of reasons (see Bird 2004; Hopkins, Cole, and Mason 1998; Kline 2004). An experiment is generally done to test if a particular independent variable has an effect on the observed responses—an interest derived from a research hypothesis (H_1) that there is a true, non-zero difference among groups. It turns out that in most cases the research hypothesis of interest cannot be tested directly because it is inexact. A research hypothesis may be that fluoride in toothpaste reduces cavities. The hypothesis simply posits a reduction without specifying the exact number of cavity reductions. The null hypothesis does specify an exact number of cavity reductions that will result from using the fluoride toothpaste, namely zero; thus H_0 can be tested directly.

The inexact nature of the research hypothesis means that it cannot be tested directly, so we do the next best thing by testing the viability of the null hypothesis, because this can be tested directly. Since the research hypothesis is not compatible with H_0, it draws support if the data and statistical test indicate that H_0 is untenable. The term "untenable" is rather vague, so this idea is expressed in terms of probabilities. Non-zero differences among groups that are very improbable if random error is the only factor operating are judged as significant, meaning it is unlikely that they are due to chance. Having thus neutralized random error statistically, in a well-designed experiment, the only remaining candidate accounting for an observed effect is the independent variable.

a. Probability and Reference Distributions

Probability and the concept of a reference distribution are important because they provide the basis for experimental decisions about the viability of null hypotheses. Simply stated, statistical hypothesis testing involves computing a relevant statistic from an experimental data sample. This calculated test statistic is then compared to a distribution of values that would be expected for the statistic if H_0 was true with only random error accounting for observed differences. If the computed statistic from the experiment falls among the extreme values in the reference distribution most distant from zero (associated with a very low probability of occurrence), then the experimental **decision** is to **reject H_0**.

The mathematics of probability can be dealt with at varying levels of complexity, from a simple question everyone should be able to answer to quite difficult problems to solve. For example, every reader should be able to answer the following question correctly: Given a fair coin, what is the probability that when tossed, the coin will land with the *heads* side up? You may respond with "half, or 50% of the time" without engaging in any arithmetic operations or analysis. This simple problem, like other probability problems, is solved by counting. Basically, two things have to be counted: the number of different ways the target result (e.g., *heads*) can occur and the total number of different results that can occur (e.g., *heads* and *tails*). Probability is the ratio of these two frequencies. If you toss a coin a single time, there is only one possible "target" result: landing with the head side up. A total of only two different results are possible: landing with the head side up or with the tail side up. Thus, the unnecessary arithmetic for this problem is $1 \div 2 = .50$ (a probability that a "head" will come up half of the time).

In the context of probability, a reference distribution is much like the descriptive label suggests. Given certain assumptions (remember, even for the coin toss we had to assume we had a "fair" coin, one that in the **long run** would come up heads and tails equally often), a distribution of probabilities for classes of outcomes for a test statistic can be generated. The specific results for the test statistic from an experiment can then be compared to this reference distribution to determine if it is among the highly probable outcomes (e.g., observing *heads* 55 times out of 100 coin tosses) according to the assumptions that generated the reference distribution, or if it is among the highly improbable extreme outcomes (e.g., observing *heads* 90 times **or more** out of 100 coin tosses). Although theoretical reference distributions for test statistics that will be considered in later chapters will be noted, we will take time in the remainder of this chapter to describe and construct a reference distribution that can be used for simple tasks (like tossing a coin) where only one of two possible responses or outcomes can occur on each observational occasion (trial). Tossing a coin is a good model for this situation; however, this illustration may be generalized to more meaningful research situations for which there are only two response possibilities on any given observation trial (e.g., a *correct* or *incorrect* response; a *left* or *right* turn for a rat traversing a two-choice T-maze; or, more generally, a "success" or a "failure").

It was noted earlier that solutions for probability questions reduce to problems of counting; however, tasks where only one of two outcomes is possible require many trials for adequate evaluation, and counting target outcomes and total outcomes by listing all possible combinations across trials in order to determine critical probability ratios quickly becomes unmanageable. For example, consider the number of possibilities if a coin is tossed as few as four times. The coin could come up heads on all four trials, it could come up tails on all four trials, and there are a variety of mixed heads and tails that could occur. Specifically, any one of the following 16 distinctly different outcome sequences could be observed (H = heads, T = tails):

Sixteen Possible Four-Trial Sequences of "Heads" (H) and "Tails" (T)

	1	2	3	4	5	6	7	8	9	10	11	12	13	14	15	16
Trial 1	H	H	H	H	H	H	H	H	T	T	T	T	T	T	T	T
Trial 2	H	H	H	H	T	T	T	T	H	H	H	H	T	T	T	T
Trial 3	H	H	T	T	H	H	T	T	H	H	T	T	H	H	T	T
Trial 4	H	T	H	T	H	T	H	T	H	T	H	T	H	T	H	T
ΣH	4H	3H	3H	2H	3H	2H	2H	1H	3H	2H	2H	1H	2H	1H	1H	0H

To create a probability distribution we need to count the number of ways given target outcomes can occur (we will use classes of results such as observing 4H, 3H, 2H, 1H, or 0H occurrences—these cover all possible outcomes as we cannot obtain an outcome of five heads when tossing a coin only four times, nor can we obtain fewer than zero heads). Of the 16 four-trial sequences, only the one in the first column meets a target outcome of 4H, thus the probability of coin coming up heads on 4 successive tosses is 1 ÷ 16 = .0625. There are 4 outcomes with 3H and 1T (columns 2, 3, 5, and 9); thus, the probability of 3 heads in four coin tosses is 4 ÷ 16 = .2500. There are 6 outcomes with 2H and 2T (columns 4, 6, 7, 10, 11, and 13), yielding a probability for this 2H and 2T of 6 ÷ 16 = .3750. There are 4 outcomes with only 1H (columns 8, 12, 14, and 15) for a probability of 4 ÷ 16 = .2500. Finally, only the sequence in the 16th column has 0H, hence the probability is 1 ÷ 16 = .0625. The sum of these proportions is 1.0. This discrete probability distribution may be summarized as follows, with p(r) indicating the probability of each specific class of r outcomes (e.g., r = 4 for four heads, r = 3 for three heads):

$$p(r = 4) = 1 \div 16 = .0625$$
$$p(r = 3) = 4 \div 16 = .2500$$
$$p(r = 2) = 6 \div 16 = .3750$$
$$p(r = 1) = 4 \div 16 = .2500$$
$$p(r = 0) = 1 \div 16 = .0625$$

b. An Illustration

No one would really do an experiment such as this with only four trials. Even with as few as 12 trials, help is needed in counting. With four trials there were $2^4 = 16$ different outcome sequences, and though somewhat tedious, each could be enumerated. With 12 trials, there are $2^{12} = 4{,}096$ different outcome sequences, and enumerating these so all classes of outcomes may be counted would be overwhelming. Fortunately, the required counting task is manageable with the help of the formidable-looking counting formula that follows:

$$p(r) = (N! \div r![N - r]!)([p]^r[q]^{[N - r]})$$

In this formula, N = number of trials, r = the target number of "successful" outcomes in question, p = probability of a "successful" outcome on any given trial, and q = probability of an "unsuccessful" outcome on any given trial. Recall from Chapter 1 that N! (read as N factorial) = (N)(N – 1)(N – 2) … . To specify a complete probability distribution you must solve p(r) for all possible values of r. If this were a coin-tossing task with 12 trials, then the possible results could be r = 12 heads, r = 11 heads, … , r = 0 heads, and p(r) must be solved for all of these 13 possibilities. The terms "successful" and "unsuccessful" are merely general designations for the two possible outcomes on any trial (e.g., heads or tails for a coin-tossing task). What is the probability that a fair coin would land with the heads side up on 12 successive trials? Applying the counting rule, we have N = 12, r = 12, p = ½, and q = ½:

$$p(r = 12) = (12! \div 12![12 - 12]!)([1/2]^{12}[1/2]^0)$$

$$p(r = 12) = (12! \div 12![0!])([1/4096][1])$$

$$p(r = 12) = (1)(1/4{,}096) = .0002 \text{ (two times in 10,000 sequences of tossing a coin 12 times – a very unlikely outcome)}$$

In calculating probabilities you need to know two important mathematical rules that involve the integer 0: zero factorial (0!) and a zero exponent (k^0). The result is 1 in both cases: 0! = 1 and $(1/2)^0 = 1$. Also, in using reference distributions we are interested in combinations of outcomes, not just a single outcome. That is, we would want to know the probability of a set of extreme outcomes, such as the probability of "heads" at least 10 times out of 12 trials **and** two or fewer times out of 12 trials (this description represents a set of extreme outcomes under the fair-coin assumption). With r = number of "heads," the solution for the probability of getting either 0, 1, 2, 10, 11, or 12 heads when tossing a fair coin 12 times is simply the sum of the separate probabilities [where r = number of heads, simply do the addition for p(r = 0) + p(r = 1) + p(r = 2) + p(r = 10) + p(r = 11) + p(r = 12)]. If you are familiar with working with fractions, you may skip this sentence; however, as a reminder to those who have been separated from fractions for a while, you need to know that when you raise a fraction to a power, you do so for both the numerator and the denominator [e.g., $(2/5)^2 = 2^2/5^2 = 4/25 = .16$]. Completing the entire probability distribution for this illustration involving tossing a coin 12 times, we have [recall that a factorial is a succession of multiplications; e.g., 12! = (12)(11)(10) …(1)]:

$$p(r = 12) = (12! \div 12![0!])([1/4096][1]) = 1 \div 4{,}096 = .0002$$

$$p(r = 11) = (12! \div 11![1!]([1/2048][1/2]) = 12 \div 4{,}096 = .0029$$

$$p(r = 10) = (12! \div 10![2!])([1/1024][1/4]) = 66 \div 4{,}096 = .0161$$

$$p(r = 9) = (12! \div 9![3!])([1/512][1/8]) = 220 \div 4{,}096 = .0537$$

$p(r = 8) = (12! \div 8![4!])([1/256][1/16]) = 495 \div 4{,}096 = .1208$

$p(r = 7) = (12! \div 7![5!])([1/128][1/32]) = 792 \div 4{,}096 = .1934$

$p(r = 6) = (12! \div 6![6!])([1/64][1/64]) = 924 \div 4{,}096 = .2256$

$p(r = 5) = (12! \div 5![7!])([1/32][1/128]) = 792 \div 4{,}096 = .1934$

$p(r = 4) = (12! \div 4![8!])([1/16][1/256]) = 495 \div 4{,}096 = .1208$

$p(r = 3) = (12! \div 3![9!])([1/8][1/512]) = 220 \div 4{,}096 = .0537$

$p(r = 2) = (12! \div 2![10!])([1/4][1/1024]) = 66 \div 4{,}096 = .0161$

$p(r = 1) = (12! \div 1![11!])([1/2][1/2048]) = 12 \div 4{,}096 = .0029$

$p(r = 0) = (12! \div 0![12!])([1][1/4096]) = 1 \div 4{,}096 = .0002$

Admittedly this is a tedious example; however, it is done for a purpose. We have now created a **reference distribution** for a small-scale problem. A reference distribution is what we *refer to* in evaluating experimental results. You can calculate a test statistic from your data and compare it to an appropriate reference distribution that associates chance probabilities of obtaining a test statistic **as extreme as, or more extreme than** the one produced by your data. For tossing a coin, the reference distribution was built on the assumption that "heads" and "tails" were equally likely outcomes ($p = q = .50$). We may proceed to test this null hypothesis [$p(H) - p(T) = 0$]; perhaps with the coin used to determine whether the NFL or AFL team receives the opening kickoff to start the annual Super Bowl football game. Since one conference won the opening coin flip an unprecedented 12 successive years counting the 2009 Super Bowl, perhaps this would not be considered a flippant example (sorry for the pun). However, you should know the probability that the same conference would win the coin toss again the following year is still $p = .50$. This claim should make sense to you if you see the humor or absurdity in a joke retold by Friedman (1995, page 36) "about the man who carries a bomb with him whenever he goes on an airplane because the odds against there being two bombs on one airplane are much higher."

In psychological research you will hear about the common .05 and .01 significance levels. These references simply mean that by *convention*, researchers discredit the null hypothesis [e.g., that this is a fair coin with $p(H) = p(T)$] for all extreme outcomes that collectively could occur by chance fewer than 5 times in 100 (or fewer than 1 time in 100 if the more conservative .01 discrediting option is used). From the 12-trial coin-tossing reference distribution, we can see that the probability of observing extreme outcomes of 10, 11, or 12 heads is equal to .0161 + .0029 + .0002 = .0192; a very unlikely class of events. However, an equally extreme set of outcomes would be 2, 1, or 0 heads in 12 tosses of a coin. This class of outcomes is also equal to .0161 + .0029 + .0002 = .0192. Collectively, the probability of tossing a coin 12 times with the result of extreme outcomes of 12, 11, 10, 2, 1, or 0 heads is .0192 + .0192 = .0384. Chance variation within this reference distribution indicates that in the long run approximately 3.8% of the time a fair coin is tossed, one would observe heads 12, 11, 10, 2, 1, or 0 times. If the .05 significance rule is used, the fair coin hypothesis (H_0) would be discredited if any of these six results had occurred. This is the way null hypothesis testing works, with the understanding that for a small proportion of research reports, H_0 will be discredited in error. That is, even a fair coin can land with the "heads" side up 12 times in a row. We will have more to say about this later.

This 12-trial sequence of two-alternative outcomes certainly puts a strain on a hand-calculation exercise, yet in terms of trials it falls far short of what would be done in an actual experiment appropriate for this model. We could advance to the next stage of complexity to see how we could use a different distribution (e.g., the z distribution) that is easier on the computational headaches for approximations with larger trial sets; however, that would add little to the main purpose of introducing the concept and application of a reference distribution. In a real-life laboratory (outside of a statistics class) if one ever had to do a meaningful statistical test of this sort, he or she would certainly let a computer do the work. Nonetheless, the z distribution is an important distribution in its own right, and it is introduced in the next section to provide a prototype illustration of "reasoning from an assumed *continuous* probability distribution."

V. The z Distribution

The probability reference distribution for a discrete two-outcome event considered in this chapter served the purpose of illustrating the concept and application of reference distributions in the statistical decision-making process. We will introduce one important reference distribution here, called the z distribution (also referred to as the *unit-normal* distribution, or *normal* distribution). Although in actual practice reference distributions developed for continuous variables are common, you will not often encounter applications of the z distribution in published research articles. The shape of the z distribution should be familiar because it is approximated in many data displays. It is schematized in Figure 4.1: the defining features are (1) a center peak, with (2) frequency of scores tailing off symmetrically in both directions from the center. Importantly, the z distribution has a **mean of 0** and a **variance (and standard deviation) of 1.0**. Thus, units along the X axis are **standard deviations** (e.g., z = ±1.0 provides information concerning the proportion of scores that are within one standard deviation of the mean, as well as information concerning scores more extreme than one standard deviation above and/or below the mean).

Although z is a **population** distribution, sample data can be converted to *z scores* in order to use z as a reference distribution for drawing statistical inferences (or approximations to the z distribution, such as the t distribution that will be discussed in future chapters). The discrete probability distribution that was generated for tossing a coin 12 times is presented graphically in Figure 4.2; you can see that in terms of general shape and symmetry, it bears a close resemblance to the z distribution depicted in Figure 4.1. Any observed number of "successes" (e.g., r = 10 heads out of 12 coin tosses) may be converted to a z score, with the probability approximated by referring to the z distribution. The transformation formula is $(r - [N][p] - .5) \div \sqrt{([N][p][q])}$. The .5 constant (called the *correction for continuity*) is subtracted in the numerator because r is larger than (N)(p) in this example (r = 10,

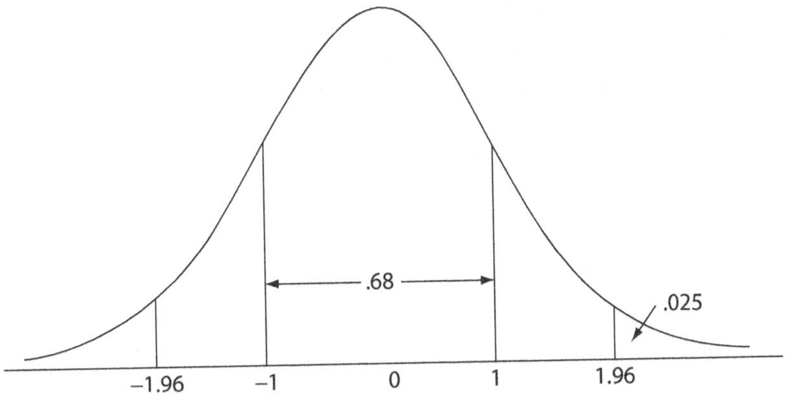

Figure 4.1: Schematic Representation of the z Distribution

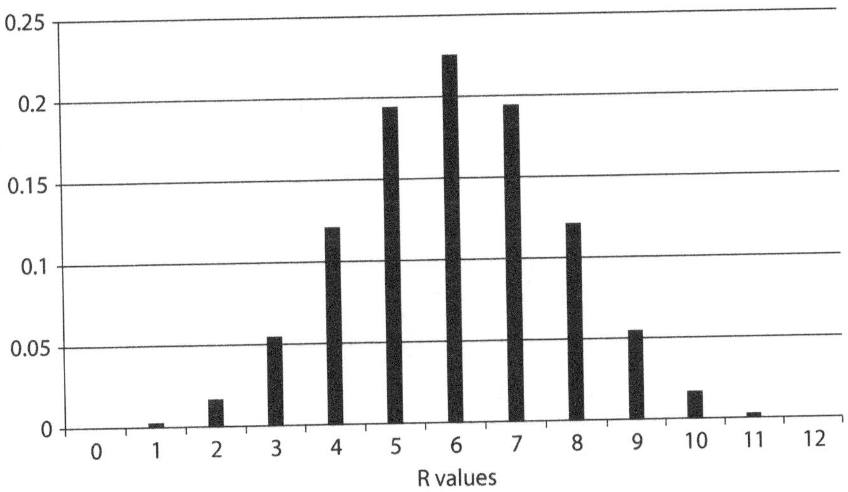

Figure 4.2: Probability Distribution for Two Outcome Experiment with N = 12, p = .5, and q = .5

N = 12, and p = .5). If r had been less than (N)(p), then the .5 constant would have been added in the numerator. These procedures are detailed in Hays (1988). Just to give you an idea about how well the approximation works even with as few as 12 trials, you may solve for z by substituting r = 10, N = 12, p = .5, and q = .5 into this z formula: the result is z = 2.02. The probability associated with a z value this large or larger is .0217 (see Hays, page 925 for a complete z reference distribution). Using the counting rule, the actual probabilities computed for r ≥ 10 were p(r = 10) = .0161, p (r = 11) = .0029, and p(r = 12) = .0002. These three exact probabilities sum to .0192; thus even for an N as small as 12, the z approximation only overestimates the probability of an outcome at least this extreme by .0217 – .0192 = .0025!

A brief digression is necessary before describing the relation between z scores and probability. This digression addresses what a z score is, and how converting key statistics to z values provides probability information. To accomplish this we will take a small **population** of five scores and make two transformations on the individual scores to see how the transformations impact the important summary statistics that measure central tendency and variability (namely, means, variances, and standard deviations). Remember, for a "population" of scores, the denominator for the variance is "n" and not the "n – 1" correction for bias when a sample variance, s^2, provides an estimate of σ^2. We will treat the small set of five scores shown in the table that follows as if these scores represented the entire population.

We will begin by considering how descriptive measures of the mean, variance, and standard deviation are affected when each score in the population is changed by adding a constant (C_1) to it. Next we will consider how these descriptive measures are affected when each score is multiplied by a constant (C_2). Finally, we will consider these two transformations in combination with specific C_1 and C_2 constants that yield a special new distribution of transformed scores ("the z distribution") that has a mean equal to 0 and both a variance and standard deviation equal to 1.0. The set of five scores we will work with, along with the C_1 and C_2 addition and multiplication constants, respectively, appear in the three columns of this table (we have selected the number 5 for the C_1 addition constant and the number 10 for the C_2 multiplication constant):

Original Y_i Scores		Y_i' Scores $(5 + Y_i)$		Y_i'' $[(10)(Y_i)]$	
	2		7		20
	5		10		50
	6		11		60
	3		8		30
	4		9		40
ΣY_i	20	$\Sigma Y_i'$	45	$\Sigma Y_i''$	200
μ_Y	4	$\mu_{Y'}$	9	$\mu_{Y''}$	40
σ_Y^2	2	$\sigma_{Y'}^2$	2	$\sigma_{Y''}^2$	200
σ_Y	1.414	$\sigma_{Y'}$	1.414	$\sigma_{Y''}$	14.142

Means for each set are determined by the following formula: $\mu = \Sigma Y_i \div n$ (e.g., for Y_i scores in Column 1, $\Sigma Y_i = 20$, and $\mu_Y = \Sigma Y_i \div n = 20 \div 5 = \mathbf{4}$). A computational formula for the variance is $\sigma^2 = (\Sigma Y_i^2 - [\Sigma Y_i]^2 \div n) \div n$. For Y_i' scores in Column 2, $\sigma_{Y'}^2 = (7^2 + 10^2 + \ldots + 9^2 - [45]^2 \div 5) \div 5 = (415 - 2{,}025 \div 5) \div 5 = (415 - 405) \div 5 = \mathbf{2}$; and for Y_i'' scores in Column 3, $\sigma_{Y''}^2 = (20^2 + 50^2 + \ldots + 40^2 - [200]^2 \div 5) \div 5 = (9{,}000 - 40{,}000 \div 5) \div 5 = (9{,}000 - 8{,}000) \div 5 = \mathbf{200}$. From this illustration with a population of only five scores you can see that adding a constant to each score increased the mean by the addition of that constant ($\mu_Y = 4$ and $\mu_{Y'} = \mu_Y + C_1 = 4 + 5 = 9$), however, it left the variance and standard deviation unchanged ($\sigma^2 = 2$ and $\sigma = 1.414$ for both the Y and Y' distributions). However, multiplying each score by a constant increased the mean by the product of the original mean and constant [$\mu_Y = 4$ and $\mu_{Y''} = (C_2)(\mu_Y) = (10)(4) = 40$], and the same was true for the standard deviation [$\sigma_Y = 1.414$, and $\sigma_{Y''} = (C_2)(\sigma_Y) = (10)(1.414)$

$= 14.14$]. The variance was increased by multiplying the original variance by the square of the constant (C_2^2) [$\sigma_Y^2 = 2$, and $\sigma_{Y''}^2 = (C_2^2)(\sigma_Y) = (100)(2) = 200$].

The final component in this exercise is to repeat the mean, variance, and standard deviation computations when C_1 and C_2 constants are applied simultaneously, and for a rather strange-looking transformation. For the addition constant we will use $C_1 = -\mu_Y$ (a minus value of the mean of the original population of Y_i scores) and for the multiplication constant we will use $C_2 = 1/\sigma_Y$ (multiplication by a fraction with the standard deviation from the original population of Y_i scores as the denominator). This new set of transformed scores are called z scores, and as you can see from the computations for the following z scores, $\mu_z = 0$, $\sigma_z = 1.0$, and $\sigma_z^2 = 1.0$.

	Original Y_i Scores		Z_i Scores $[(Y_i - \mu_Y)(1/\sigma_Y)]$
	2		$(2 - 4)(1/1.414) =$ -1.414
	5		$(5 - 4)(1/1.414) =$ +.707
	6		$(6 - 4)(1/1.414) =$ +1.414
	3		$(3 - 4)(1/1.414) =$ -.707
	4		$(4 - 4)(1/1.414) =$ 0
ΣY_i	20	ΣZ_i	0
μ_Y	4	μ_z	0
σ_Y^2	2	σ_z^2	1.0
σ_Y	1.414	σ_z	1.0

The calculations of the mean (μ_z) and variance (σ_z^2) for the transformed z scores are

$$\mu_z = \Sigma Z_i \div n = 0 \div 5 = \mathbf{0}$$

$$\sigma_z^2 = (Z_i^2 - [\Sigma Z_i]^2/n) \div n = ([-1.414]^2 + .707^2 + 1.414^2 + [-.707]^2 + 0^2 - [0^2/5]) \div 5$$

$$\sigma_z^2 = (5.0 - 0) \div 5 = \mathbf{1.0}$$

Given this minimal z distribution, we can proceed to determine a probability for any given score. For this population of five scores we probably should restrict our question to scores ranging from 2 to 6; however, for illustration purposes we will ask the following question: "What is the probability of observing a score of **8**?" To answer this question, we begin by converting the raw score of 8 to a z score using the z score transformation formula: $z_i = (Y_i - \mu_Y)(1/\sigma_Y)$. Substituting $Y_i = 8$, $\mu_Y = 4$ and $\sigma_Y = 1.414$ into the formula, we have $z = (8 - 4)(1/1.414) = 4 \div 1.414 = \mathbf{2.83}$. The center (mean) of the z distribution is 0, and the probability of obtaining a z score as far removed from the center as 2.83 standard deviation units or more can be determined by consulting a z table. Relevant portions of a z table are given in Table 4.5, and a z value this large or larger (2.83) is very unlikely (only .0023 of the total area encompassed by a z distribution lies between +2.83 and $+\infty$). Combining the two extreme tails of the z distribution (areas from -2.83 to $-\infty$ and +2.83 to $+\infty$), the probability of such extreme z scores is only .0046 (.0023 + .0023 = .0046). Since a score of "8" from this population of five scores has such a low probability of occurrence, "statistical reasoning" would lead us to discredit the notion that it came from this population. Since the entire population of five scores was presented, it is obvious that a score of 8 is not in the population of scores; however, in actual experimental research, statistical decisions involve drawing inferences without the benefit of having information about all scores in a population.

Table 4.5: Sample Values of z and Associated Probabilities

z score	proportion from z to ∞	z score	proportion from z to ∞
1.00	.1587	2.00	.0028
- - -		- - -	
1.64	.0505	2.32	.0102
1.65	.0495	2.33	.0099
- - -		- - -	
1.95	.0256	2.57	.0051
1.96	.0250	2.58	.0049
- - -		- - -	
		2.83	.0023

The z values sampled in Table 4.5 provide the proportion of the total distribution that corresponds to (1) conventional "critical cutoff points" for reference distributions (5% and 1% of the total area beyond one tail of the distribution and beyond both tails of the distribution), (2) proportions that enable us to determine how much of the total area of the z distribution is within one standard deviation of the mean ($z = \pm1.00$), and how much is within two standard deviations of the mean ($z = \pm 2.00$), and (3) the specific z score ($z = 2.83$) that corresponds to $Y_i = 8$.

Statistical decisions are most often made using either a 5% or 1% cutoff point, which means that special status is given to values this far removed from the range of likely expected values in the reference distribution. The values reported in this limited selection from the z table indicate that 5% of z scores may be expected to be about 1.645 or larger (a 5% cutoff point appears to be halfway between the tabled values of 1.64 and 1.65), and more generally for what are referred to as "two-tailed" tests, 5% may be expected to be either from -1.96 to $-\infty$ (2.5% of the scores on the extreme negative side of μ_z) or +1.96 to $+\infty$ (2.5% of the scores on the extreme positive side of μ_z). Similarly, 1% of z scores may be expected to be about 2.325 or larger (a 1% cutoff point appears to be halfway between the tabled values of 2.32 and 2.33); with the percentage split between the extreme negative and positive scores (0.5% at each extreme), 1% may be expected to be either from -2.575 to $-\infty$ or +2.575 to $+\infty$ (again, the values of ±2.575 splits the difference between $z \geq 2.57$ for a .0051 proportion and $z \geq 2.58$ for a .0049 proportion).

The z values of 1.00 and 2.00 in Table 4.5 are interesting benchmarks because they provide information regarding how much of a distribution is more than one or two standard deviations beyond the mean, respectively. From Table 4.5 you can see that the proportion of the distribution at $z \geq 1.00$ is

.1587. Similarly, the proportion of the distribution at $z \leq -1.00$ is .1587. Thus, .1587 + .1587 = 31.74% of the distribution is outside of ±1.0 standard deviation from the mean (or, if we subtract this value from 100%, you can see that 68.26% of the distribution is within one standard deviation of the mean). At two or more standard deviations beyond the mean (either larger than or smaller than the mean) we have only 4.56% of the distribution (.0228 + .0228 = .0456 for z scores of ±2.00). Subtracting these values from 100% indicates that 95.44% of the z distribution lies within the area bounded by two standard deviation units. Finally, for the small population of five scores in this illustration, 0.23% of z scores are equal to or larger than 2.83; and 0.46% are either between -2.83 and -∞ or +2.83 and +∞. Although our focus with inferential statistics will be with evaluations of **sample means rather than individual scores,** we will not develop an extension of the z distribution to sample means [sample mean minus population mean divided by the standard error of the mean: $z = (\bar{Y} - \mu) \div s_E$], as psychology research rarely involves comparing a sample mean with a known population mean. However, it is important to recognize that the z distribution serves as a model for sampling distributions we will consider with specific statistical tests introduced in future chapters (e.g., the t distribution that may be used as a reference distribution with experiments involving two different groups or conditions provides an approximation of the z distribution with limited sample sizes), and the z reference distribution serves as an introduction to hypothesis testing based on probability and the concept of reference distributions.

VI. Summary

The primary virtue of experimental research is that all things except one are equivalent across experimental conditions. Equating groups is achieved through a combination of procedures. Chief among these is randomization. With random assignment, the effects of other variables (known and unknown) that could potentially confound the experiment are distributed randomly (i.e., "by the luck of the draw") across groups. Randomization guarantees group equivalence in the long run; however, random error is likely to contribute to observed performance differences among conditions in the short run. Inferential statistics draws on probability theory in assessing whether the independent variable in an experiment should be credited in accounting for observed differences among groups or whether all of the observed differences among groups fall reasonably within the confines of random variation.

Comparing different sources of variances among experimental groups lends itself nicely to parceling out the contributions to observed variations in performance (from the independent variable and random error). All subjects within a group are treated alike, leaving random error as the only candidate for the source of variation among the scores. However, subjects in different groups experience different treatment conditions as defined by the independent variable. Therefore, variation among representative scores (e.g., means) from different groups may be attributed to both random error and the independent variable. One statistical test, the F test from the Analysis of Variance, makes use of this reasoning by comparing the ratio of these two measures of variance (between-group to within-group variances) to a distribution with random variation around an expected value of 1.0 (the expected reference distribution value of the F statistic if the effect of the independent variable was zero—the H_0 assumption).

The development and use of a specific reference distribution for experiments involving only one of two possible outcomes for each observation trial was illustrated. Classes of outcomes were presented in a probability reference distribution for two contrived examples. The first example had only four trials, which actually enabled us to list and count every possible sequence of trial outcomes. Probability of a class of outcomes (e.g., getting 2 heads out of 4 coin tosses) was determined by counting the number of 2H sequences and counting the total number of distinctly different 4-trial sequences. The probability of 2H occurring was the former divided by the latter, or as illustrated, 6 ÷ 16 = .375. The second illustration expanded the experiment to a level that made a solution by listing all possible outcome sequences and then counting them impractical. The expansion tripled the number of trials to 12, hence, a listing-counting-classifying-based solution would require listing and grouping 4,096 distinctly different sequences. A counting rule was introduced that makes this task somewhat manageable through application of the following formula: $p(r) = (N! \div r![N - r]!)(p)^r(q)^{(N - r)}$. Applying this counting rule successively across each class of outcomes (12H, 11H, 10H, etc.) produced the appropriate probability reference distribution. The chapter concluded by moving from a discrete reference distribution to a continuous reference distribution, specifically by describing the z distribution that has $\mu = 0$ and both σ and $\sigma^2 = 1.0$.

4.1: Model Problem 1. The examples in the text considered common cases where it was reasonable to assume that the two possible outcomes on any given trial were equally probable: $p = q = ½$. The probability distribution for two-outcome experiments can be adapted to accommodate specific alternative assumptions concerning probabilities about the two outcome possibilities, such as $p = ¼$ and $q = ¾$. For example, assume each trial involves a display of four equal quadrants projected on a screen, with the quadrants labeled A, B, C, and D. On each trial a subject must identify the quadrant in which a "+" sign was flashed. To make the task somewhat challenging, the + sign appeared too briefly and too faintly to be consciously perceived (i.e., a "subliminal" presentation). Nonetheless, on each trial subjects had to indicate (guess) whether the + sign appeared in Quadrant A, Quadrant B, Quadrant C, or Quadrant D. For this problem, create two probability reference distributions: (A) one for an experiment involving four trials, and (B) one for an experiment involving 12 trials.

The starting assumption is that a chance guessing model leads to the expectation that a subject will make a correct guess one-fourth of the time and an incorrect guess three-fourths of the time. Thus, $p = ¼$ and $q = ¾$. The only change required from the two similar examples in the text is to change $p = ½$ to $p = ¼$ and to change $q = ½$ to $q = ¾$.

Part A: The four-trial probability distribution (now using the counting rule) appears below:

$p(r = 4)$: $(4! \div 4![0]!)(1/4)^4(3/4)^0 = (1)(1) \div 256 = .0039$

$p(r = 3)$: $(4! \div 3![1]!)(1/4)^3(3/4)^1 = (4)(3) \div 256 = .0469$

$p(r = 2)$: $(4! \div 2![2]!)(1/4)^2(3/4)^2 = (6)(9) \div 256 = .2109$

$p(r = 1)$: $(4! \div 1![3]!)(1/4)^1(3/4)^3 = (4)(27) \div 256 = .4219$

$p(r = 0)$: $(4! \div 0![4]!)(1/4)^0(3/4)^4 = (1)(81) \div 256 = .3164$

Part B: The 12-trial probability distribution (note: $4^{12} = 16{,}777{,}216$, giving us a pretty large denominator throughout):

$p(r = 12) = (12! \div 12![0!])([1/4]^{12}[3/4]^0] = (1)(1) \div 16{,}777{,}216 = .0000$

$p(r = 11) = (12! \div 11![1!])([1/4]^{11}[3/4]^1) = (12)(3) \div 16{,}777{,}216 = .0000$

$p(r = 10) = (12! \div 10![2!])([1/4]^{10}[3/4]^2) = (66)(9) \div 16{,}777{,}216 = .0000$

$p(r = 9) = (12! \div 9![3!])([1/4]^9[3/4]^3) = (220)(27) \div 16{,}777{,}216 = .0004$

$p(r = 8) = (12! \div 8![4!])([1/4]^8[3/4]^4) = (495)(81) \div 16{,}777{,}216 = .0024$

$p(r = 7) = (12! \div 7![5!])([1/4]^7[3/4]^5) = (792)(243) \div 16{,}777{,}216 = .0115$

$p(r = 6) = (12! \div 6![6!])([1/4]^6[3/4]^6) = (924)(729) \div 16{,}777{,}216 = .0401$

$p(r = 5) = (12! \div 5![7!])([1/4]^5[3/4]^7) = (792)(2{,}187) \div 16{,}777{,}216 = .1032$

$p(r = 4) = (12! \div 4![8!])([1/4]^4[3/4]^8) = (495)(6{,}561) \div 16{,}777{,}216 = .1936$

$p(r = 3) = (12! \div 3![9!])([1/4]^3[3/4]^9) = (220)(19{,}683) \div 16{,}777{,}216 = .2581$

$p(r = 2) = (12! \div 2![10!])([1/4]^2[3/4]^{10}) = (66)(59{,}049) \div 16{,}777{,}216 = .2324$

$p(r = 1) = (12! \div 1![11!])([1/4]^1[3/4]^{11}) = (12)(177{,}147) \div 16{,}777{,}216 = .1267$

$p(r = 0) = (12! \div 0![12!])([1/4]^0[3/4]^{12}) = (1)(531{,}441) \div 16{,}777{,}216 = .0317$

4.2: Model Problem 2. For a 12-trial experiment with p(correct guess) = ¼ and q(incorrect guess) = ¾, what is the probability of observing **at least** 7 correct guesses?

As obvious as it may sound, it is essential that you understand what must be counted to satisfy the conditions set by the question. In this case, the phrase "at least" must be correctly interpreted. A criterion of at least 7 correct guesses includes probabilities for outcomes of 7, 8, 9, 10, 11, and 12 correct guesses. These six possible outcomes are relevant for this question because each one meets the criterion of being "at least 7." Thus, we need to sum the separate probabilities for these six outcomes that meet the stated criterion:

$$p(r = \text{"at least 7"}) = p(r = 7) + p(r = 8) + p(r = 9) + p(r = 10) + p(r = 11) + p(r = 12)$$

From the table in Model Problem 4.1, Part B, p(r = "at least 7") = .0115 + .0024 + .0004 + .0000 + .0000 + .0000 = **.0143** (a little over 1 chance in 100 that one of these six outcomes would occur).

4.3: Practice Problem 1. A T-maze is a narrow, straight-walled alley ending in two alternative pathways: a 90° turn to the right or a 90° turn to the left. Given the assumption that turning right and turning left are equally probable, (A) determine the chance probability distribution for all possible outcomes in an eight-trial experiment (N = 8). (B) What is the probability that a rat placed in this maze would go down the right-hand path on at least six of the eight trials?

4.4: Practice Problem 2. To complicate the life of a laboratory rat, a third choice point was added—now the rat could choose to go left, center, or right, and one of the three alternatives had a food reward at the end while the other two alternatives did not have a food reward at the end. Repeat Parts A and B from Practice Problem 4.3 with the new assumption that you now have a maze with three equally probable choice points, one of which has food at the end, and the other two alternatives do not have food at the end. (A) Determine the chance probability distribution for all possible outcomes in an eight-trial experiment (N = 8), where p = the probability of a successful (food) choice and q = the probability of an unsuccessful (no food) choice. (B) What is the probability by chance alone that a rat placed in this maze would select the rewarded pathway on at least six of the eight trials?

4.5: Practice Problem 3. Twin sisters claim that each one "feels" what the other twin experiences. Suppose the following test is conducted. One sister is in Reno, Nevada, and her twin is in Monterey Bay, California. An experimenter is with each sister, and at pre-designated times, the Reno sister is given a mild electrical shock either to her (1) left shoulder, (2) right shoulder, (3) left hand, (4) right hand, (5) left foot, or (6) right foot. Positions of the shock are determined at random, and of course, are not known in advance by either sister. The Monterey Bay twin is only informed that her sister will be given a shock at one of these six locations on her body, and her task is to indicate where she was shocked. If 10 shock tests are given, what is the chance probability that the Monterey Bay twin's response will correctly identify the place on her sister's body where the shock was given on at least five of these 10 trials?

4.6: Practice Problem 4. Given a decision criterion set at the .05 level (the probability of the extreme outcomes must be .05 or smaller to be considered as "significant" or to justify rejecting H_0), what is the minimum number of trials that must be given in Practice Problem 2 in order to test the H_0 that rats will choose the arm of the maze leading to the food reward with a greater than chance frequency? For example, if only one trial were given, a rat would be expected by chance alone to select the "food path" one-third of the time, thus more than one trial would be necessary to test H_0 because there are no outcomes with a probability of occurrence as rare or extreme as .05.

4.7: Practice Problem 5. What is the minimum number of trials that must be given in Practice Problem 3 in order to test H_0 at the .05 level? If the decision criterion was set at the .01 level, what is the minimum number of trials that would have to be given?

4.8: Practice Problem 6. The 16 scores from Practice Problem 3 in Chapter 3 are copied below:

12, 4, 7, 7, 1, 0, 6, 11, 14, 3, 5, 12, 2, 8, 9, 7

A. What is the z score for the first subject who had a Y score of 12?

B. What is the z score for the second subject who had a Y score of 4?

C. Assume these 16 scores were randomly drawn for a population with $\mu = 8.5$. Convert the sample mean to a z score using the following formula: $(\bar{Y} - \mu) \div s_E = ?$

Chapter 5

Getting Ready for the Analysis of Variance

It is easy to lie with statistics, but it is easier to lie without them.
—Moesteller

I. The Language of Analysis of Variance

The test statistic for the analysis of variance is the **F ratio**, which is simply an estimate of one variance from a data set divided by an estimate of another variance from the same data set. In the course of estimating the two variances, new labels, new computational organizations, and new formats are introduced. Since computations done on a hand calculator involve several steps, it is helpful to follow a well-organized system for completing all of the required steps (especially for experiments with several groups and large sample sizes). The system we will use for this purpose follows the action plan described by Keppel and his associates (Keppel and Saufley 1980; Keppel, Saufley, and Tokunaga 1992; Keppel and Wickens 2004).

We will start this section with some "name-calling." In analysis of variance (referred to as **ANOVA**) terminology, a variance (s^2) is called a "**Mean Square**" (**MS**). Subscripts are used with abbreviations denoting the source of the variance [e.g., MS_{BG} or MS_A is used to designate a variance estimate **between** group means (e.g., between the A groups); MS_{WG} or $MS_{S/A}$ is used to designate a variance estimate among subjects (S) **within** the A groups]. Uppercase letters, beginning with A, are used to identify the treatment factor or independent variable, often referred to as **Factor A** (Factors A and B for designs we will consider later that have two independent variables). Numerical subscripts that accompany lower case letter codes denote different levels or specific treatment conditions; e.g., from the experiment by Loftus and Palmer (1974) referred to in Chapter 1, $\mathbf{a_1}$ could refer to a misinformation condition with the verb "smashed," $\mathbf{a_2}$ to a misinformation condition with the verb "hit," and $\mathbf{a_3}$ to the control group that was not asked a question about vehicle speed. The F ratio is formed by taking an estimate of the variance between group means and dividing it by an estimate of the variance within groups. A general form of this F ratio is $\mathbf{MS_{BG} \div MS_{WG}}$; a more specific version for individual experiments is $\mathbf{MS_A \div MS_{S/A}}$, with the subscript A representing a variance estimate **between the A groups** and the subscript S/A ("S slash A") representing a variance estimate for the **subjects within the A groups** (S for subjects, the slash mark for "within," and A for groups).

Calculating variances to determine an F ratio represents the final product we are after. The additional labels noted in this paragraph identify the component parts in the computations. Recall that the basic computational form of the variance had two parts in the numerator [ΣY_i^2 and $(\Sigma Y_i)^2/n$—note the "slash" mark in this context denotes a division operation as $(\Sigma Y_i)^2$ is divided by n] and one part in the denominator (n – 1). Both parts have names commonly used in the analysis of variance: the numerator is called **sum of squares (SS)**, and the denominator is called **degrees of freedom (df)**. One additional, less common ANOVA breakdown involves the two components of the numerator for the variance. In the earlier editions of the texts cited in the first paragraph in this section, Keppel and Saufley (1980), and Keppel et al. (1992) suggested doing the SS calculations in two separate steps, referring to the required component parts as "basic ratios." For a total variance estimate based on all scores in a data set, the computational formula is $s^2 = (\Sigma Y_i^2 - [\Sigma Y_i]^2/n) \div (n - 1)$. For ANOVA, this would be written as $\mathbf{MS = SS \div df}$. The two component parts of the sum of squares may be expressed in terms of bracketed coding letters [Y] and [T] (called *basic ratios*), where the basic ratio $\mathbf{[Y] = \Sigma Y_i^2}$ and the basic ratio $\mathbf{[T] = (\Sigma Y_i)^2/n}$; thus, $\mathbf{SS = [Y] - [T]}$ (the sum of squares based on all scores in a data set is referred to as the total sum of squares and coded as SS_{Total} or simply SS_T). Putting these various parts together, you have the following computational path to an estimate of the **total** variance for a set of numbers: First, $[Y] = \Sigma Y_i^2$ and $[T] = (\Sigma Y_i)^2/n$; Second, $SS_T = [Y] - [T]$; Third, $df_T = n - 1$; Fourth, $MS_T = SS_T \div df_T$.

II. Identifying Three Variances

The variance is a distance score, and as such, it shares certain properties of distance measures. Importantly, a distance traveled can be meaningfully partitioned into component parts, and this is what is done when identifying three variances within a data set. If a "complete" data set can be organized into different subsets, then we have the necessary conditions for partitioning a total variance (based on distances of individual scores from the mean calculated from all scores) into two component parts. The total variance and its two component variances constitute the three variances for an ANOVA.

The two component variances represent (1) a restricted variance based on subsets of the data, and (2) an expanded concept of variance to include consideration of how different **group means** vary about a common reference point. The first of these is perhaps a little easier to deal with at the conceptual level.

It simply requires calculating variances separately for each group in an experiment. For example, if an experiment had six groups with 10 subjects in each group, then in addition to a total variance based on all 60 subjects, six separate variances can be calculated, one for each group of 10 subjects. An average of these six group variances represents one component part of the total variance (the within-group variance). A second component part refers to the variability of individual group means, with the overall mean based on all individual scores serving as the reference point (the between-group variance).

Sums of squares for these two component parts (based on the average of individual distance scores from the means of their respective groups, and the distance scores of group means from the grand mean based on all scores), when added together, equal the total sum of squares based on all scores considered without regard to group distinctions. Recall the basic ingredient in a variance is the individual distance score, or how far away an individual score is from a mean. This distance may be covered in two steps: in getting from point A to point C, one may travel to some intermediate point B, then travel from point B to point C. If you considered point A as the individual, point B as the group reference point (the group mean), and point C as the final destination or reference point (the grand mean based on all scores in a distribution), then the total distance A to C (individual to grand mean) is covered by the distance from A to B (individual to respective group mean) plus the distance from B to C (the respective group mean to the grand mean). This simplified description is intended to show that the total sum of squares (SS_T) is divided into two parts: a within-group sum of squares (SS_{WG}) and a between-group sum of squares (SS_{BG}). The two component parts are important in the analysis of variance because the F test statistic is formed by dividing the between-group estimate of variance (the representation of the B to C distances) by the within-group estimate of variance (the representation of the A to B distances).

III. $F = MS_{BG} \div MS_{WG}$: A Relevant Test of H_0

In order to calculate an F statistic for an analysis of variance, at least two variances must be computed from the data collected in an experiment. Remember, an experiment requires a minimum of two groups, thus at least two group means can be determined, which allows for an estimation of a variance between group means (MS_{BG}) in reference to a total mean reference point based on all scores. Also, having distinct groups allows for estimating variances of individual scores within a group (MS_{WG}) with the respective group means as reference points. For the F test statistic, the ratio involves a between-group variance estimate in the numerator and a within-group variance estimate in the denominator ($F = MS_{BG} \div MS_{WG}$, or $F = MS_A \div MS_{S/A}$).

Now that we have the two variances for F, the next issue to address concerns the rationale for F as a test statistic for H_0. If the null hypothesis is true, the independent variable (Factor A) does not affect performance, thus it is likely that any observed differences between groups are the result of random error. First, consider the causes of variation among the scores from individuals within the same group. All participants in a given group have experienced the same level of Factor A (i.e., the same treatment); hence performance differences among these subjects cannot be attributed to the independent variable (in effect, Factor A is held constant for subjects within the same group). All other potential factors (known and unknown) have been distributed randomly in the experiment, thus we cannot specify the causes of observed differences, so they are attributed to the catchall category of *random error*. Therefore, within-group variance ($MS_{S/A}$) provides an estimate of random error variability.

Second, consider the causes of variation among the means of different groups in the experiment. Certainly random error is at work here just as it was for the within-group variance estimate since each of the groups is composed of different randomly assigned subjects. However, in addition to random error, individuals in the different groups received different treatment experiences (levels of Factor A). If the independent variable has an effect on performance, it will be reflected by differences between the group means. Therefore, between-group variance (MS_A) provides an estimate of random error **and** an estimate of variability caused by the independent variable. Hopefully, the illustrations of random assignment in Chapter 4 are helpful in justifying this assertion.

The F test reasoning follows directly. The numerator in the F ratio provides an estimate of variance influenced by random error and potentially influenced by the independent variable. The denominator in the F ratio also provides an estimate of variance influenced by random error. However, since all subjects within any given group receive the same treatment, unlike the variance estimate in the numerator (MS_{BG}), the independent variable does not contribute to the denominator variance estimate (MS_{WG}). Given these contributing sources for variance estimates in the numerator and denominator of the F ratio, there is an obvious result to expect for situations for which H_0 is true. If the independent variable has no effect on the performance being measured, then the expected value of F (long run average) is 1.0, as we would expect both the numerator and denominator simply to provide estimates of random error (if the numerator and denominator of a ratio are the same, the value of the ratio is 1.0). Of course, in the short run the two estimates are not likely to be identical as sometimes random error as estimated by variation between group means (the numerator in the F ratio) may be a little larger than random error as estimated by variation among individuals within the groups (the denominator in the F ratio). And, sometimes the opposite will be true, resulting in a larger estimate of random error within groups compared to the between-group estimate.

A schematic reference distribution for F is shown in Figure 5.1 based on the H_0 assumption that the independent variable does not contribute to the variability observed between groups. Since variances can only be positive numbers, **F ratios cannot be negative**. The reference distribution is expected to peak around a value of 1.0, and it will be positively skewed, meaning F values 3, 4, or more units larger than an expected value of 1.0 are possible. However, F values 3, 4, or more units smaller than an expected value of 1.0 are not possible (e.g., if the numerator variance estimate is four times larger than the denominator variance estimate, $F = 4 \div 1 = +4.00$; however, if the denominator variance estimate is four times larger than the numerator variance estimate, $F = 1 \div 4 = +.25$).

Computed values for F from experiments are compared to the appropriate F reference distribution, much like the illustrations involving the creation of reference distributions for the two outcome experiments (e.g., flipping a coin and recording whether it landed with the "heads" side up or the "tails" side up) described in the last chapter. However, F reference distributions (yes, there are many with each determined by the number of groups being compared and the number of subjects within the groups) are theoretical, continuous distributions and critical information about these distributions are presented in tables that identify key points in the distributions. The critical information needed for statistical decision-making involves a comparison between an F value calculated from a data set and the F value in the appropriate reference distribution at a standard cutoff point. Thus, rather than presenting all of the complete F reference distributions, only values of F at major cutoff points are presented (e.g., the number

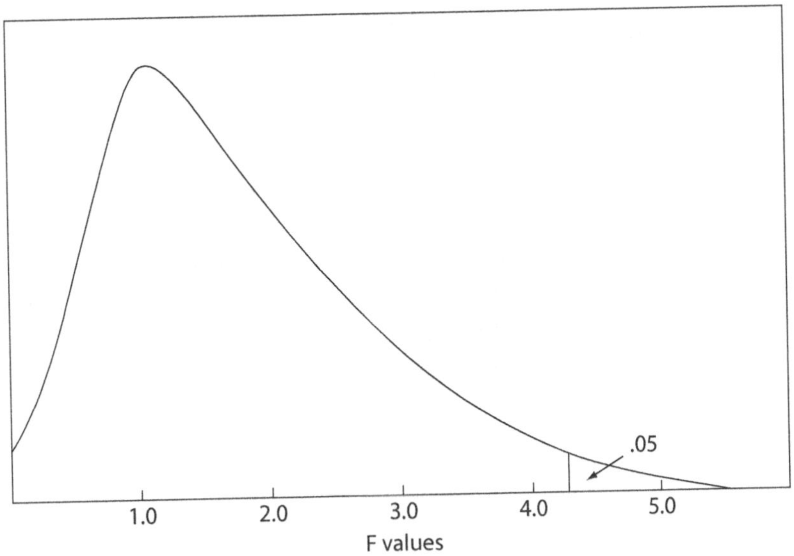

Figure 5.1: Schematic Illustration of a Reference Distribution for F with a = 2 and n = 10

that "cuts off" the most extreme 5% of F values and the most extreme 1% of F values). We will see later that the critical tabled F value for an experiment involving two groups with 10 subjects in each group is F = 4.41 at the .05 probability level and F = 8.29 at the .01 probability level (the specific F reference distribution for $df_{BG} = 1$ and $df_{WG} = 18$). The .05 and .01 probability levels simply identify conventional decision criteria. The interpretation is as follows: if random error is the only factor contributing to the two estimates of variance, then an $F \geq 4.41$ would be expected to occur no more than 5% of the time, and $F \geq 8.29$ would be expected to occur no more than 1% of the time.

If the experimenter decided in advance to use the .05 cutoff for rejecting H_0 (the hypothesis that the only source for group differences was random error), and the F calculated from experimental data was equal to 10.25, then H_0 would be rejected. In the somewhat convoluted language of statistical hypothesis testing, this means that the experimental decision is that the independent variable (Factor A) is **regarded** as having produced an effect on performance (with the effect of Factor A described as *significant*). Had the calculated F been only 3.50 (below the critical F = 4.41 at the .05 level), the experiment would have failed to reject H_0, (Factor A described as *non-significant*) and the experimenter would not be in a position to mount an effective argument that the independent variable made a difference in performance.

IV. Components of Variances

Given these basics for calculating variances "ANOVA style," we are almost ready to apply the operations to a fictitious data set. One more parallel operation must be described before moving on to this application. The parallel operation involves treating group means like we treated individual scores in computing a basic ratio. Bracket A [A] replaces [Y] that was used for computing the total SS. As you can see from the following formula, [A] is computed by squaring the sum or total score for each group, then summing these squared totals and dividing by n (the number of subjects in each group): $[A] = \Sigma A_j^2 \div n$, where A_j = the sum of all the individual scores in Group a_j. Using bracketed terms (basic ratios) as the starting points for the necessary calculations, the formulas for SS_{BG} or SS_A and SS_{WG} or $SS_{S/A}$ are as follows: $SS_A = [A] - [T]$; $SS_{S/A} = [Y] - [A]$.

The numerator for the third variance that could be calculated is $SS_T = [Y] - [T]$; however, since the F ratio involves only MS_A and $MS_{S/A}$, it is not necessary to calculate SS_T. We did make use of the relationship that SS_T was divided into two component parts, SS_A and $SS_{S/A}$, in deriving the bracketed terms for $SS_{S/A}$. If $SS_T = SS_A + SS_{S/A}$, then in bracketed terms, this is equivalent to $[Y] - [T] = [A] - [T] + SS_{S/A}$. Solving for $SS_{S/A}$, we have $SS_{S/A} = [Y] - [T] - [A] + [T]$. The minus [T] and plus [T] cancel out, hence we are left with $SS_{S/A} = [Y] - [A]$.

The variance estimates (mean squares) are the sums of squares divided by their respective degrees of freedom. For SS_A, degrees of freedom correspond to the number of groups minus 1 ($df_A = a - 1$). For $SS_{S/A}$, degrees of freedom are determined by subtracting 1 from the number of subjects in one group, then multiplying this number by the number of groups in the experiment [$df_{S/A} = (a)(n - 1)$]. The formulas for between-group and within-group variance estimates are $MS_A = SS_A \div df_A$ and $MS_{S/A} = SS_{S/A} \div df_{S/A}$, respectively.

To illustrate the computational steps for calculating a between-group variance and a within-group variance, we will use the fictitious data set presented in Table 5.1. Assume that the scores in this table were obtained under either one of two drug conditions (a_1 or a_2). There are two groups ($a = 2$) and 20 subjects in each group ($n = 20$). The total number of scores is determined by $(a)(n) = (2)(20) = 40$. The total variance for this set of 40 scores is given by $s^2 = MS_T = [\Sigma Y_i^2 - (\Sigma Y_i)^2/n] \div [(a)(n) - 1)]$. Squaring each of the 40 individual scores and summing these squared scores, we have: $\Sigma Y_i^2 = 16{,}393$. The sum of the 40 scores is $\Sigma Y_i = 799$; and this sum squared and then divided by 40 is $(\Sigma Y_i)^2/40 = 15{,}960.02$. The total sum of squares is 16,393 – 15,960.02 = **432.98**. The SS_T can be partitioned into an SS_{BG} (SS_A) and an SS_{WG} ($SS_{S/A}$), with $SS_T = SS_{BG} + SS_{WG}$; thus, we now know one thing about SS_A and $SS_{S/A}$, namely that $SS_A + SS_{S/A} = 432.98$.

Table 5.1: Scores from Subjects Tested Following Administration of Drug A_1 or Drug A_2

Drug A_1 (a_1)		Drug A_2 (a_2)	
18	16	21	17
19	21	19	19
21	22	15	21
26	18	17	15
20	16	18	18
20	24	12	22
22	19	21	16
23	26	18	24

15	25	25	23
21	22	21	23

The steps for calculating MS_A and $MS_{S/A}$ are as follows:

Step 1. The Basic Ratios:

$$[Y] = \Sigma Y_i^2 = 18^2 + 19^2 + \ldots + 23^2 = \mathbf{16{,}393.00}$$

$$[T] = (\Sigma Y_i)^2 \div (a)(n) = (799)^2 \div 40 = \mathbf{15{,}960.02}$$

$$[A] = \Sigma A_j^2 \div n = (414^2 + 385^2) \div 20 = \mathbf{15{,}981.05}$$

Step 2. The Sums of Squares (SS):

$$SS_A = [A] - [T] = 15{,}981.05 - 15{,}960.02 = \mathbf{21.03}$$

$$SS_{S/A} = [Y] - [A] = 16{,}393.00 - 15{,}981.05 = \mathbf{411.95}$$

[Note: Earlier we calculated $SS_T = 432.98$; and here we see that $SS_A + SS_{S/A} = 21.03 + 411.95 = 432.98$]

Step 3. Degrees of Freedom (df):

$$df_A = a - 1 = 2 - 1 = \mathbf{1}$$

$$df_{S/A} = (a)(n - 1) = (2)(20 - 1) = (2)(19) = \mathbf{38}$$

[Note: $df_A + df_{S/A} = df_T$. With a total of 40 scores, $df_T = (a)(n) - 1 = (2)(20) - 1 = 40 - 1 = 39$; $df_A = (a - 1) = (2 - 1) = 1$, $df_{S/A} = (a)(n - 1) = (2)(20 - 1) = (2)(19) = 38$; and $df_A + df_{S/A} = 1 + 38 = 39$.]

Step 4. The Variances (MS):

$$MS_A = SS_A \div df_A = 21.03 \div 1 = 21.03 = \mathbf{21.03}$$

$$MS_{S/A} = SS_{S/A} \div df_{S/A} = 411.95 \div 38 = \mathbf{10.84}$$

Calculating $MS_{S/A}$ using the basic ratio formulas is a shorthand way for determining within-group variance. Within-group variance is simply **the average of the variances calculated separately for each group**. For this two group example problem, we could also have calculated $MS_{S/A}$ by using the following formula: $MS_{S/A} = (s^2_{A1} + s^2_{A2}) \div 2$. The advantage of the shorthand formula may be more apparent with experiments involving several groups: for example, with six groups, the basic ratio procedure still involves solving $SS_{S/A} = [Y] - [A]$, and then dividing this result by $df_{S/A}$: $MS_{S/A} = SS_{S/A} \div df_{S/A}$. The same result could be found with the more time-consuming procedure of calculating six separate variances and then taking the average of these six variances: $MS_{S/A} = (s^2_{A1} + s^2_{A2} + s^2_{A3} + s^2_{A4} + s^2_{A5} + s^2_{A6}) \div 6$.

The sum of squares for the 20 A_1 scores is $SS_{A1} = [Y_{A1}] - [T_{A1}]$, where Y_{A1} and T_{A1} refer to the 20 scores in the first two columns of Table 5.1.

$$[Y_{A1}] = 18^2 + 19^2 + \ldots + 22^2 = 8{,}764.00$$

$$[T_{A1}] = 414^2 \div 20 = 171{,}396 \div 20 = 8{,}569.80$$

$$SS_{A1} = [Y_{A1}] - [T_{A1}] = 194.20$$

$$s^2_{A1} = MS_{A1} = SS_{A1} \div df_{A1} = 194.20 \div (n - 1) = 194.20 \div 19 = \mathbf{10.22}$$

Repeat the computations for the 20 A_2 scores (columns 3 and 4):

$$[Y_{A2}] = 21^2 + 19^2 + \ldots + 23^2 = 7{,}629.00$$

$$[T_{A2}] = 385^2 \div 20 = 148{,}225 \div 20 = 7{,}411.25$$

$$SS_{A2} = [Y_{A2}] - [T_{A2}] = 217.75$$

$$s^2_{A2} = MS_{A2} = SS_{A2} \div df_{A2} = 217.75 \div (n - 1) = 217.75 \div 19 = \mathbf{11.46}$$

These calculations confirm the fact that the within-group mean square calculated earlier [$MS_{S/A} = SS_{S/A} \div df_{S/A} = 411.95 \div 38 = \mathbf{10.84}$] is simply the average of the individual group variances: $MS_{S/A} = (s^2_{A1} + s^2_{A2}) \div a = (10.22 + 11.46) \div 2 = \mathbf{10.84}$. Since $F = MS_A \div MS_{S/A}$, $F = 21.03 \div 10.84 = 1.94$.

V. Summary

The analysis of variance is a statistical procedure for testing a null hypothesis (H_0). Mathematically it requires computation of at least two variances, as the F statistic from the ANOVA is a ratio of two variance estimates. Since computations on a hand calculator for ANOVA can be complicated and tedious, an organized step-wise approach in completing these operations is recommended. Working backwards, the final computational step (the F ratio) involves dividing the between-group variance estimate by the within-group variance estimate ($F = MS_A \div MS_{S/A}$). The numerator portion of the between-group variance estimate is the sum of squares between groups (SS_A) and the denominator is the degrees of freedom between groups (df_A): $MS_A = SS_A \div df_A$. Similarly, the numerator portion of the within-group variance estimate is the sum of squares within groups ($SS_{S/A}$) and the denominator is the degrees of freedom within groups ($df_{S/A}$): $MS_{S/A} = SS_{S/A} \div df_{S/A}$. Degrees of freedom between- and within-groups are easily determined: df_A is the number of groups (levels of Factor A) minus 1, or $df_A = (a - 1)$; $df_{S/A}$ is the number of subjects within a group minus 1 $(n - 1)$ multiplied by the number of groups in the experiment, or $df_{S/A} = (a)(n - 1)$. Each sum of squares has two component parts, expressed as bracketed terms or basic ratios. The two basic ratios for SS_A are [A] and [T]. The two basic ratios for $SS_{S/A}$ are [Y] and [A]. The computational work begins with computing basic ratios. The basic ratio for A is obtained by squaring each group sum, adding the squared sums, and dividing this sum by the number of subjects in each group (n). The basic ratio for T is obtained by summing all scores in the data set, squaring the resulting sum, and dividing this sum by the total number of scores [(a)(n)]. The basic ratio for Y is obtained by squaring each individual score and then summing the squared scores.

Once an F ratio for a data set has been computed, a decision is made about whether or not the independent variable had an effect on performance in the experiment by comparing the calculated F to the appropriate reference distribution of F values. The reference distribution is built on the H_0 assumption that the only thing affecting performance of the different groups in the experiment is random error. If this were true, then hypothetical F values from thousands of tests would hover around a value of 1.0, with some lower (but never less than 0) and some higher. The question is, does the F ratio calculated from the experimental data fit in the extreme portion of the appropriate reference distribution (e.g., mingled amongst the rarest 5% of the reference distribution F values)? If an F calculated from an experiment is in this extreme 5% region and out of the main stream of reference distribution F values, then the experimenter decides (knowing there is a small error risk of this being an incorrect decision; referred to as an α or Type I error) that the independent variable had an effect on performance.

5.1: Model Problem. Answer the following questions based on the data in the table below:

Groups	a_1	a_2	a_3	a_4	a_5
	4	6	2	3	0
	4	9	5	4	4
	7	7	4	5	2
	5	6	6	5	3
	3	5	5	7	6
	6	9	3	4	3
ΣA_j	**29**	**42**	**25**	**28**	**18**

A. What does a = ?
The lower case letter code "a" is used to identify the number of groups in the experiment. In this case the data are organized into five groups, or **a = 5.**

B. What does n = ?
The lower case letter code "n" is used to identify the number of subjects in a group. There are six scores in each group column, or **n = 6.**

C. What does df_A = ?
The degrees of freedom between groups is determined by subtracting 1 from the number of groups in the experiment, or $\mathbf{df_A = a - 1 = 5 - 1 = 4}$.

D. What does $df_{S/A}$ = ?
The degrees of freedom within groups is determined by subtracting 1 from the number of subjects in a group, then multiplying the degrees of freedom for one group by the number of groups (a) in the experiment, or $\mathbf{df_{S/A} = (a)(n - 1) = (5)(6 - 1) = (5)(5) = 25}$.

E. What does [Y] = ? What does [T] = ? What does [A] = ?

$[Y] = \Sigma Y_i^2 = (4^2 + 7^2 + \ldots + 3^2) = \mathbf{788.00}$
$[T] = (\Sigma Y_i)^2 \div (a)(n) = (4 + 7 + \ldots + 3)^2 \div (5)(6) = (142)^2 \div 30 = 20{,}164 \div 30 = \mathbf{672.13}$
$[A] = \Sigma A_j^2 \div n = (29^2 + 42^2 + 25^2 + 28^2 + 18^2) \div 6 = 4{,}338 \div 6 = \mathbf{723.00}$

F. What does SS_A = ?

$SS_A = [A] - [T] = 723.00 - 672.13 = \mathbf{50.87}$

G. What does $SS_{S/A}$ = ?

$SS_{S/A} = [Y] - [A] = 788.00 - 723.00 = \mathbf{65.00}$

H. What does $MS_A = ?$

$MS_A = SS_A \div df_A = 50.87 \div 4 = \mathbf{12.72}$

I. What does $MS_{S/A} = ?$

$MS_{S/A} = SS_{S/A} \div df_{S/A} = 65.00 \div 25 = \mathbf{2.60}$

5.2: Practice Problem 1. Using the data in the following table, provide the information called for in Parts A through I identified in the Model Problem (i.e., for this data set, a = ?, n = ?, $MS_{S/A}$ = ?, etc.)

Groups	a_1	a_2	a_3
	10	14	21
	8	15	15
	12	10	14
	5	9	11
	15	10	12
	9	12	20
	11	11	16

5.3: Practice Problem 2. Calculate the between-group variance (MS_A) and the within-group variance ($MS_{S/A}$) for the scores listed in the following table:

Groups	a_1	a_2	a_3	a_4
	25	33	29	40
	16	20	28	29
	21	19	35	35
	14	22	25	30
	24	21	25	32
	28	30	32	36
	17	15	26	44
	25	30	28	27
	30	29	25	24
	20	25	22	24

5.4: Practice Problem 3. An experimenter measures running speed (measured in seconds) of albino rats placed in a 5-meter straight alley. There are 6 groups in his experiment (a = 6), with 8 subjects in each group (n = 8). The independent variable (Factor A) is the number of small food pellets at the end of the alley (1, 3, 5, 7, 9, or 11). Calculate the F ratio for comparing the running speeds of these 6 groups of rats.

Data Matrix: Running Speed in Seconds

	Number of Food Pellet Rewards					
Albino Rats	a_1(1)	a_2(3)	a_3(5)	a_4(7)	a_5(9)	a_6(11)
	18	16	23	19	25	20
	29	22	19	20	15	16
	36	25	24	25	18	21
	30	30	28	22	16	18
	32	29	30	24	27	20
	25	26	22	20	19	13
	20	21	18	18	20	17
	33	25	22	20	17	18

5.5: Practice Problem 4. The data from Practice Problem 2 in Chapter 3 were presented in a two-group format (Group I vs. Group II), and a variance was calculated for each group: $s_1^2 = 18.54$ and $s_2^2 = 10.04$. The average s^2 for these two groups is $(18.54 + 10.04) \div 2 = 14.29$. Report values for [Y], [A], $SS_{S/A}$, and $MS_{S/A}$ for this data set (given on pages 46 and 47). Since $MS_{S/A}$ is actually the average of the individual group variances, you know that $MS_{S/A}$ must equal 14.29 (within rounding error).

VII. Practice Test 1: Chapters 1–5

(answers given in Appendix B)

I. Multiple choice: Circle the letter or number for the best answer to each question

1. A student did a "pilot" study to test whether accurate recognition of faces was related to Race. Specifically, she collected a large number of photographs of Asian women. Her experiment involved showing 10 randomly selected "target" faces, and allowing research participants 30 seconds to study the faces. She then had participants complete a survey (a task that took 15 minutes to complete). After the survey participants were shown 50 photos of Asian women and asked to select the 10 they saw earlier. There were three groups of research subjects, with 25 volunteers in each group. She had one group of White males, one group of Black males, and one group of Asian males. Her theory was that the "In-Group" (Asian males) would be better at recognizing the Asian faces (average number of correct recognitions) than the other two groups. If the pilot study confirmed expectations, she planned to repeat the experiment with photos of White and Black women.

   A. The independent variable in this experiment is:
      1. The 50 photos of Asian women
      2. The 15-minute survey task
      3. Racial classification of male participants
      4. Number of faces correctly recognized

   B. The dependent variable in this experiment is:
      1. The 50 photos of Asian women
      2. The 15-minute survey task
      3. Racial classification of male participants
      4. Number of faces correctly recognized

2. In an experiment, subject mortality (loss of subjects from one or more groups) is a particular concern because it
   a. may destroy initial group equivalence.
   b. may redefine the independent variable in the experiment.
   c. may eliminate homogeneity of variance.
   d. may require a change in the research hypothesis.

3. The statistic developed as an index of variability was based on the concept of taking an 'average' of squared deviations from the mean. However, the variance ended up having n−1 in the denominator instead of n. This modification was necessary
   a. to avoid always having a result of 0.
   b. to insure that the sum of the squared deviation scores would be the smallest possible.

c. to correct a bias because sample variance systematically underestimated population variance.
d. all of the above.

4. A measure of central tendency that divides a distribution of scores in half (an equal number of scores above and below this point) is called
   a. mean
   b. median
   c. range
   d. standard deviation

5. Assume a mean for a distribution of scores is 6 and two variances are calculated. For one variance the mean is used as the reference point and for the other variance, 0 is used as the reference point. Which of the following statements is true?
   a. Variance about the mean = variance about 0.
   b. Variance about the mean < variance about 0.
   c. Variance about the mean > variance about 0.
   d. Cannot be determined without knowing the scores in the distribution.

6. The mean and variance on a 50-point quiz in an introductory psychology class were $\bar{Y} = 25$ and $s^2 = 16$. There were 100 students in the class. To get a better idea of where he stood in relation to the rest of the class, a student who earned a score of 30 on the quiz converted it to a z score, which was
   a. $z = -1.00$
   b. $z = +.31$
   c. $z = +1.00$
   d. $z = +1.25$

7. Who is credited with 'inventing' the Analysis of Variance?
   a. Freud
   b. Loftus
   c. Pearson
   d. Fisher

8. The reading speeds of three groups of 3rd grade children were compared: a public-school group, a private-school group, and a home-school group. Average reading speeds in terms of words read per minute were 133, 130, and 120, respectively. The null hypothesis states that the reading scores for the three groups should
   a. all be 0.
   b. be 133, 130, and 120, respectively.
   c. all be the same.
   d. all be different.

9. For an inferential statistical test, a statistic is computed from data that have been collected, and the value of the statistic is compared to a *reference* distribution of possible values. The reference distribution is determined based on assumptions regarding how the statistic would most likely be distributed if the ______________________.
   a. null hypothesis (H_0) was true
   b. null hypothesis (H_0) was false
   c. research hypothesis (H_1) was true
   d. research hypothesis (H_1) was false

10. Assume an experiment involves three groups from a prison population: one taking a low-key anger-management class; one taking an aggressive anger-management class; and a no-treatment control group. Of course, there is variability both between groups and within groups. Random error (RE) and the independent variable (IV: anger-management class condition) may account for between- and within-group variability as described below.
   a. RE for within-group and IV for between-group variability.
   b. IV for within-group and RE for between-group variability.
   c. RE for within-group and both RE and IV for between-group variability.
   d. RE and IV for both within and between-group variability.

II. Re-organize the following 25 scores, presenting them in a **grouped frequency distribution, setting 35–38 as the highest interval**:
23, 33, 37, 35, 18, 21, 25, 16, 30, 29, 30, 14, 20, 16, 32, 23, 24, 26, 25, 21, 19, 16, 36, 20, 28

INTERVAL	FREQUENCY
35–38	

III. Below are 16 scores. Do the necessary computations to determine (1) the **mean** ($\bar{Y}$), (2) the **median** (Mdn), (3) the **sum of squares** (SS), and (4) the **standard error of the mean** (abbreviated s_E or s_M).

SCORES

15
9
21
25
12

(1) $\bar{Y}$ =

10	(2) **Mdn** =
10	
16	
11	
12	(3) **SS** =
20	
18	
10	
8	
15	
22	(4) $\mathbf{s_E}$ =

IV. An experimenter developed an apparatus connecting a pacifier to a stereo system. She was interested in determining if infants display preferences for different styles of music. Sucking on the pacifier resulted in a 30-second musical presentation through the stereo system. Her study involved 21 infants tested when they were 6 months of age. She randomly assigned the infants to one of three groups ($a = 3$), such that there were 7 in each group ($n = 7$). Each infant was placed in a room for 30 minutes and given a pacifier. For infants in Group C, when an infant sucked on the pacifier classical music was played. For infants in Group F, when an infant sucked on the pacifier folk music was played. For infants in Group J, when an infant sucked on the pacifier jazz music was played. For each infant, the total time (in minutes) that music was played during the 30-minute session was recorded, and the data are presented in the table below. What is the value of the F statistic for these data?

Number of Minutes Infants "Played" Music

Group C	Group F	Group J
11.5	12.0	6.0
15.5	11.5	9.5
10.0	7.0	10.5
16.0	9.0	12.0
20.5	14.0	11.0
8.5	10.0	8.0
17.0	13.5	13.0

V. Although there are only two columns of numbers below, three different sums of squares can be calculated (a between-group sum of squares, a within-group sum of squares, and a total sum of squares). Calculate the values for these three sums of squares.

Group I	Group II
4	5
3	8
5	4
5	6
2	4
6	7
1	3

Chapter 6

The Single-Factor ANOVA

To be a statistician is great! You never have to be absolutely sure. …
Being reasonably certain is enough.
—Guarisma

I. Inferential Analysis of a Two-Group Experiment

In the last chapter, a step-wise procedure was described for calculating variances. A number set of 40 scores was used for this illustration, with the numbers organized as if they were scores from a two-group experiment (comparing Drug A_1 with Drug A_2) with 20 individuals in each group (a = 2, n = 20). That illustration ended with the calculation of two variances: (1) a between-group variance, $\mathbf{MS_{BG}}$ **= 21.03**; and (2) a within-group variance, $\mathbf{MS_{WG}}$ **= 10.84**. The example problem concluded with most of the major computational work involved for an analysis of variance completed. There is one remaining simple calculation left to perform, and one minor coding change that will be used hereafter. The independent variable is referred to as Factor A, hence from now on we will replace the BG subscript for between groups with A, and we will replace the WG subscript for within groups with S/A (read as "subjects within the A groups"). The test statistic in the analysis of variance is F (named in honor of Sir Ronald A. Fisher, a noted statistician and evolutionary biologist, 1890–1962), and F = MS_A ÷

Table 6.1: A Selection of Critical F Values from Table A-1

			df_A		
$df_{S/A}$	1	2	3	4	5
10	**4.96**	**4.10**	**3.71**	**3.48**	**3.33**
	10.0	7.56	6.55	5.99	5.64
20	**4.35**	**3.49**	**3.10**	**2.87**	**2.71**
	8.10	5.85	4.94	4.43	4.10
30	***4.17***	**3.32**	**2.92**	**2.69**	**2.53**
	7.56	5.39	4.51	4.02	3.70
40	**4.08**	**3.23**	**2.84**	**2.61**	**2.45**
	7.31	5.18	4.31	3.83	3.51
60	**4.00**	**3.15**	**2.76**	**2.53**	**2.37**
	7.08	4.98	4.13	3.65	3.34

$MS_{S/A}$. Therefore, we can complete the analysis of variance for the data set used in the illustration from Chapter 5 by performing this last calculation: $F = MS_A \div MS_{S/A} = 21.03 \div 10.84 = 1.94$.

At this point all that remains to be done is the interpretation of the ANOVA results (the F test). Interpreting ANOVA results requires a decision regarding whether or not the independent variable (Factor A, which for this fictitious data set was the specific drug treatment) made a difference in the observed scores (i.e., was the drug effect *significant*?). The conventional procedure for this decision involves a comparison of the F obtained from the ANOVA with an appropriate reference distribution.

A copy of a portion of an F table is presented in Table 6.1, with df_A values for columns and $df_{S/A}$ values for rows. Two critical cutoff points are shown for each reference distribution represented: one in bold for the .05 cutoff point (which means that if group means were **exactly equal** as specified by the null hypothesis, then observing an F statistic as large as or larger than the tabled value at this .05 cutoff should occur no more than 5% of the time) and the second tabled value is for F at the .01 cutoff point (an F value this large or larger should occur no more than 1% of the time if the true difference between group means is 0 as specified by H_0). These critical cutoff points are identified as either α (referring to a Type I or α error risk) ≤ .05 and .01 respectively, or p (referring to probability) ≤ .05 and .01. Each row by column intersection identifies critical F (F_{crit}) values for different reference distributions. For the example problem we considered, a = 2 and n = 20; therefore $df_A = a - 1 = 1$ and $df_{S/A} = (a)(n - 1) = (2)(19) = 38$. The F_{crit} values for this reference distribution will be in the column labeled "1" and the row labeled "38" (shown in italics). The F table from which this excerpt was taken (see Appendix A, Table A-1) skips critical F values for within group degrees of freedom of 31–39; so we will approximate F_{crit} by using $df_{S/A} = 30$. This F_{crit} value at p ≤ .05 is 4.17. The ANOVA computations for our fictitious data set resulted in F = 1.94, which does not exceed F_{crit}, meaning that the group difference was not so extreme as to render it as highly improbable based on random error only). You may note that $F_{crit} = 4.08$ for $df_{S/A} = 40$. Thus, if F from the ANOVA had been greater than 4.08, but less than 4.17, we would need to go to a more detailed table for the significance test. If exact degrees of freedom are not specified in the F table, we recommend that you use the smaller df value: if the F you calculate from the data is significant with the smaller $df_{S/A}$ (e.g., 30), it will certainly be significant with the actual $df_{S/A}$ (e.g., 38). When using a computer program for analysis of variance, exact probabilities are provided, so limitations of different F tables are not a problem.

A second reporting issue involves the format used to convey ANOVA results. It is helpful to present computational results in a detailed Analysis of Variance Summary Table—this is the format commonly encountered in statistics textbooks. The standard analysis of variance summary table identifies the sources of variance, sums of squares, degrees of freedom, mean squares, and F values. The comparison F_{crit} may also be reported in the table, or an alternative indicator of the significance decision may appear below the table or next to the F value (e.g., $p \leq .05$ for a significant F value and ns or non-sig for a non-significant F). The complete ANOVA summary table for the problem at hand appears in Table 6.2.

Table 6.2: Analysis of Variance Summary Table

Source	SS	df	MS	F	F_{crit}
A (between-groups)	21.03	1	21.03	1.94	4.17
S/A (within-groups)	411.95	38	10.84		

A briefer format for reporting ANOVA results is common for journal publications. The F value, degrees of freedom, error variance, significance value, and often one additional index that we will introduce later (an estimate of effect size) are reported following a verbal description of the effect. For example, a recent study reported the decision to use $p \leq .05$ as the critical rejection region for all statistical tests, and the following description of the results for one of several statistical tests reported: "Proportion of false recognitions were higher for experimental (M = .35) than for control lures (M = .26), $F(1, 92) = 16.49$, MSE = .05" (Halcomb, Taylor, DeSouza, and Wallace 2008, p. 453). In this shorthand statistical phrase, the observed F was reported (16.49), degrees of freedom were given in parentheses next to F ($df_A = 1$, $df_{S/A} = 92$), the error variance estimate was given (the within-group variance, referred to as MSE [for "mean square error"] = .05), and in addition, means for the groups involved in the contrast (coded as M = .35 and M = .26) were reported.

In the remainder of this section we will do three statistical tests on a small data set ($a = 2$, $n = 10$). The three statistical analyses involve (1) an analysis of variance (the test procedure just considered), (2) the t test statistic (known as Student's t developed by W. S. Gossett, 1876–1937), and (3) a confidence interval test. The ANOVA is the most versatile of the three test procedures, as Student's t and confidence interval applications are restricted to comparisons between two groups or conditions. It is fair to say that most published experiments today involve more than two groups or conditions; consequently, a justification for emphasizing ANOVA procedures in this book is based on their greater utility.

At the risk of spoiling the surprise concerning the relation among these three statistical tests, you should know all three procedures produce identical conclusions regarding H_0. Indeed, this must be the case since $F = t^2$ whenever $df_A = 1$. A $CI_{.95}$ will include 0 in the range of highly probable differences between group means whenever $F_{.05}$ and $t_{.05}$ values indicate a non-significant effect, because in such cases a zero difference between groups cannot be ruled out. A $CI_{.95}$ will not include 0 in the range of highly probable differences between group means whenever F and t values indicate a significant effect.

So does it matter which procedure is used? Using F or t really does not matter here; however, since t is limited to comparisons between two groups and F is appropriate for experiments with two or more groups, our focus will be on F. The t test is introduced primarily because it is used sometimes and reported in published experiments, and because it is commonly introduced in introductory statistics textbooks in psychology when developing H_0 testing from simple two-group designs to complex designs involving multiple groups. For confidence intervals, one can certainly make a good case for preferring the CI approach (see Bird 2004; Kline 2004) to t and F tests when an experiment involves comparisons between two groups, because CIs not only provide the same evaluation of H_0 as t and F, but, as we will see, a CI is richer in information as it also identifies a range of scores wherein lies the likely true difference between group means. The data set that will be worked, and reworked to illustrate F, t, and CI appears in Table 6.3.

Table 6.3: Fictitious Data for a Two-Group Experiment to Illustrate F, t, and CI Procedures

Group A_1	Group A_2
3	7
5	5
2	9
4	6
7	6
5	9
1	8
4	4
6	5
2	5

a. Two-Group Analysis of Variance (F Test)

Nothing is new here except the data set with a = 2 and n = 10. The step-wise computations necessary for calculating F are as follows:

Step 1. The Basic Ratios:

$$[Y] = \Sigma Y_i^2 = 3^2 + 5^2 + \ldots + 5^2 = \mathbf{623.00}$$

$$[T] = (\Sigma Y_i)^2 \div (a)(n) = (103)^2 \div 20 = \mathbf{530.45}$$

$$[A] = \Sigma A_j^2 \div n = (39^2 + 64^2) \div 10 = \mathbf{561.70}$$

Step 2. The Sums of Squares (SS):

$$SS_A = [A] - [T] = 561.70 - 530.45 = \mathbf{31.25}$$

$$SS_{S/A} = [Y] - [A] = 623.00 - 561.70 = \mathbf{61.30}$$

Step 3. Degrees of Freedom (df):

$$df_A = a - 1 = 2 - 1 = \mathbf{1};\ df_{S/A} = (a)(n - 1) = (2)(10 - 1) = (2)(9) = \mathbf{18}$$

Step 4. The Variances (MS):

$$MS_A = SS_A \div df_A = 31.25 \div 1 = \mathbf{31.25}$$

$$MS_{S/A} = SS_{S/A} \div df_{S/A} = 61.30 \div 18 = \mathbf{3.41}$$

Step 5. The F ratio:

$$F = MS_A \div MS_{S/A} = 31.25 \div 3.41 = \mathbf{9.16}$$

Step 6. The Reference Distribution Comparison: From Table A-1 in Appendix A at the back of the book, we see that $F_{crit}(1, 18) = 4.41$. Since the F obtained in this example ($F = 9.16$) exceeds F_{crit}, the difference between groups is significant at the .05 level. The summary of this analysis of variance is presented in the following table.

Table 6.4: Analysis of Variance Summary Table

Source	SS	df	MS	F	F_{crit}
A	31.25	1	31.25	9.16 $p < .05$	4.41
S/A	61.30	18	3.41		

b. t-Test Alternative for Two-Group Comparisons

Since $F = t^2$, we already know that within rounding error, t will be the square root of F: $t = \sqrt{9.16} = \pm 3.03$; and further, that this value of t will be significant at $p \leq .05$. The t statistic is the difference between group means divided by an estimate of error variance. Two equivalent formulas for t are given: one that might be considered a "pre-ANOVA" version and the one we will use that might be considered a "post-ANOVA" version since it expresses error variance in terms of $MS_{S/A}$. Our choice of this latter version makes the computational tasks for this problem somewhat easier since $MS_{S/A}$ has already been calculated ($MS_{S/A}$ = 3.41). The formula for t is $t = (\bar{Y}_{A2} - \bar{Y}_{A1}) \div s_{diff}$, where s_{diff} is the standard error of the difference. In textbooks that introduce the t test before moving on to the analysis of variance, the formula for s_{diff} may appear as: $s_{diff} = \sqrt{([s^2_{A2} + s^2_{A1}] \div n)}$. This formula is equivalent to the following formula that expresses the estimate of error variance with the ANOVA calculation of $MS_{S/A}$: $s_{diff} = \sqrt{([2][MS_{S/A}] \div n)}$.

For the so-called post-ANOVA formula, we need to calculate $\bar{Y}_{A2}$ and $\bar{Y}_{A1}$. We know from the ANOVA calculations that $MS_{S/A} = 3.41$, and determining means should by now be a relatively easy exercise: $\bar{Y}_{A1} = 39 \div 10 = 3.90$ and $\bar{Y}_{A2} = 64 \div 10 = 6.40$. Therefore, $t = (6.40 - 3.90) \div \sqrt{([2][3.41 \div 10])} = 2.5 \div \sqrt{([2][.341])} = 2.5 \div \pm .826 = \pm 3.03$.

The t reference distributions are approximations of the z distribution, thus they are symmetrical (although slightly flatter) with the high point at t = 0. Different t distributions are determined by the degrees of freedom parameter. The t test is appropriate for comparisons between two groups, and with two groups, degrees of freedom is simply the sum of degrees of freedom for each group. If there are n_1 subjects in Group 1 and n_2 subjects in Group 2, then $df = (n_1 - 1) + (n_2 - 1)$. With a "large" sample size, a t reference distribution comes very close to just overlapping the z reference distribution (e.g., at the $p \leq .05$ cutoff point, t and z both equal ± 1.96 when $n = \infty$; statistical tables of critical t values often do not provide probability information beyond n = 60 or n = 120; e.g., critical .05 cutoff points are t = ±2.00 with df = 60; t = ±1.98 with df = 120; and t = ±1.96 with df = ∞).

Figure 6.1 provides a schematic representation of the z reference distribution, with a specific approximation t reference distribution having 18 degrees of freedom superimposed (represented by the dashed lines). The flatter reference distribution for t means that slightly more of the area under the curve is out at the extremes. That is, you must go further out along the baseline of t values in the t distribution before you cut off 5% of the total area under the curve (±2.10) than you do for z values in the z distribution (for z, the 5% cutoff point occurs at ±1.96).

Table A-2 in Appendix A in the list of statistical tables identifies critical cutoff points for selected t reference distributions having different $df_{S/A}$ values. You should regard each critical cutoff value listed in this table as either a positive or negative number. For example, we would expect to find $t \geq +2.10$ with df = 18 approximately 2.5% of the time by chance alone; we would expect to find $t \leq -2.10$ approximately

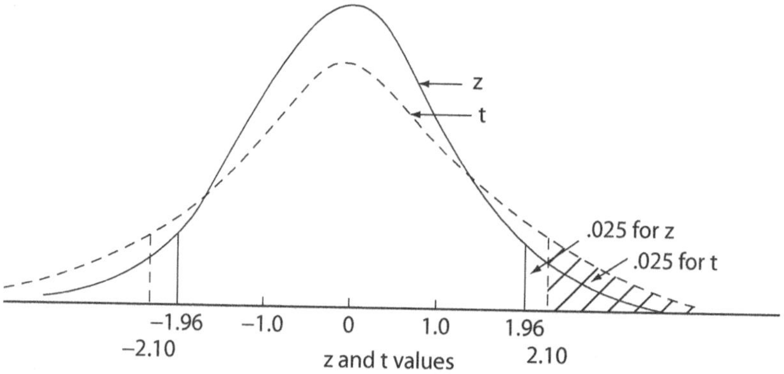

Figure 6.1: The z Reference Distribution with Critical Cutoff Points of ±1.96 for p ≤.05 and a t Reference Distribution with df = 18 and Critical Cutoff Points of ±2.10 for p ≤ .05

2.5% of the time by chance alone (the combined probability for either of these two extreme outcomes is .025 + .025 = .05, with equally extreme values of t lying in both tails of the reference distribution; hence the typical test is called a *two-tailed* test of H_0). Recall for a comparison between two groups, $t^2 = F$, thus $t_{crit}^2 = F_{crit}$. A quick check of the statistical tables in Appendix A indicates that for F (Table A-1) with $df_A = 1$ and $df_{S/A} = 18$, F_{crit} at $p \leq .05$ is 4.41. From Table A-2 we see that t_{crit} at $p \leq .05$ with df = 18 is 2.10, and $2.10^2 = 4.41$. For this data set, we found t = 3.03, which exceeds $t_{crit}(18) = 2.10$; hence t is significant at $p < .05$.

c. A Confidence Interval (CI) for the Difference Between Group Means

Before considering the CI approach to testing the difference between **two** group means, it will be necessary to digress briefly to explain a CI in the context of estimating a population parameter such as the mean (μ) based on a statistic estimating μ from a sample ($\bar{Y}$). A random sample of cases is often examined for the purpose of understanding something about the population from which the sample was drawn. The sample mean certainly provides a reasonable estimate of the mean for the population; however, in most cases the population mean and sample mean will not be identical. So if the sample mean is 10.0, it may not be very impressive to say to someone that my best guess is that the population mean is 10.0, but I am quite sure that is wrong! Wouldn't it be better to create a reasonable margin of error so that even though the response about the population mean is inexact, a narrow band within which it lies can be estimated with a high degree of confidence? Of course, to be 100% certain, you could say that "I am absolutely sure that the population mean is somewhere between $-\infty$ and $+\infty$"; a totally useless response, and one that does not require sampling from a population to justify. So we are looking for a compromise: a response that is somewhat inexact, allowing for a range of possible values for μ that can be expressed with a high degree of confidence (e.g., with a 95% or 99% degree of confidence).

In developing a confidence interval we will center it around the mean of the sample since that is the best point information we have about the location of the population mean. A range of values below and above this sample mean starting point will be determined based on estimates of variability due to random sampling error. It should be emphasized that with CIs the interest is in the main stream of scores—that is, the objective is to set lower and upper boundaries in such a way that if you were to do this 100 times for 100 different samples, you would expect the true population mean to be included in 95% or 99% of the intervals so constructed. That is, 'If we were to repeat the experiment over and over, then 95% of the time the confidence intervals contain the true mean' (Hoekstra, Morey, and Rouder 2014, p. 1160).

To determine confidence intervals, we sort of turn the tables on the H_0 testing procedures, making use of critical cutoff values at the .05 level (for a 95% confidence interval) or the .01 level (for a 99% confidence interval). We will illustrate a single-sample CI estimation by using the Group a_1 data and a 95% confidence interval ($CI_{.95}$). The formula for CIs is composed of recognizable elements: $CI_{.95}$: $\bar{Y}_{A1} - (t_{.05})\sqrt{(s^2_{A1} \div n)} \leq \mu \leq \bar{Y}_{A1} + (t_{.05})\sqrt{(s^2_{A1} \div n)}$; we know $\bar{Y}_{A1} = 3.9$, n = 10, and from Table A-2 in Appendix A, we see that $t_{.05}(9) = 2.26$ (for a t test comparing a single sample mean with a population mean, df = n – 1), and $\sqrt{(s^2_{A1} \div n)}$ is the *standard error of the mean* (commonly coded as s_E; the above formula is equivalent to the standard deviation divided by the square root of the sample size: $s_E = s_{A1} \div \sqrt{n}$). Carrying out the computations for this $CI_{.95}$ we have:

$s^2_{A1} = [(3^2 + 5^2 + \ldots + 2^2) - 39^2/10] \div 9 = (185 - 152.1) \div 9 = 32.9 \div 9 = 3.66;$

and the standard error of the mean is $s_E = \sqrt{(s^2_{A1} \div n)} = \sqrt{(3.66 \div 10)} = .60$

$CI_{.95}$: $3.9 - (2.26)(.60) \leq \mu \leq 3.9 + (2.26)(.60) = 3.9 - 1.36 \leq \mu \leq 3.9 + 1.36$

$CI_{.95}$: $2.54 \leq \mu \leq 5.26$

What this CI means is that while we cannot say for certain what the true mean in the population is, we can state with 95% confidence that μ is at least 2.54 and no greater than 5.26. The extension relevant to testing H_0 changes the estimation of a population mean based on a sample mean to the estimation of the difference between two population means based on the difference between two sample means. In determining the difference between sample means it does not matter if $\bar{Y}_{A1}$ is subtracted from $\bar{Y}_{A2}$, or vice versa. In order to have a positive difference score, we will continue to subtract the smaller $\bar{Y}_{A1}$ mean from the larger $\bar{Y}_{A2}$ mean.

Before considering the modified formula for CIs for the difference between two group means, let's look at this in the context of H_0 testing. What does it mean if a t and F fail to reject H_0? It means that we cannot rule out the possibility that the true difference between group means is 0, or more generally, that the actual numerical difference in the experiment is merely a chance variation from this true 0 difference. So if a non-significant F means that a difference of 0 between group means is viable, then a corresponding CI must specify an interval that includes 0 as a possibility (i.e., a negative difference between sample means at the lower boundary and a positive difference at the upper boundary). Thus, based on the $F_{.05}$ and $t_{.05}$ tests, we know something about $CI_{.95}$ (the 95% confidence interval) without doing any calculations. The F and t tests in the example problem were significant, resulting in rejection of H_0. If H_0 (the hypothesis that the difference between groups is 0) is rejected at the .05 probability level, then the interval for the 95% confidence level must exclude 0 (either both lower and upper boundaries will be negative numbers or both will be positive numbers). Let's check to make sure this is the case.

The CI formula for the difference between two group means is an expansion of the formula for a CI estimating a population mean from a single sample: t is found the same way; the difference between two means $(\bar{Y}_{A2} - \bar{Y}_{A1})$ replaces the single sample mean in the formula; the *standard error of the difference* $[\sqrt{(s^2_{A2} + s^2_{A1}) \div n}]$ replaces the standard error of the mean $[\sqrt{(s^2_{A1} \div n)}]$, and degrees of freedom for finding the $t_{.05}$ value from Table A-2 is determined by summing the degrees of freedom for the two groups $(n_1 - 1 + n_2 - 1)$. Thus, the formula for the 95% confidence interval for the difference between means is given by the following formula:

$$CI_{.95}: (\bar{Y}_{A2} - \bar{Y}_{A1}) - (t_{.05})\sqrt{([s^2_{A2} + s^2_{A1}] \div n)} \leq \mu_2 - \mu_1 \leq (\bar{Y}_{A2} - \bar{Y}_{A1}) + (t_{.05})\sqrt{([s^2_{A2} + s^2_{A1}] \div n)}.$$

The variance for Group a_1 was calculated for the single sample CI just completed: $s^2_{A1} = \mathbf{3.66}$. However, we need to repeat this operation for Group a_2: $s^2_{A2} = [(7^2 + 5^2 + \ldots + 5^2) - 64^2/10] \div 9 = (438 - 409.6) \div 9 = 28.4 \div 9 = \mathbf{3.16}$. Substituting these two variance estimates into the formula for the standard error of the difference (s_{diff}) we have: $s_{diff} = \sqrt{([3.66 + 3.16] \div 10)} = \sqrt{.682} = .826$. For the two-group comparison, degrees of freedom are determined by the sum of the degrees of freedom for each group, or $n_1 - 1 + n_2 - 1$, which is $10 - 1 + 10 - 1 = 18$, and $t_{.05}(18) = 2.10$. The respective means for Group a_2 and Group a_1 are 6.4 and 3.9. Substituting $\bar{Y}_{A2}$, $\bar{Y}_{A1}$, $t_{.05}$, and s_{diff} in the $CI_{.95}$ formula, we have:

$$CI_{.95}: (6.4 - 3.9) - (2.10)(\surd.682) \leq \mu_2 - \mu_1 \leq (6.4 - 3.9) + (2.10)(\surd.682)$$

$$CI_{.95}: 2.5 - (2.10)(.826) \leq \mu_2 - \mu_1 \leq 2.5 + (2.10)(.826)$$

$$CI_{.95}: 2.5 - 1.73 \leq \mu_2 - \mu_1 \leq 2.5 + 1.73 = \mathbf{+.77 \leq \mu_2 - \mu_1 \leq +4.23}$$

Alternatively, the standard error of the difference could be calculated using $MS_{S/A}$ from the ANOVA, and for this two-group example, we would obtain the same lower and upper boundaries for the 95% confidence interval: $s_{diff} = \surd([2][MS_{S/A}] \div n)$; and since $MS_{S/A}$ is given in the ANOVA summary table (From Table 6.4, $MS_{S/A} = 3.41$), there is a computational advantage in using this "post-ANOVA" formula.

$$CI_{.95}: (6.4 - 3.9) - (2.10)\surd([2][3.41] \div 10) \leq \mu_2 - \mu_1 \leq (6.4 - 3.9) + (2.10)\surd([2][(3.41] \div 10)$$

$$CI_{.95}: 2.5 - (2.10)\surd(6.82 \div 10) \leq \mu_2 - \mu_1 \leq 2.5 + (2.10)\surd(6.82 \div 10)$$

$$CI_{.95}: 2.5 - (2.10)(.826) \leq \mu_2 - \mu_1 \leq 2.5 + (2.10)(.826)$$

$$CI_{.95}: 2.5 - 1.73 \leq \mu_2 - \mu_1 \leq 2.5 + 1.73 = \mathbf{+.77 \leq \mu_2 - \mu_1 \leq +4.23}$$

Consistent with results of F and t, zero is not included in this $CI_{.95}$ interval. As you can see, the 95% confidence interval not only informs us that a zero difference (assumed by H_0) between group means is unlikely ($p \leq .05$), it also informs us about the limits for the true difference between group means (i.e., it tells us something about what the true difference between group means is, as it can be stated confidently that $\mu_2 - \mu_1$ lies somewhere in the interval from +.77 to +4.23: t and F tests merely tell us what the true difference between group means likely is not—i.e., it is not 0).

Further consideration of confidence intervals reveals a serious limitation of the null hypothesis testing approach. A **non-significant t or F** justifies only a weak conclusion that group means do not differ. A corresponding CI when F is not significant would include 0 in the interval (e.g., for a $CI_{.95}$ of $-1.25 \leq \mu_1 - \mu_2 \leq +3.00$, then an F computed from the same data set would not be significant at $p \leq .05$). Since 0 is within the interval from -1.25 to +3.00, a zero difference between groups (i.e., the groups are "equal") cannot be ruled out. However, there are many values other than 0 which similarly cannot be ruled out. For example, a true difference between group means of -1.00 is just as reasonable as a true difference of 0 or +2.50, etc. A mere non-significant F does not differentiate among the many possible mean differences within the interval from -1.25 to +3.00. Consequently, interpreting a non-significant F as indicating that group means are equal (a zero difference) is at best considered a **weak** conclusion. A strong case has been made for preferring CIs to t and F when comparing pairs of group means (Bird 2004; Kline 2004).

For experiments with three or more groups or conditions, an ANOVA is preferred to conducting several two-group comparisons with t tests. The two illustrations that follow are appropriate for designs with a single independent variable. The data in Table 6.5 simply repeat the data in Table 6.3 with the addition of a new third column of scores to accommodate the addition of Group a_3.

Table 6.5: Data for a Three-Group Experiment

Group a_1	Group a_2	Group a_3
3	7	6
5	5	7
2	9	11
4	6	10
7	6	8
5	9	6
1	8	5
4	4	9
6	5	9
2	5	10

The computational steps remain unchanged for this problem; however, details in the computations have some alterations to accommodate the additional group (a = 3).The step-wise computations necessary for calculating F are as follows:

Step 1. The Basic Ratios:

$$[Y] = \Sigma Y_i^2 = 3^2 + 5^2 + \ldots + 10^2 = \mathbf{1,316.00}$$

$$[T] = (\Sigma Y_i)^2 \div (a)(n) = (184)^2 \div 30 = \mathbf{1,128.53}$$

$$[A] = \Sigma A_j^2 \div n = (39^2 + 64^2 + 81^2) \div 10 = \mathbf{1,217.80}$$

Step 2. The Sums of Squares (SS):

$$SS_A = [A] - [T] = 1,217.80 - 1,128.53 = \mathbf{89.27}$$

$$SS_{S/A} = [Y] - [A] = 1,316.00 - 1,217.80 = \mathbf{98.20}$$

Step 3. Degrees of Freedom (df):

$$df_A = 3 - 1 = \mathbf{2}$$

$$df_{S/A} = (a)(n - 1) = (3)(10 - 1) = (3)(9) = \mathbf{27}$$

Step 4. The Variances (MS):

$$MS_A = SS_A \div df_A = 89.27/2 = \mathbf{44.64}$$

$$MS_{S/A} = SS_{S/A} \div df_{S/A} = 98.20/27 = \mathbf{3.64}$$

Step 5. The F ratio:

$$F = MS_A \div MS_{S/A} = 44.64 \div 3.64 = \mathbf{12.26}$$

Step 6. The Reference Distribution Comparison: From Table A-1 in Appendix A at the back of the book, we can locate the critical F ratio where the second column ($df_A = 2$) intersects the 27th row ($df_{S/A} = 27$). Table A-1 indicates that $F_{crit}(2, 25) = 3.39$, and $F_{crit}(2, 30) = 3.32$. Although the actual $F_{crit}(2, 27)$ is not given, we know that it lies between the values of 3.39 and 3.32. Since the F obtained in this example ($F = 17.27$) exceeds both of these critical F values, it would certainly exceed the actual F_{crit} with 2 and 27 degrees of freedom. Thus, the difference among the three groups is significant. The ANOVA results are presented in Table 6.6:

Table 6.6: Analysis of Variance Summary Table

Source	SS	df	MS	F	
A	89.27	2	44.64	12.26	$p < .05$
S/A	98.20	27	3.64		

In order to provide an ANOVA illustration for a four-group experiment, we will need to add one more column of data, and this has been done in Table 6.7:

Table 6.7: Data for a Four-Group Experiment

Group a_1	Group a_2	Group a_3	Group a_4
3	7	6	7
5	5	7	4
2	9	11	8
4	6	10	8
7	6	8	5
5	9	6	3
1	8	5	6
4	4	9	7
6	5	9	6
2	5	10	8

The new computations to accommodate the addition of Group a_4 are as follows:

Step 1. The Basic Ratios:

$$[Y] = \Sigma Y_i^2 = 3^2 + 5^2 + \ldots + 8^2 = \mathbf{1{,}728.00}$$

$$[T] = (\Sigma Y_i)^2 \div (a)(n) = (246)^2 \div 40 = \mathbf{1{,}512.90}$$

$$[A] = \Sigma A_j^2 \div n = (39^2 + 64^2 + 81^2 + 62^2) \div 10 = \mathbf{1{,}602.20}$$

Step 2. The Sums of Squares (SS):

$$SS_A = [A] - [T] = 1{,}602.20 - 1{,}512.90 = \mathbf{89.30}$$

$$SS_{S/A} = [Y] - [A] = 1{,}728.00 - 1{,}602.20 = \mathbf{125.80}$$

Step 3. Degrees of Freedom (df):

$$df_A = 4 - 1 = \mathbf{3}\text{; and } df_{S/A} = (a)(n - 1) = (4)(10 - 1) = (4)(9) = \mathbf{36}$$

Step 4. The Variances (MS):

$$MS_A = SS_A \div df_A = 89.30 \div 3 = \mathbf{29.77}$$

$$MS_{S/A} = SS_{S/A} \div df_{S/A} = 125.80 \div 36 = \mathbf{3.49}$$

Step 5. The F ratio:

$$F = MS_A \div MS_{S/A} = 29.77 \div 3.49 = \mathbf{8.53}$$

Step 6. The Reference Distribution Comparison: From Table A-1 in the Appendix at the back of the book, $F_{crit} = 2.92$ (this is actually F_{crit} with 3 and 30 degrees of freedom). Since the F calculated in this example ($F = 8.53$) would most certainly exceed $F_{crit}(3, 36)$, the difference among the four groups is significant. The ANOVA summary table for this illustration is presented in Table 6.8:

Table 6.8: Analysis of Variance Summary Table

Source	SS	df	MS	F	
A	89.30	3	29.77	8.53	$p < .05$
S/A	125.80	36	3.49		

III. ANOVA for Groups with Unequal Numbers of Subjects

Most often experiments are designed with the intention of having equal numbers of participants in each experimental group. Occasionally they are planned differently, and occasionally the best laid plans go awry and an experiment may end up with unequal numbers of subjects in each group. A number of complications may enter the picture with unequal sample sizes (see Winer, Brown, and Michels 1991), and care should be taken to make sure that basic assumptions underlying the statistical model are still justified (e.g., random groups). For now, we will assume it is okay to proceed with a simple modification of the computational procedures used for calculating F. The modification only involves MS_A, which of course has a "trickle-up" impact on the corresponding sum of squares (SS_A) and basic ratio ([A]). Happily, the necessary modification is a very logical one. In calculating [A], a total score for each group was squared, and these squared group totals were summed and divided by n. An equivalent procedure would have been to square the total score for Group a_1, then divide it by n. Do the same for Group a_2, Group a_3, etc., and after each group total has been squared and divided by n, simply add up the results. That is, $[A] = \Sigma A_j^2 \div n = (A_1^2 \div n) + (A_2^2 \div n) + (A_3^2 \div n) \ldots$. Once this equivalence is recognized, then the simple solution for calculating [A] with unequal numbers of subjects in each of k groups follows from the second formula: just divide each squared group total (A_j^2) by its respective n_j: $[A] = (A_1^2 \div n_1) + (A_2^2 \div n_2) + \ldots + (A_k^2 \div n_k)$.

To illustrate an ANOVA for an experiment with unequal numbers of subjects per group, we will take the data from Table 6.5 and arbitrarily exclude the last score in Group a_1 ($n_1 = 9$), the final two scores in Group a_2 ($n_2 = 8$), and the final four scores in Group a_3 ($n_3 = 6$). The data for this illustration are presented in Table 6.9.

Table 6.9: Data for a Three-Group Experiment with Unequal n

Group a_1	Group a_2	Group a_3
3	7	6
5	5	7
2	9	11
4	6	10
7	6	8
5	9	6
1	8	
4	4	
6		

The step-wise computations necessary for calculating F are as follows:

Step 1. The Basic Ratios:

$[Y] = \Sigma Y_i^2 = 3^2 + 5^2 + \ldots + 6^2 = \mathbf{975.00}$

$[T] = (\Sigma Y_i)^2 \div (n_1 + n_2 + n_3) = (139)^2 \div 23 = \mathbf{840.04}$

$[A] = \Sigma(A_j^2 \div n_j) = (37^2 \div 9) + (54^2 \div 8) + (48^2 \div 6) = 152.11 + 364.50 + 384.00 = \mathbf{900.61}$

Step 2. The Sums of Squares (SS):

$SS_A = [A] - [T] = 900.61 - 840.04 = \mathbf{60.57}$

$SS_{S/A} = [Y] - [A] = 975.00 - 900.61 = \mathbf{74.39}$

Step 3. Degrees of Freedom (df):

$df_A = 3 - 1 = \mathbf{2}$

$df_{S/A} = a(n - 1)$, which for three groups is simply $(n - 1) + (n - 1) + (n - 1)$

[For unequal n, $df_{S/A} = (n_1 - 1) + (n_2 - 1) + (n_3 - 1) = 8 + 7 + 5 = \mathbf{20}$]

Step 4. The Variances (MS):

$$MS_A = SS_A \div df_A = 60.57 \div 2 = \mathbf{30.28}$$

$$MS_{S/A} = SS_{S/A} \div df_{S/A} = 74.39 \div 20 = \mathbf{3.72}$$

Step 5. The F ratio:

$$F(2, 20) = MS_A \div MS_{S/A} = 30.28 \div 3.72 = \mathbf{8.14}$$

Step 6. The Reference Distribution Comparison: From Table A-1 in Appendix A, $F_{crit}(2, 20) = 3.49$. Since the F obtained in this example (F = 8.14) exceeds F_{crit}, the difference among groups is significant.

Table 6.10: Analysis of Variance Summary Table

Source	SS	df	MS	F	
A	60.57	2	30.28	8.14	$p < .05$
S/A	74.39	20	3.72		

IV. Summary

Statistical analyses for experiments involving a single independent variable and two, three, and four groups were illustrated in this chapter. In addition, the analysis of variance was adapted for experiments with unequal numbers of subjects in various groups. For all designs, the analysis of variance is an appropriate statistical procedure. In addition, two alternative testing procedures were introduced for experiments (or follow-up *analytical comparisons*, a consideration that will be described in later chapters) that involve only two groups. The three test procedures (F, t, and CI) yield identical decisions about rejecting or failing to reject H_0. Since ANOVA is applicable regardless of number of groups, while t and use of confidence intervals (CI) are limited to two-group contrasts, a reasonable question might be, "Why worry beginning statistics students about these latter two alternatives?"

While the following responses may or may not serve as satisfactory justifications for everyone, they represent the main reasons these procedures were included in this chapter. Two group comparisons using a t test are reported sometimes in the literature in component analyses of more complex experiments; and traditional introductory statistics textbooks usually introduce hypothesis testing with a prototype two-group contrast and the t test, building up to more complex experiments and the F test. Thus, t is a hypothesis-testing technique that is currently used often enough to justify including it in an introductory textbook, even if not as a focal interest.

The rationale for introducing confidence intervals in this context is a little different. At present, there is not a widespread use of CIs for testing whether or not data indicate that two groups differ. However, a case can be made that perhaps journal editors should insist that authors report CIs any time they do an H_0 test of the difference between two groups. Confidence intervals provide the same "yes or no" decision about a null hypothesis for a given contrast that t and F provide, and in addition, CIs provide information about the lower and upper limits of the interval that most likely reveals the magnitude of the discrepancy between two population means.

It is fair to say that most experiments in psychology today involve more than two groups or conditions. Computational illustrations were provided for experiments with three groups and for experiments with four groups. It should not be necessary to keep adding illustrations to cover experiments with five groups, six groups, seven groups, etc. A reasonable assumption from an instructor's perspective is that if a student can successfully do an analysis of variance for an experiment involving two, three, and four groups, adding one or more groups to a design should not pose a serious problem. Most experiments are designed for groups with equal numbers of subjects; however, there can be exceptions. The final problem presented in this chapter illustrated computational steps for the analysis of variance adapted to a three-group, unequal-n experiment.

Model problems are not presented for this chapter since several problems were worked out in detail in the main body of the text.

6.1: Practice Problem 1. Consider the data in Table 6.7 (page 106) from a four-group experiment. Statistical tests for six different two-group contrasts are possible with this data set (a_1 vs. a_2; a_1 vs. a_3, a_1 vs. a_4; a_2 vs. a_3, a_2 vs. a_4; a_3 vs. a_4), and given that three different test statistics (F, t, and CI) were introduced for two-group comparisons, this means that this little set of data could be used for 18 different exercises—Oh my! So there really isn't any need to have multiple practice problems with two-group experiments. You can do as many or as few of these 18 comparisons as you or your instructor desire. Some of the comparisons are reported in the text, and answers to three of the possible 18 are provided below:

A. A t test comparing Groups a_1 and a_2: **$t(18) = 3.03$, $p < .05$** (actually this test was reported after the data for the a_1 and a_2 groups were presented in Table 6.3—and both F and $CI_{.95}$ were reported); however, even if the F and $CI_{.95}$ work had not been done, you would know what F(1, 18) equals and whether $CI_{.95}$ does or does not span 0.

B. An ANOVA comparing Groups a_2 and a_3: **$F = 3.98$, $F_{crit} = 4.41$**, therefore you know what t equals and whether $CI_{.95}$ does or does not span 0. What does t equal? Does $CI_{.95}$ include 0?

C. For the difference between the means for Groups a_3 and a_4: **$CI_{.95}$: $.12 \leq \mu_3 - \mu_4 \leq 3.68$**, therefore you at least know whether or not t and F would be significant at $p \leq .05$. Are they significant?

6.2: Practice Problem 2. Scores from a 4-group experiment are presented in the following table (**$a = 4$, $n = 6$**):

a_1	a_2	a_3	a_4
5	0	6	9
5	2	5	8
7	1	7	10
3	5	7	12
1	3	9	9
6	7	5	6

Complete the ANOVA summary table below, and indicate the critical F value that must be equaled or exceeded for the differences among groups to be significant at $p \leq .05$:

Source	**SS**	**df**	**MS**	**F**	**$F_{(crit)}$ (for $p \leq .05$)**
A					
S/A					

6.3: Practice Problem 3. Is there a significant difference ($p \leq .05$) among the six groups below?

Groups

a_1	a_2	a_3	a_4	a_5	a_6
5	8	12	10	15	9
5	6	10	7	10	12
7	7	6	8	8	10
4	7		5	10	12
6	5		11		9
8					7

6.4: Practice Problem 4. An experiment involved 5 groups, and the experimenter was interested in determining if participants would be confused about whether a "new" (not studied) test word had been presented earlier **if** a word related to it had been on the study list. For one group (a_1), respective study and new test words were semantically related (e.g., study word "hammer," new test word "nail"). For a second group (a_2), respective study and new test words were related in terms of both sight and sound (e.g., study word "hamster"; test word "hammer"). For the third group (a_3), the two words were connected via a missing "meditational" link (e.g., study word "hamster"; test word "nail," with *hammer* the presumed missing mediational link). The remaining two groups were controls in that the respective related words were not presented on the study list. Subjects in one control group (a_4) were tested on the same words as subjects in Group a_1 and Group a_3 (e.g., nail) and subjects in the other control group (a_5) were tested on the same words as subjects in Group a_2 (e.g., hammer). Fictitious data are presented in the following table:

Number of False Recognitions:

a_1	a_2	a_3	a_4	a_5
3	3	1	0	1
2	3	2	2	1
3	5	2	1	0
2	5	1	1	3
3	3	0	0	1
4	3	1	1	0
3	1	2	1	1
1	0	0	0	0

A. Do an analysis of variance to determine whether differences among these 5 groups are significant ($p \leq .05$), reporting the results of your analysis in a standard analysis of variance summary table.

B. Do a t test using only the data from Groups a_1 and a_3. Is t significant at $p \leq .05$? Would a 95% confidence interval for $\mu_1 - \mu_3$ (based on the sample means for Groups a_1 and a_3) include 0? Confirm your answer to this question by determining $CI_{.95}$ for the difference between $\bar{Y}_{A1}$ and $\bar{Y}_{A3}$.

Chapter 7

Additional Analyses for More Detailed Information

If I had only one day to live, I'd live it in my statistics class—it would seem so much longer.
—Unknown (obviously a student)

I. Keeping a Proper Perspective: What Do ANOVA Results Reveal?

You should understand by now that a significant F ratio **does not prove** that H_0 is false, and that the difference between groups is something other than 0. The inference drawn from the statistical test is a probabilistic one, enabling the researcher to have the odds substantially in his or her favor when the data call for rejecting H_0. The term "significant" in the statistical context simply means that it is unlikely (e.g., five or fewer times per hundred) that a difference between groups as large as or larger than what was found in the experiment could be merely the result of random variation. Recognition of this uncertainty, *viz.* that the decision process involves a small risk for error, represents one limitation of H_0 testing that needs to be kept in perspective. A second limitation, and the one emphasized in this chapter, has to do with the number of groups involved when results of a statistical test indicate that the difference among groups is significant. If only two groups are involved, then the interpretation of a significant difference is straightforward: Scores for Group a_1 are either higher or lower than scores for Group a_2, and we only need to look at group means to know the direction of the significant difference. The case is more complex for experiments involving three or more groups.

Three ANOVA problems with three or more groups ($a \geq 3$) were completed in the last chapter: two with $a = 3$ (one had unequal numbers of subjects per group) and one with $a = 4$. In all three examples, the obtained F was significant at the .05 level. So what did these statistical tests reveal? Did these F tests provide statistical information about whether any given group differed significantly from any other group (e.g., in the four-group example, could the researcher conclude that a Group a_2 mean equal to 6.4 was significantly larger than a Group a_1 mean equal to 3.9)? If this question is asked in class, the answer given should be a confident "**NO**." For the four-group example problem worked out in the text of the last chapter, the group means were as follows: $\bar{Y}_{A1} = 3.9$, $\bar{Y}_{A2} = 6.4$, $\bar{Y}_{A3} = 8.1$, and $\bar{Y}_{A4} = 6.2$. The H_0 assumption tested was that means of the four groups were equal: $\bar{Y}_{A1} = \bar{Y}_{A2} = \bar{Y}_{A3} = \bar{Y}_{A4}$. A significant F result implies that at least one inequality existed among this set of four groups. Of course, we can "eyeball" the means and see that in order of magnitude $\bar{Y}_{A3} > \bar{Y}_{A2} > \bar{Y}_{A4} > \bar{Y}_{A1}$. However, if it is important to know whether $\bar{Y}_{A2}$ was *significantly* larger than $\bar{Y}_{A1}$, then the argument here is that the difference needs to be tested statistically. The purpose of this chapter is to describe follow-up comparisons (referred to as *analytical comparisons*) that **may** be considered if more detailed information about differences among specific group means is sought. Given that a significant F with $df_A > 2$ only indicates that the likelihood that the means for all groups are not equal, it certainly seems reasonable to test the statistical significance of more focused comparisons (see Rosenthal, Rosnow, and Rubin 2000).

II. Classes of Analytical Comparisons

a. A Second Kind of α Error Risk (α_{FW})

By convention, most of the time a statistical test of the null hypothesis is performed, the significance level is set at $\alpha \leq .05$. A decision criterion is associated with a risk of making a false positive error, i.e., incorrectly attributing observed performance differences to the independent variable instead of random error or just chance variation. Setting $\alpha \leq .05$ presumably serves to keep this type of decision error to an acceptably low percentage. When multiple H_0 tests are performed on a given data set, such as all pair-wise contrasts in a four-group experiment, there is a second, related type of α error risk to consider. In the context of performing multiple statistical tests comparing two groups at a time from a set of four groups, six such comparisons are possible: a_1 vs. a_2, a_1 vs. a_3, a_1 vs. a_4, a_2 vs. a_3, a_2 vs. a_4, and a_3 vs. a_4. If the critical cutoff point for any statistical comparison between two groups is set at the .05 level (per comparison error risk; α_{PC}), then there is a second error risk for the experiment as a whole which may be of concern. The probability for this type of error is referred to as the *family-wise* error rate (α_{FW}) for a set or family of follow-up comparisons; α_{FW} will always be larger than the per comparison error rate (α_{PC}).

The per-comparison error rate (α_{PC}) refers to the error risk we have considered thus far; in the context of analytical comparisons, it is the error risk or α level set for a statistical test comparing Group a_1 with Group a_2, a_1 with a_3, etc. After completing several statistical pair-wise comparisons, you may now consider the entire collection of comparisons and ask, "What is the error risk or α level for **one or more comparisons** within the set of pair-wise comparisons to be significant by chance alone?" The α level associated with this question that considers a set of statistical tests is referred to as the family-wise error risk or α_{FW}.

In order to understand the distinction between α_{PC} and α_{FW}, consider four sets of scores drawn randomly from a table of random numbers. Thus, differences between group means would be the product of random error. Setting α_{PC} at .05 for each two-group comparison, we could test the difference between one or more pairs of groups. Thus, for each statistical test, the error risk of incorrectly rejecting H_0 is .05. However, for the bigger picture of looking at the collection of follow-up comparisons, we may ask: If two or more statistical tests between different pairs of groups are conducted, what is the probability that at least one of the tests will result in a significant difference (a family-wise error rate question, i.e., α_{FW})? To answer this question exactly, we could return to the binomial illustration described in Chapter 4 for tossing a coin, where $p(r) = [N! \div r!(N - r)!](p)^r(q)^{(N - r)}$. However, in the coin-tossing example, it was assumed that "heads" and "tails" were equally likely outcomes, with p (probability of "heads") = .50 and q (probability of "tails") = .50. The only thing that changes in applying this counting formula to H_0 testing when multiple tests are conducted is the value of p and the value of q. If p is the **chance probability** for rejecting H_0 and q is the probability of failing to reject H_0, then given H_0 is true, p = .05 and q = .95. The remaining terms have similar meanings as they did in the coin-tossing example: N is the number comparisons made or statistical tests conducted (trials) and r is the number of successes (for the purpose of this illustration, consider a significant difference rejecting H_0 as a "successful" outcome). If both a_1 vs. a_2 and a_3 vs. a_4 are tested statistically with $\alpha_{PC} = .05$, then this formula can be applied to determine the probability (the α_{FW} error risk) that at least one of the two tests would come out significant even when differences between pairs of groups tested are due entirely to chance. Three outcomes are possible in this simple situation: (1) both of the tests could result in significant differences; (2) one test could result in a significant difference and the other test could fail to reject H_0; or (3) both of the tests could result in a failure to reject H_0. With N = 2, p = .05, and q = .95, we have the following probabilities for r = 2 successes (significant differences), r = 1 success and r = 0 successes:

p(r = 2) (probability of 2 successes): $(2! \div 2![2 - 2]!)(.05)^2(.95)^0 = (1)(.0025)(1) =$ **.0025**

p(r = 1): $(2! \div 1![2 - 1]!)(.05)^1(.95)^1 = (2)(.05)(.95) =$ **.0950**

p(r = 0): $(2! \div 0![2 - 0]!)(.05)^0(.95)^2 = (1)(1)(.9025) =$ **.9025**

The family-wise error risk concerns whether either one (r = 1) or both (r = 2) comparisons are significant when H_0 is true (the "chance" probability), and it is readily apparent that α_{FW} presents an error risk greater than .05. For this example, the probability that at least one pair-wise contrast among selected group means will be significant by chance when two follow-up comparisons are made, even though each comparison is tested at the .05 level, is determined by taking the sum of the probabilities for the two outcome possibilities that include a significant result: $\boldsymbol{\alpha_{FW} = p(r = 2) + p(r = 1) = .0025 + .0950 = .0975}$. A quick way to **estimate** α_{FW} without going through the exact binomial calculations is to multiply α_{PC} by the number of pairs of groups tested: $\boldsymbol{\alpha_{FW} \approx (C)(\alpha_{PC})}$, where C is the number of statistical tests conducted (actually or potentially). With two tests at $\alpha_{PC} = .05$, the exact family-wise error risk illustrated in this two-comparison example was $\alpha_{FW} = .0975$. Using the quick estimate procedure, we have $\alpha_{FW} \approx (2)(.05) = .10$. You can save considerable time by using the estimation procedure for determining α_{FW}, assuming that you can tolerate a small degree of imperfection (.0025 in this case). Actually, most researchers would not worry about an inflated α_{FW} resulting from a second statistical comparison between groups; however, adding just a few more comparisons could cause concerns. It was mentioned earlier that six comparisons were possible among pairs of intact groups in a four-group experiment. Using the estimation procedure, if all six comparisons are tested, $\alpha_{FW} \approx (6)(.05) = .30$ (i.e., at least one α error could be expected nearly one-third of the time!).

b. Limiting Analytical Comparisons to a Meaningful Subset

When multiple follow-up comparisons are contemplated, the researcher faces a dilemma. How does one hold α_{FW} in check without suffering the consequences of undesirable side effects? After all, running a 30% or higher risk that at least one statistical inference of rejecting H_0 is incorrect seems excessive. Considering the formula for estimating α_{FW} $[(\alpha_{FW} \approx C(\alpha_{PC})]$, we can see that there are only two components that we can work with in trying to control α_{FW}: they are α_{PC} and the number of comparisons, C. Since α_{FW} increases directly as a function of C, one can keep α_{FW} relatively low by **planning** only a small number of *post hoc* comparisons. Alternatively, one can simply adopt a lower (compromise) α_{PC} level. For example, if we decided to use a more conservative α level (e.g., $\alpha_{PC} = .01$), testing all six comparisons in a four-group experiment would result in a family-wise error risk of approximately $\alpha_{FW} \approx (6)(.01) = .06$. This latter adjustment appears to be a sensible solution; however, it has a downside of its own. Lowering α_{PC} to .01 or lower in order to maintain α_{FW} close to .05, increases the risk of another type of inferential error: a **failure** to decide that groups differ when in reality they actually do differ (this second type of error risk is referred to as a Type II or β error, and it will be considered in more detail in Chapter 8).

For our purposes, and consistent with what is often reported in research publications, we will adopt the following game plan: If only a small number of follow-up comparisons are conducted, and if the comparisons in this small set are theoretically meaningful and planned in advance of data collection, then researchers in effect ignore α_{FW} and proceed doing these "planned" analytical comparisons, with α_{PC} for each comparison set at .05. As a rule of thumb, we can define "small number of comparisons" as a number equal to or less than df_A.

Any time decisions about analytical comparisons are driven by the data (e.g., "I think I'll test the difference between Group a_1 and Group a_3 because the difference between these two group means is larger than the difference between any other pair of group means."), α_{PC} adjustments should be made. In the next chapter we will consider two examples of standard α_{PC} adjustment procedures

(Scheffé and Tukey tests); in the Appendix to Chapter 8, two additional α_{PC} adjustment procedures will be introduced. Since the F reference distributions in Table A-1 of Appendix A provide critical F values for both the .05 and .01 levels, at this stage if it is necessary to prevent α_{FW} from becoming too large, we can proceed by setting $\alpha_{PC} = .01$ in determining the critical cutoff (F_{crit}) for each individual comparison.

III. Comparing Pairs of Intact Groups

For experiments involving three or more groups, a significant F indicates that all groups are not equal. If more detailed information about specific pairs of groups is important, then additional statistical tests (analytical comparisons) will need to be carried out. Computational options for comparing two groups, as described in the last chapter, may be used for this purpose. Since F and t tests are equivalent, we will use only F for analytical comparisons, although t tests for this purpose are sometimes reported in the literature. It is true that reporting confidence intervals also reveals the same H_0 information as ANOVA for two-group analytical comparisons; even though F is reported more often in the literature, an illustration for confidence intervals with selected pairs of intact groups (not new combinations of groups) will be included if for no other reason than to encourage wider adoption of this method of testing H_0 (for CI applications for complex pair-wise comparisons, see Bird 2004).

a. The F Test

In the next section a new computational procedure will be introduced for calculating sums of squares when conducting analytical comparisons. To demonstrate that this new procedure (using coefficients to identify and weight appropriately groups involved in a comparison) is comparable to a limited analysis of variance on the two groups being compared, we will begin with an example that works through the analytical comparison procedures by simply extracting the data for the two groups being compared and doing a two-group analysis of variance. The advantage of the new coefficient procedure that will be introduced next is that it facilitates analyses of complex comparisons that go beyond merely testing a difference between two existing groups (i.e., it is useful for comparing a combination of groups such as Groups a_1 and a_2 combined vs. Group a_3 and for testing complex functional relations across groups such as a test for a linear or higher order "trend," a special class of analytical comparisons introduced at the end of this chapter).

For illustrative purposes we will need a data set. Assume 40 students are randomly assigned to one of four groups, such that there are 10 students in each group. Each individual was given two reasoning tests, with the 2nd test occurring 30 hours after the 1st test. Following completion of the 1st test, subjects in Group a_1 were not allowed to sleep for 15 hours (it is probably fair to say that going without sleep for only 15 hours actually represents a no-sleep-deprivation condition), with subjects in the remaining groups experiencing increasing levels of sleep deprivation. Subjects in Group a_2 were not allowed to sleep for 20 hours (perhaps only a slight sleep-deprivation condition), subjects in Group a_3 were not allowed to sleep for 25 hours, and subjects in Group a_4 were not allowed to sleep for 30 hours. The dependent variable was the difference between scores on the 1st and 2nd test, with larger numbers indicating relatively poorer performance on the 2nd test. Fictitious data appear in Table 7.1.

Table 7.1: Sleep Deprivation and Performance on Reasoning Tests

	Hours of Sleep Deprivation			
	15 hrs. (a_1)	20 hrs. (a_2)	25 hrs. (a_3)	30 hrs. (a_4)
	7.0	9.5	8.3	10.0
	6.5	9.0	7.5	9.6
	5.9	8.4	7.5	9.5
	5.5	8.0	7.1	9.4
	5.0	7.6	6.2	8.0
	4.7	7.5	6.2	7.8
	4.0	6.0	5.0	7.8
	3.8	5.9	4.8	7.2
	3.1	4.0	4.0	7.2
	2.7	3.8	3.9	7.1
ΣY_i	**48.2**	**69.7**	**60.5**	**83.6**
ΣY_i^2	**250.74**	**521.07**	**387.93**	**710.54**
$\bar{Y}$	**4.82**	**6.97**	**6.05**	**8.36**

Analytical comparisons should be considered only if the overall analysis of variance resulted in a significant F ratio. The justification for this ground rule is that a non-significant F is interpreted as indicating that the means for all groups do not differ beyond what is reasonably expected by chance or random variation. Therefore any follow-up question that basically asks if specific pairs of groups are equal has already been answered by a non-significant F ratio in the overall analysis of variance. Thus, we need to report the four-group ANOVA, and if F is significant, we may consider doing analytical comparisons. There is a second reason we need the calculations from the ANOVA summary table. For testing analytical comparisons in designs with different subjects in each group, it is acceptable to use $MS_{S/A}$ as the denominator for each $F_{comparison}$, rather than doing the extra work to calculate a new $MS_{S/A}$ based only on the two groups being compared. Using $MS_{S/A}$ from the completed ANOVA is not only a labor-saving practice; it is also quite reasonable. Stability and accuracy of estimation should be a function of 'a' (number of groups contributing to the estimation); thus an estimate of random error based on information from four groups should be better than a comparable estimate based on information from only two groups. Table 7.2 presents results of the overall ANOVA—you may wish to confirm these results by doing the calculations for [Y], [A], [T], SS_A, $SS_{S/A}$, etc.

Table 7.2: ANOVA Summary Table

Source	SS	df	MS	F	F_{crit}
A	66.95	3	22.32	9.22 $p < .05$	2.92
S/A	87.23	36	2.42		

i. A Two-Group ANOVA

Before doing analytical comparisons in a new way by using weighting coefficients, we will first do a two-group ANOVA as if only the 15- and 30-hr. sleep-deprivation groups had been included in the experiment (a_1 vs. a_4). The data and calculations for SS_A appear below:

	15 hrs. (a_1)	30 hrs. (a_4)
	7.0	10.0
	6.5	9.6
	5.9	9.5
	5.5	9.4
	5.0	8.0
	4.7	7.8
	4.0	7.8
	3.8	7.2
	3.1	7.2
	2.7	7.1
ΣY_i	**48.2**	**83.6**

Since we may use $\mathbf{MS_{S/A} = 2.42}$ from the overall ANOVA for the estimate of random error, all we need to calculate from this two-group data set is the between-group variance estimate for the specific Group a_1 vs. Group a_4 comparison.

$[A_{comp}] = (48.2^2 + 83.6^2) \div 10 = 9{,}312.20 \div 10 = 931.22$
$[T_{comp}] = 131.8^2 \div 20 = 17{,}371.24 = 868.56$
$SS_{Acomp} = 931.20 - 868.56 = \mathbf{62.66}$; and since $df_{Acomp} = 1$, $\mathbf{MS_{Acomp} = 62.66}$
$F_{Acomp}(1, 36) = 62.66 \div 2.42 = \mathbf{25.89}$, which is significant at $p < .05$.

ii. The Coefficient Method

We began with a simple analytical comparison between two of the original four groups in the experiment: 15 hours (a_1) vs. 30 hours (a_4) of sleep deprivation. We will now repeat this analytical comparison with the *coefficient method*, and in the next section we will consider a second complex comparison that involves treating Groups a_2, a_3, and a_4 as if they were a single group (e.g., conditions that probably involved some sleep deprivation) to compare with Group a_1 (a group with possibly little or no sleep deprivation). With the coefficient method, a sum of squares for a comparison of interest is determined by multiplying group means by specific weighting coefficients (c_i) defining the comparison of interest, with the following formula for the between-groups sum of squares: $SS_{Acomp} = n(\Sigma c_i \bar{Y}_{Ai})^2 \div \Sigma c_i^2$.

Coefficients selected for each specific comparison may be any numbers provided the following conditions are satisfied: (1) Positive numbers are used for one side of the comparison and negative numbers are used for the other side of the comparison. (2) Both sides of a comparison must be equally weighted, thus the sum of the positive weighting coefficients must equal the sum of the negative weighting coefficients ($\Sigma c_i = 0$). (3) With the exception of the special case of analysis of *trend*, if more than one group contributes to one side of a comparison, each of the contributing group means must have identical coefficients assigned. Perhaps these rules will make more sense once we work through specific example problems. Since each group mean is multiplied by its respective coefficient, a table of group means and coefficients provides the necessary starting point for calculating SS_{Acomp}.

Table of Group Means and Weighting Coefficients for the a_1 vs. a_4 Comparison

Group Means:	<u>4.82(a_1)</u>	<u>6.97(a_2)</u>	<u>6.05(a_3)</u>	<u>8.36(a_4)</u>
Weighting Coefficients (c_i):	+1	0	0	-1

The comparison of interest is between Group a_1 ($c_1 = +1$) and Group a_4 ($c_4 = -1$) (15 hours of sleep deprivation vs. 30 hours of sleep deprivation). Given the equal weighting for both sides of the contrast, could different numbers have been selected for the coefficients? The answer is, "yes." For example, for this comparison we could have used +½ and -½ for c_1 and c_4, respectively; or +2 and -2, etc. Notice that one side of the comparison has a positive-signed coefficient and the other side has a negative-signed coefficient, and that the sum of the positive and negative coefficients equals zero. Also, notice that Group a_2 and Group a_3 are not involved in the contrast of interest, thus coefficients of 0 are used for these two groups. Multiplying the Group a_2 mean by 0 and multiplying the Group a_3 mean by 0 results in both products equaling 0, thus eliminating these two groups from further consideration.

Comparison 1 (a_1 vs. a_4):

$$SS_{Acomp1} = n(\Sigma c_i \bar{Y}_{Ai})^2 \div \Sigma c_i^2 = 10([+1][4.82] + [-1][8.36])^2 \div (+1^2 + -1^2)$$

$$SS_{Acomp1} = 10(-3.54)^2 \div 2 = 10(12.53) \div 2 = \mathbf{62.66};$$

Since $df_{Acomp1} = 1$, $MS_{Acomp1} = \mathbf{62.66}$

Using $MS_{S/A}$ from the original ANOVA, we have $F_{Acomp1} = MS_{Acomp1} \div MS_{S/A} = 62.66 \div 2.42 = \mathbf{25.89}$.

From Table A-1 in Appendix A at the back of the book, we see that $F_{crit}\ (1, 36) = 4.17$, thus, F_{Acomp1} is significant at $p < .05$ ($F_{Acomp1} = 25.89$ using this coefficient procedure is the same as the F_{Acomp1} computed earlier by simply extracting Group a_1 and Group a_4 from the original data set and doing a two-group analysis of variance).

b. Using Confidence Intervals

We will do one more repetition involving this data set; however, this time we will compare Group a_1 with Group a_4 by using confidence intervals. Although we do not know what the lower and upper limits of the $CI_{.95}$ will be for this comparison, we do know that by virtue of the fact that F_{Acomp1} (1, 36) was significant at p = .05, that 0 will not be in this $CI_{.95}$ interval. However, a caveat is in order here. We can only be certain of this claim if the same error term ($MS_{S/A}$) that was used for F_{Acomp1} is also used for $CI_{.95}$. The 95% confidence interval for the difference between means for Group a_1 ($\bar{Y}_{A1}$ = 4.82) and Group a_4 ($\bar{Y}_{A4}$ = 8.36) is determined by the following formula:

$$CI_{.95}: (\bar{Y}_{A4} - \bar{Y}_{A1}) - (t_{.05})\sqrt{(2[MS_{S/A}] \div n)} \leq \mu_4 - \mu_1 \leq (\bar{Y}_{A4} - \bar{Y}_{A1}) + (t_{.05})\sqrt{(2[MS_{S/A}] \div n)}$$

In the above formula for $CI_{.95}$, the standard error of the difference is calculated using $MS_{S/A}$ rather than s_4^2 and s_1^2 [$s_{diff} = \sqrt{(2[MS_{S/A}] \div n)}$, since we want to use the same random error estimate as was used for F_{Acomp1} ($MS_{S/A}$ = 2.42), and $t_{.05}(36)$ = 2.03 (a t value from Table A-2 in Appendix A that is midway between t values with df = 30 and df = 40). If s_{diff} had been estimated by taking the square root of the sum of s_1^2 and s_4^2 divided by n, then the appropriate t value from the table would have been $t_{.05}(18)$ = 2.10, since the estimates of random error would have been based only on the 10 subjects in Group a_1 and the 10 subjects in Group a_4. With a = 4, $MS_{S/A}$ is simply the average of the four group variances: $MS_{S/A} = (s_1^2 + s_2^2 + s_3^2 + s_4^2) \div a$. Each a_i group provides an estimate of random error; using the average of these four random error estimates provides a more stable estimate of $\sigma^2_{S/A}$ than an estimate provided by only two groups.

The argument for using $MS_{S/A}$ rather than s_1^2 and s_4^2 in the CI formula is that in the long run, an estimate of random error based on data from four groups ought to be more accurate than an estimate of random error based on data from only two groups. In addition, for this example problem, $MS_{S/A}$ has 36 degrees of freedom, whereas df = 18 for an estimate of random error restricted to s_1^2 and s_4^2. We will see later that using $MS_{S/A}$ with its greater degrees of freedom has the added advantage of increasing *power* or sensitivity for detecting the presence of a difference between means. Determining $CI_{.95}$ for the difference between means for Group a_1 and Group a_4, we have:

$$CI_{.95}: (8.36 - 4.82) - (2.03)\sqrt{(2[2.42] \div 10)} \leq \mu_4 - \mu_1 \leq (8.36 - 4.82) + (2.03)\sqrt{(2[2.42] \div 10)}$$

$$CI_{.95}: 3.54 - (2.03)\sqrt{.484} \leq \mu_4 - \mu_1 \leq 3.54 + (2.03)\sqrt{.484}$$

$$CI_{.95}: 3.54 - (2.03)(.696) \leq \mu_4 - \mu_1 \leq 3.54 + (2.03)(.696)$$

$$CI_{.95}: 3.54 - 1.41 \leq \mu_4 - \mu_1 \leq 3.54 + 1.41$$

$$\mathbf{CI_{.95}: 2.13 \leq \mu_4 - \mu_1 \leq 4.95}$$

The second comparison illustrated is a complex comparison contrasting Group a_1 with the combination of Groups a_2, a_3, and a_4 treated as if they represented a single group (possibly a no-sleep-deprivation group vs. groups with varying degrees of sleep deprivation). This contrast also has equally weighted positive (+3) and negative [(-1) + (-1) + (-1) = -3] coefficients. As was true for the earlier comparison between two original groups, different numbers could have been selected for the coefficients. For example, for this complex comparison we could have used +1 for a_1 and -1/3 for a_2, -1/3 for a_3, and -1/3 for a_4; or -6 for a_1 and +2 for a_2, +2 for a_3, and +2 for a_4. Again, you can see that the sum of the coefficients is 0. An appropriate table of group means and weighting coefficients is given in the table that follows:

Table of Group Means and Weighting Coefficients for the a_1 vs. $a_2 + a_3 + a_4$ Comparison

Group Means:	4.82(a_1)	6.97(a_2)	6.05(a_3)	8.36(a_4)
Weighting Coefficients (c_2):	+3	-1	-1	-1

Once an appropriate set of weighting coefficients has been determined and assigned to each group, the calculations for a sum of squares for a complex analytical comparison proceed by applying the same formula that was used for a direct analytical comparison between two existing groups from the experiment: $SS_{Acomp2} = n(\Sigma c_j \bar{Y}_{Aj})^2 \div \Sigma c_j^2$.

$$SS_{Acomp2} = 10([+3][4.82] + [-1][6.97] + [-1][6.05] + [-1][8.36])^2 \div (+3^2 + -1^2 + -1^2 + -1^2)$$

$$SS_{Acomp2} = 10(-6.92)^2 \div 12 = 10(47.87) \div 12 = \mathbf{39.89}$$

Even though all four groups contribute to this comparison, it constitutes a two-group contrast, as the original groups are reorganized into one group consisting of scores with positive-signed coefficients and the other group consisting of scores with negative-signed coefficients. Therefore, this $df_{Acomp2} = 1$, and $MS_{Acomp2} = SS_{Acomp2} \div df_{Acomp2} =$ **39.89**. Using $MS_{S/A}$ from the original ANOVA, we have $F_{Acomp2} = 39.89 \div 2.42 =$ **16.48**; and $F_{crit}(1, 36) = 4.17$, therefore F_{Acomp2} is significant at $p < .05$.

a. Analysis of Trend: A Special Class of Complex Comparisons

Analytical comparisons involving tests for functional relations across conditions of an experiment may represent both an increase and a decrease in difficulty and complexity. On the conceptual level, testing for a trend or pattern involves more abstract relations among groups. On the computational side, doing a trend analysis ought to be one step easier than other analytical comparisons. For a trend analysis it is not necessary to create coefficients that identify comparisons of interest. Coefficients for linear, quadratic, and cubic trends are given in Table A-3 in Appendix A at the end of the book.

If treatment conditions that define different groups in an experiment represent **ordered differences along a quantitative dimension**, then a trend analysis as a follow-up to an analysis of variance may be relevant. The fictitious sleep-deprivation data in this chapter were created deliberately because the independent variable meets these criteria: time measured in terms of hours of sleep deprivation is a quantitative variable, and the four groups were separated by equal distances (5 hours) along this dimension. Thus, a trend analysis may be considered (actually, meeting the equal distance criterion enables us to use the coefficients available in Table A-3; otherwise coefficients would have to be customized to fit distances of different magnitude—see Keppel and Wickens 2004 for computational procedures for customizing coefficients).

A test of trend is designed to determine if there is a systematic pattern in a data set related to the systematic differences along the dimension defining treatment differences. Basically, the patterns in a data set identified may be a straight-line (*linear* function), with scores either consistently increasing or decreasing across the ordered conditions of the experiment. If there is one change in direction in the pattern (e.g., scores first increase and then decrease, or vice-versa), the trend is *quadratic*. A *cubic* trend reflects two changes in direction (e.g., scores may first increase, then decrease, and then increase again across treatment conditions); regardless of how many groups may be involved in an experiment, we will not consider trend components beyond linear, quadratic, and cubic (well, maybe one little exception at the end of the practice problems). However, you should know that the number of possible trends that can be tested for a given data set equals df_A.

Each trend component has **one degree of freedom**. For the four-group experiment in this section, it is possible to assess three trend components: linear, quadratic, and cubic, and $df_{Linear} = 1$, $df_{Quadratic} = 1$ and $df_{Cubic} = 1$. A table is given after this paragraph listing the means for the four groups from the sleep-deprivation example and their respective coefficients from Table A-3 from Appendix A for testing linear, quadratic, and cubic trends. Calculations will be shown for linear and quadratic trends, and although the SS_{Cubic} will be given, the direct computation of the cubic trend will not be done. You can see by the positive and negative signs that the coefficients for the linear trend reflect no change in direction; the coefficients for the quadratic trend reflect a single change in direction (decreasing from a_1 to a_2, then increasing from a_3 to a_4); and the coefficients for the cubic trend reflect two changes in direction (increasing from a_1 to a_2, then decreasing from a_2 to a_3, then increasing again from a_3 to a_4). Also, as was true for other comparisons using coefficients, the sum of the positive and negative coefficients equals 0.

Table of Group Means and Weighting Coefficients for Linear, Quadratic, and Cubic Trends:

	Group Means			
Trend Comparisons	4.82(a_1)	6.97(a_2)	6.05(a_3)	8.36(a_4)
c_{Linear}	-3	-1	+1	+3
$c_{Quadratic}$	+1	-1	-1	+1
c_{Cubic}	-1	+3	-3	+1

The mean square error from the overall analysis of variance (for this data set, $MS_{S/A} = 2.42$) may again be used as the denominator for each $F_{\text{“trend”}}$, therefore the primary calculation required is a sum of squares for each trend ($SS_{ALinear}$, $SS_{AQuadratic}$, and SS_{ACubic}). The computational formula for the sum of squares for trend components is the same as previously used for other analytical comparisons using the coefficient method: e.g., $SS_{ALinear} = n(\Sigma c_{Linear}\bar{Y}_{Ai})^2 \div \Sigma c_{Linear}^{\ 2}$.

1. $SS_{ALinear} = 10([-3][4.82] + [-1][6.97] + [1][6.05] + [3][8.36])^2 \div (-3^2 + -1^2 + 1^2 + 3^2)$
   $SS_{ALinear} = 10(9.7)^2 \div 20 = 940.9 \div 20 = \underline{47.04}$; as indicated earlier, $df_{ALinear} = 1$, therefore $MS_{ALinear} = 47.04$ and **$F_{ALinear}(1, 36) = 47.04 \div 2.42 = 19.44$; $p < .05$.**

2. $SS_{AQuadratic} = 10([1][4.82] + [-1][6.97] + [-1][6.05] + [1][8.36])^2 \div (1^2 + -1^2 - 1^2 + 1^{2)}$
   $SS_{AQuadratic} = 10(.16)^2 \div 4 = .26 \div 4 = .06$; $MS_{AQuadratic} = .06$ and **$F_{AQuadratic}(1, 36) = .06 \div 2.42 = .02$; clearly a non-significant F.**

3. $df_A = 3$, thus only three trends can be computed; therefore, $SS_{ACubic} =$ **19.85**! You can check this by calculating SS_{ACubic} directly; however, since $SS_{ALinear} + SS_{AQuadratic} + SS_{ACubic} = SS_A$, it follows that $SS_{ACubic} = SS_A - SS_{ALinear} - SS_{AQuadratic}$; thus, $SS_{ACubic} = 66.95 - 47.04 - .06 = 19.85$. **$F_{ACubic}(1, 36) = 19.85 \div 2.42 = 8.20$, $p < .05$.**

The conclusion from the trend analysis is that both a linear and a cubic trend are present across the sleep-deprivation conditions.

V. Summary

This chapter began with a cautionary note about recognizing limitations accompanying the report of a statistically significant difference among groups in an experiment that involves more than two groups. The significant F ratio only calls into question the H_0 assumption that performance scores for all groups are equal. Additional testing is required for statistical justification for more detailed information (e.g., given statistical support that three or more groups differ, is the difference between a selected pair of groups significant?). Accepting the position that additional analytical comparisons may be important, we have two things to address: (1) consideration of the α error risk for a family or set of additional comparisons (α_{FW}); and (2) the operations involved in conducting additional statistical tests. Formal procedures for making adjustments in α_{PC} in order to control α_{FW} will be addressed in the next chapter.

A simple way to estimate α_{FW} is to multiply the number of analytical comparisons one makes or would be willing to make (C) by the α level set for each comparison (α_{PC}). If the number of meaningful comparisons is small (e.g., $C \leq df_A$), then the researcher often just "lives" with the inflated α_{FW}. Corrective procedures that involve reducing α_{PC} are recommended if you are conducting a large number of planned analytical comparisons or whenever you are conducting two or more exploratory or unplanned follow-up comparisons.

Following an ANOVA with a significant effect for groups, if two specific groups are targeted for comparison, one could simply extract the groups from the original data set and compute t or F. However, either of two alternative test procedures was recommended. The first option involved calculating SS_{Acomp} by using weighting coefficients, because we will apply this procedure for determining F_{Acomp} for complex comparisons that involve combining groups and for a trend analysis. The second option involved determining a confidence interval (e.g., $CI_{.95}$) for the difference between the two group means. A $CI_{.95}$ that does not include 0 is equivalent to a significant F ratio, and it also provides additional information about lower and upper limits for the difference between means. An example problem illustrating each methodological approach was presented.

Complex comparisons involving combinations of experimental groups may also be of interest (e.g., comparing data from three groups that are combined and treated as if they came from a single group with data from a fourth group). The F test using the method of weighting coefficients is best suited for complex comparisons. One special case of a complex comparison that may be considered for quantitative independent variables is the analysis of trend, and coefficients for specific trend components are provided in Table A-3 in Appendix A at the back of the book. For experimental designs with different individuals randomly assigned to each group in the experiment, it is appropriate to use $MS_{S/A}$ as the estimate of random error variance for all procedures, which simplifies the computational steps since $MS_{S/A}$ is reported in the original analysis of variance.

7.1: Model Problem. Although computational procedures were illustrated in the text, a three-part problem using the method of weighting coefficients with complex comparisons is presented here because we will have occasion to refer to it again in the context of a future chapter. The three parts involve the following three complex comparisons: (1) treating a_1 and a_2 as a single group to compare with a_3 and a_4 treated as a single group; (2) $a_1 + a_3$ vs. $a_2 + a_4$; and (3) $a_1 + a_4$ vs. $a_2 + a_3$. The formula for calculating the sum of squares using weighting coefficients is $SS_{Acomp} = n(\Sigma c_j \bar{Y}_{Aj})^2 \div \Sigma c_j^2$. The data set is the sleep-deprivation illustration in the text (a = 4, n = 10), and the denominator for each F_{Acomp} is $MS_{S/A} = 2.42$ from the overall ANOVA. Group means and weighting coefficients for each comparison are presented in the table below:

Table of Group Means and Weighting Coefficients for Three Complex Comparisons:

	Group Means			
Comparisons	4.82(a_1)	6.97(a_2)	6.05(a_3)	8.36(a_4)
A_{comp1}	+1	+1	-1	-1
A_{comp2}	+1	-1	+1	-1
A_{comp3}	+1	-1	-1	+1

Comp 1: $SS_{Acomp1} = n(\Sigma c_j \bar{Y}_{Aj})^2 \div \Sigma c_j^2 = 10(4.82 + 6.97 - 6.05 - 8.36)^2 \div (1^2 + 1^2 + [-1]^2 + [-1]^2)$
$SS_{Acomp1} = 10(-2.62)^2 \div 4 = 10(6.86) \div 4 = 68.60 \div 4 = \mathbf{17.16}$; $df_A = 1$; $df_{S/A} = 36$; and $MS_{S/A} = 2.42$; therefore, $\mathbf{F_{Acomp1}(1, 36) = 17.16 \div 2.42 = 7.09, p < .05}$.

Comp 2: $SS_{Acomp2} = 10(4.82 - 6.97 + 6.05 - 8.36)^2 \div (1^2 + [-1]^2 + 1^2 + [-1]^2)$
$SS_{Acomp2} = 10(-4.46)^2 \div 4 = 10(19.89) \div 4 = 198.90 \div 4 = \mathbf{49.73}$; $df_A = 1$; $df_{S/A} = 36$; and $MS_{S/A} = 2.42$; therefore, $\mathbf{F_{Acomp2}(1, 36) = 49.73 \div 2.42 = 20.55, p < .05}$.

Comp 3: $SS_{Acomp3} = 10(4.82 - 6.97 - 6.05 + 8.36)^2 \div (1^2 + [-1]^2 + [-1]^2 + 1^2)$
$SS_{Acomp3} = 10(.16)^2 \div 4 = 10(.03) \div 4 = .30 \div 4 = \mathbf{.08}$; $df_A = 1$; $df_{S/A} = 36$; and $MS_{S/A} = 2.42$; therefore, $\mathbf{F_{Acomp3}(1, 36) = .08 \div 2.42 < 1.0}$, **which is not significant**. Coincidentally, the coefficients for Comparison 3 are identical to the coefficients used for assessing the quadratic trend illustrated in the text.

7.2: Practice Problem 1. As can be seen from the table of group means and coefficients in the Model Problem, means for Group a_2 (6.97) and Group a_3 (6.05) are quite close numerically. By inspection, we might well expect that they do not differ significantly. If that is the case, then 0 would fall between the lower and upper limits for $CI_{.95}$. By the same token, the Group a_4 mean (8.36) is noticeably larger than the Group a_3 mean (6.05), thus it is possible that these two group means are significantly different. If these two groups differ significantly (at $\alpha < .05$), then 0 would not be within the upper and lower limits for the 95% confidence interval. Determine $CI_{.95}$ for each of these contrasts (the difference between means for Group a_2 and Group a_3, and the difference between means for Group a_3 and Group a_4).

7.3: Practice Problem 2. Assume the scores in the data matrix for this problem were obtained from three groups (a = 3), with 11 subjects in each group (n = 11).

Groups	a_1	a_2	a_3
	15	12	20
	16	18	20
	22	25	18
	14	16	23
	11	24	27
	17	16	30
	12	14	32
	24	20	24
	20	21	20
	18	25	22
	10	19	15

A. Do an analysis of variance on these data and indicate whether the difference among these three groups is significant.
B. Regardless of whether F in Part A is significant, for practice purposes do the following analytical comparisons: (1) a_1 vs. a_3 and (2) $a_1 + a_2$ vs. a_3.
C. Assume Factor A is a quantitative independent variable with equally spaced intervals. Is there a significant linear trend? Is there a significant quadratic trend? (Did you have a déjà vu reaction when you were calculating $SS_{ALinear}$?)

7.4: Practice Problem 3. The data from Table 6.7 (page 106) are copied below. For these data $F(3, 36) = 8.53$, $MS_{S/A} = 3.49$, $p < .05$. Your task is to calculate F for the following analytical comparisons: (1) F_{Acomp1} contrasting $a_1 + a_4$ vs. $a_2 + a_3$; (2) F_{Acomp2} contrasting a_1 vs. $a_2 + a_3 + a_4$ as a single group, and (3) F_{Acomp3} contrasting $a_1 + a_2$ as a single group vs. a_4. Would any of these comparisons be significant if we set a more stringent $\alpha_{PC} = .01$ for determining the critical cutoff point?

a_1	a_2	a_3	a_4
3	7	6	7
5	5	7	4
2	9	11	8
4	6	10	8
7	6	8	5
5	9	6	3
1	8	5	6
4	4	9	7
6	5	9	6
2	5	10	8

7.5: Practice Problem 4. An experimenter has 9th grade boys ($n = 4$), shooting a basketball 50 times from a distance away from the basket of 6 feet for boys randomly assigned to Group a_1, a distance of 8 feet for boys randomly assigned to Group a_2, a distance of 10 feet for boys randomly assigned to Group a_3, or a distance of 12 feet for boys randomly assigned to Group a_4. A table showing number of missed shots out of 50 and the ANOVA summary table follow. As you can see, the results of the overall analysis of variance indicate that the four groups differed in number of missed shots, $F(3, 12) = 6.44$, $p \leq .05$. Factor A represents equally spaced intervals along a quantitative dimension (distance in feet from the basket). Do the appropriate analytical comparisons to determine if there is a significant linear trend, a significant quadratic trend, and a significant cubic trend relating distance from the basket and shot accuracy.

Number of Missed Shots

	a_1	a_2	a_3	a_4
	8	9	16	20
	11	13	20	15
	12	18	22	25
	5	8	15	17
$\bar{Y}$	**9.00**	**12.00**	**18.25**	**19.25**

ANOVA Summary Table

Source	SS	df	MS	F
A	292.25	3	97.42	6.44
S/A	181.50	12	15.12	

7.6: Practice Problem 5. An elementary school principal compared five different spelling books marketed for second-grade children. Her elementary school had two second-grade classes, each with 25 students. The principal assigned spelling books to students at random, with the restriction that 5 students in each class use a different one of the five spelling books. One book was the "standard" (S) book currently used at the school, and the remaining four books were new competitors, designated by the respective publishers' initials. For this study, $a = 5$ and $n = 10$. Percent correct scores on a national spelling test given at the end of the school year are presented in the following table:

S	RH	HM	PH	AW
60	70	80	80	70
55	55	60	80	90
75	65	55	70	90
90	95	85	80	75
70	80	80	80	70
65	75	70	75	65
40	90	85	70	75
80	50	60	75	85
80	85	80	90	80
70	100	70	75	85

A. The primary purpose for doing this study was to compare the new spelling books collectively and individually with the standard book. However, before doing this, an overall analysis of variance is in order to determine if there is a significant difference at $p \leq .05$ among the five spelling books. You need to do the ANOVA to test if the different spelling books had an effect on final test performance and to determine the value of $MS_{S/A}$, which you need for Parts B and C of this problem.

B. Regardless of the results of the ANOVA, do the appropriate analytical comparison to test if this collection of new spelling books, as a group, resulted in significantly different spelling scores at the end of the year, compared to the standard spelling book (S) that had been in use at this school.

C. The book the principal would like to adopt for future use is the one published by RH. Thus, she was particularly interested in whether test scores at the end of the year were higher for the 10 students who used this book compared to the 10 students in Group S who used the standard book. Is the difference between these two groups significant at the .05 level?

7.7: Practice Problem 6. As a final exercise for this chapter, do a complete trend analysis for the fictitious data accompanying this problem (you can see from the completed ANOVA summary table that Factor A is significant). Since there are five groups in this experiment, you could actually evaluate four different trend components: Linear, Quadratic, Cubic, and a trend that was not mentioned in the text, *viz.*, a Quartic trend (an assessment of a "zig-zag" pattern of changes across the five group means (e.g., a trend that might indicate that there was an increase as $\bar{Y}_{A2}$ is larger than $\bar{Y}_{A1}$, followed by a decrease as $\bar{Y}_{A3}$ is smaller than $\bar{Y}_{A2}$, followed by an increase as $\bar{Y}_{A4}$ is larger than $\bar{Y}_{A3}$, and finally another decrease as $\bar{Y}_{A5}$ is smaller than $\bar{Y}_{A4}$). However, you will encounter one small problem. Table A-3 in Appendix A only provides coefficients for assessing linear, quadratic, and cubic trends; hence a quartic trend cannot be evaluated directly by using weighting coefficients. However, you can overcome this problem by making use of the relationship that $SS_A = SS_{Alinear} + SS_{Aquadratic} + SS_{Acubic} + SS_{Aquartic}$.

a_1	a_2	a_3	a_4	a_5
3	4	5	3	6
4	4	7	8	4
2	6	8	8	6
1	5	9	7	2

Source	SS	df	MS	F
A	55.30	4	13.82	4.65 $p < .05$
S/A	44.50	15	2.97	

A. What does $F_{Alinear}$ equal, and is it significant?
B. What does $F_{Aquadratic}$ equal, and is it significant?
C. What does F_{Acubic} equal, and is it significant?
D. What does $F_{Aquartic}$ equal, and is it significant?

Chapter 8

Assumptions and Refinements for ANOVA

An approximate answer to the right question is worth a good deal more than an exact answer to an approximate question.
—Tukey

I. Adjustment Procedures with Analytical Comparisons for Controlling α_{FW}

a. Using a More Conservative Per-Comparison (α_{PC}) for F_{crit}

A problem exists when multiple t or F tests are conducted because even though an acceptable error risk (e.g., $\alpha_{PC} = .05$) is set for each test, the probability that at least one test in a series of pair-wise comparisons will be significant may become unacceptably high (e.g., $\alpha_{FW} \approx .30$ for a series of six follow-up comparisons). That is, while each individual test may be conducted with a 5% error risk, there is a different error probability associated with the entire collection or family of tests. The probability that at least one among several statistical tests conducted will be significant by chance alone will be considerably larger than the per-comparison $\alpha_{PC} \leq .05$. A procedure was described in the last chapter for calculating this second, family-wise α_{FW} error risk, as well as a simple procedure for estimating it [$\alpha_{FW} \approx C(\alpha_{PC})$]. Thus far we have only considered controlling the family-wise error risk by simply setting a more conservative critical value of F for each pair-wise comparison (e.g., for six pair-wise comparisons, setting $\alpha_{PC} = .01$ will result in a more acceptable $\alpha_{FW} \approx .06$).

As you can see, holding α_{FW} to an acceptable level is done at the expense of α_{PC}, as each pair-wise contrast is tested with a more conservative (lower) α_{PC}, thus increasing the difficulty of detecting the presence of a real effect. When specific analytical comparisons are meaningful (e.g., theoretically relevant), researchers may be reluctant to make sacrifices in α_{PC} in order to keep α_{FW} in check. However, when experimenters run a number of exploratory *post hoc* statistical tests just to see what turns up, it is certainly appropriate to adopt a more stringent α_{PC} level to guard against an unacceptably high α_{FW} error risk. Benazzi, Horner, and Good (2006) justified this distinction rather nicely in a study that encompassed both concerns: some analytical comparisons were "theory-driven" and tested with α_{PC} set at .05, while others were *post hoc* exploratory comparisons tested at a more conservative α_{PC} level in order to maintain an acceptable family-wise error rate ($\alpha_{FW} = .05$).

In this chapter, two formal α_{PC} correction procedures will be illustrated for a contrived experiment with a = 4 and n = 8. There are six possible pair-wise comparisons among the original four groups that could be tested; if an experimenter tested all six contrasts (a_1 vs. a_2; a_1 vs. a_3, a_1 vs. a_4; a_2 vs. a_3, a_2 vs. a_4; a_3 vs. a_4) at $\alpha_{PC} = .05$, then the critical F with 1 and 28 degrees of freedom for each pair-wise test is 4.20. However, with $\alpha_{PC} = .05$ and six analytical comparisons, $\alpha_{FW} \approx 6(.05) = .30$. The F table at the back of the book (Table A-1) provides critical cutoff values (F_{crit}) at both the .05 and .01 levels. The experimenter could bring α_{FW} down to a more acceptable error risk level simply by setting $\alpha_{PC} = .01$ [which would result in $\alpha_{FW} \approx 6(.01) = .06$]. Setting the critical region at a .01 level we have $F_{crit}(1, 28) = 7.64$ for each two-group comparison. The two formal procedures introduced in this chapter (the Scheffé test and the Tukey test) maintain α_{FW} at a desired level (e.g., $\alpha_{FW} = .05$) by adjusting α_{PC} appropriately for the conditions of the experiment and number of analytical comparisons. The Scheffé test is the more conservative of the two (requiring a larger difference between groups being compared in order to be regarded as significant, hence a smaller α_{PC}), and it should be used when virtually all possible comparisons between groups and combinations of groups are under consideration.

b. The Scheffé Test

A Scheffé test (Scheffé 1953) simply involves conducting a set of analytical comparisons that are held to a strict testing standard. As such, doing a Scheffé test mostly involves doing analytical comparisons by calculating several $F_{comparisons}$ (F_{comps}) following procedures introduced in the last chapter. The only new feature required for a Scheffé test involves determining an appropriate F_{crit} for comparison with the F_{comps} calculated from the data. We will illustrate the procedure with a fictitious data set, and most of the work illustrated involves familiar operations in doing an analysis of variance and calculating F_{comps}. The only new requirement occurs at the final stage in converting the critical comparison F_{crit} from Table A-1 to a new critical F (F_S standing for a critical Scheffé F). The formula for this conversion is $\mathbf{F_S = df_A(F_{crit})\{df_A, df_{S/A}\}}$, where F_{crit} is from Table A-1 using the df_A column, not the df_{Acomp} column!

Consider a pretend experiment with a = 5 and n = 9. The pretend experiment examined the effect of different drug treatments on anxiety scores on a questionnaire given to 45 individuals in therapy. The 45 clients were randomly assigned to one of five conditions (a no-drug control or one of four different drug treatment groups). Assume the analysis of variance indicates that the effect of the drug conditions was significant. Analytical comparisons may then be tested using the Scheffé procedure, and if desired, setting a conventional error risk for α_{FW} of $p \leq .05$. Rather than presenting a data matrix and doing an analysis of variance, we will simply accept the ANOVA summary results for these data depicted in Table 8.1.

Table 8.1: ANOVA Summary Table for an Anxiety Drug Pretend Experiment

Source	SS	df	MS	F	
A	69.20	4	17.30	4.87	p < .05
S/A	142.00	40	3.55		

Given this significant F ratio, if we wanted to consider testing all possible pair-wise comparisons among different pairs of the original five groups, and among various combinations of groups (e.g., a_1 vs. the four drug treatment groups combined), we would certainly want to control α_{FW}. Using the Scheffé procedure implies that many such comparisons would be tested; however, for this illustration we will only follow through testing two possible analytical comparisons. Calculating an F_{comp} for a Scheffé test is no different from calculating an F_{comp} without an α_{PC} adjustment. Using the coefficients method, we need to know the mean for each group and the coefficients that define and weight the comparisons of interest. For this illustration involving two analytical comparisons, A_{comp1} will contrast Group a_1 (no-drug) with Group a_3 (one of the drug conditions), and A_{comp2} will contrast Group a_1 with Group a_5 (another one of the drug conditions). Assume the five group means were as follows: $\bar{Y}_{A1} = 3.11$, $\bar{Y}_{A2} = 4.78$, $\bar{Y}_{A3} = 5.33$, $\bar{Y}_{A4} = 6.00$, and $\bar{Y}_{A5} = 6.78$. The group means and corresponding weighting coefficients for the two comparisons of interest are given in the table that follows. The calculations for SS, MS, and F for the two comparisons of interest appear following the table.

Groups and Means:	a_1(3.11)	a_2(4.78)	a_3(5.33)	a_4(6.00)	a_5(6.78)
c_is for A_{comp1}	+1	0	-1	0	0
c_is for A_{comp2}	+1	0	0	0	-1

$SS_{Acomp1} = (n)(\Sigma c_i\bar{Y}_i)^2 \div \Sigma c_i^2 = 9(3.11 - 5.33)^2 \div 2 = 9(-2.22)^2 \div 2 = 22.18$, and since $df_{Acomp1} = 1$, $MS_{Acomp1} = 22.18$.

From the ANOVA summary table, $MS_{S/A} = 3.55$; therefore $F_{Acomp1}(1, 40) = 22.18 \div 3.55 = 6.25$.

Inspection of Table A–1 in Appendix A indicates that $F_{crit}(1, 40) = 4.08$. If α_{FW} had been ignored, the statistical decision would have been that the difference between Group a_1 and Group a_3 is significant ($p < .05$). Similarly, $SS_{Acomp2} = 9(3.11 - 6.78)^2 \div 2 = 60.61$, and since $df_{Acomp2} = 1$, $MS_{Acomp2} = 60.61$ and $F_{Acomp2}(1, 40) = 60.61 \div 3.55 = 17.07$, also a significant difference.

The remaining new step for completing a Scheffé test is to determine the appropriate F_S by solving $F_S = (df_A)(F_{crit})\{df_A, df_{S/A}\}$. The "squiggly" brackets ({}) do not represent a multiplication instruction; rather, here they merely indicate between- and within-group degrees of freedom for MS_A (**not for MS_{Acomp}**) and $MS_{S/A}$—that is, they direct you where to look in Table A-1 in Appendix A in order to find the F_{crit} in the formula. From the overall ANOVA, $df_A = 4$, $df_{S/A} = 40$, and from Table A-1 we see that $F_{crit}(4, 40) = 2.61$. The formula for the Scheffé test only requires a simple multiplication: $\mathbf{F_S = (4)(2.61) = 10.44}$. This F_S calculation is all that is new in carrying out a Scheffé test. In order for a tested comparison to be significant with a Scheffé test, the corresponding F_{Acomp} must equal or exceed $F_S = 10.44$. The specific $F_{Acomp1} = \mathbf{6.25}$ in this example is not larger than F_S, therefore the difference between Group a_1 and Group a_3 is not significant with a Scheffé test. However, the specific $F_{Acomp2} = 17.07$ is larger than F_S, therefore the difference between Group a_1 and Group a_5 is significant by a Scheffé test. You can see from Table A-1 that α_{PC} for a Scheffé test for this experiment sets the error risk even lower than .01, as $F_{crit}(1, 40)$ at $p \leq .01$ is only 7.31, compared to $F_S = 10.44$.

c. The Tukey Test

The Tukey test (a 1953 unpublished manuscript cited in Keppel et al. 1992) is not as stringent as the Scheffé test, meaning that the per comparison F_{crit} for a Tukey test (F_T) will be more conservative than the unadjusted $F_{crit} = 4.08$ and less conservative than $F_S = 10.44$. The Tukey procedure for controlling α_{FW} would be appropriate for a restricted but large number of follow-up comparisons, such as all pair-wise tests for differences between means of the original groups (e.g., Group a_1 vs. Group a_2; a_1 vs. a_3; … a_4 vs. a_5); however, additional comparisons between new combinations of groups would not be under consideration. A Tukey test is conducted much like a Scheffé test in that the only new operation occurs at the final stage. That is, using the weighting coefficient method, F_{Acomp} values for various analytical comparisons are computed following an overall ANOVA that resulted in a significant F for Factor A. For a Tukey test, an adjusted F_{crit} (coded as F_T) is calculated from the following formula: $F_T = q_T^2 \div 2$. Each F_{Acomp} is significant if it equals or exceeds F_T. The arithmetic required for calculating F_T is simple in that it only involves a q_T number that is squared and then divided by 2. Further, no computations are required for finding q_T, as q_T values are given in a table that looks very similar to Table A-1, the reference distributions for F. Table A-4 in Appendix A presents q_T values organized in rows and columns, with rows corresponding to $df_{S/A}$ (just as they do in the F table, Table A-1) and columns corresponding to **a = number of groups** (the organization used for columns in this table poses a potential source of confusion because the numbered columns in Table A-1 correspond to $\mathbf{df_A}$, and the numbered columns in Table A-4 correspond to **number of groups**). Since these two tables are similar in appearance, you need to be careful not refer to Table A-4 to find critical F values for an analysis of variance, and not refer to Table A-1 to find q_T values for a Tukey test.

To illustrate application of the Tukey test we will use the same data and the same analytical comparisons that were just evaluated with the Scheffé test. Thus, all the calculations for the $F_{comparison}$ values, F_{Acomp1} (a_1 vs. a_3) and F_{Acomp2} (a_1 vs. a_5), have been completed. All that remains to be done is to determine if these analytical comparisons are significant when F_T provides the critical comparison F. For these tests we need to compute $F_T = q_T^2 \div 2$; from Table A-4 in Appendix A, with a = 5 (the column with "5" at the top) and $df_{S/A} = 40$ (the row labeled "40"), we see that $q_T = 4.04$ for controlling α_{FW} at the .05 level. Thus, $\mathbf{F_T = 4.04^2 \div 2 = 16.32 \div 2 = 8.16}$. The specific $F_{Acomp1} = 6.25$ is not larger than F_T, therefore the conclusion based on the Tukey test is that the difference between the a_1 and a_3 groups is not significant. The specific $F_{Acomp2} = 17.07$ is clearly larger than F_T, hence the difference between Group a_1 and Group a_5 is significant with the Tukey test. The finding that F_{Acomp2} is significant should come as no surprise since this latter analytical comparison had already been shown to be significant with the more conservative Scheffé test procedure. Note that F_T is also higher than the critical F at the .01 level ($F_T = 8.16$ and $F_{crit} = 7.31$ for $p \leq .01$), which means that holding the α_{FW} error risk to p = .05 results in α_{PC} error risks that are also more conservative than p = .01.

II. Errors in Statistical Decisions: Type I (α) and Type II (β) Errors

In the last chapter and continuing into this chapter we discussed two kinds of α error risks when multiple analytical comparisons are done: one for each comparison between pairs of group means (α_{PC}) and one for the experiment as a whole (α_{FW}). In this section we will consider a second type of decision error that concerns researchers. A Type I or α error involves a decision based on observed differences among groups, *viz.*, that the H_0 assumption is incorrect and should be rejected (an error if it were known that H_0 was true). Since "truth" is not really known, experimenters live with some degree of uncertainty. Basically, as was shown in Figure 5.1 on page 79, H_0 is rejected if an F ratio calculated from experimental data is a relatively poor fit in the theoretical reference distribution of F values based on the assumption that H_0 is true (i.e., if H_0 is true, the expected value of F is 1.0). If H_0 were actually true, then a Type I or α decision error would occur on those rare occasions (e.g., 5% of the time) that the data result in an F_A that is larger than F_{crit} (the extreme right-hand side of the F reference distribution).

A Type II or β error involves a decision that differences among groups can readily be accounted for in terms of random variation (based on the results of the experiment, the H_0 assumption is **not** rejected) when in truth H_0 is incorrect. The distinction between Type I and Type II errors is summarized in the *truth table* shown in Table 8.2. This table indicates "truth" along one dimension, which we unfortunately do not know, and decisions about H_0 along the other dimension. Of course, it is only when truth and decision are not in agreement that an error occurs: an α error when H_0 is true and the decision is to reject H_0, and a β error when H_0 is false and the statistical test failed to reject H_0.

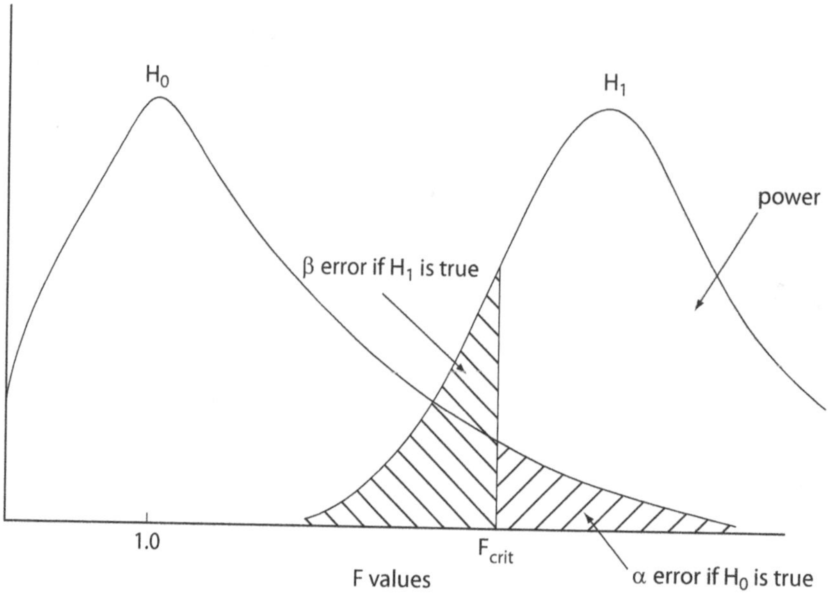

Figure 8.1: Hypothetical H_0 and H_1 Reference Distributions for F

Table 8.2: True Status and Experimental Decision Regarding H_0

	The Experimental Decision	
Truth about H_0	Reject H_0	Accept H_0
H_0 is True	Type I or α Error	Correct Decision
H_0 is False	Correct Decision	Type II or β error

Type II error risks are more difficult to deal with than Type I error risks because we do not have a specific reference distribution based on an assumption that groups differ. That is, many reference distributions are compatible with a general hypothesis that group means differ. Figure 8.1 provides an illustration of one specific reference distribution (labeled H_1) added to the graph containing the H_0 reference distribution. Remember, the decision point (F_{crit}) is set on the H_0 distribution; groups differ significantly if the F value calculated from the data is to the right of F_{crit} (larger than F_{crit}) **on the H_0 distribution**, and differences among groups are not significant if the calculated F is to the left of F_{crit} (smaller than F_{crit}). In other words, **F_{effect} values in the 95% portion of the H_0 distribution (to the left of the F_{crit}) result in a failure to reject H_0**. So how likely is it that a failure to reject H_0 will occur if a specific alternative assumption (e.g., H_1) is correct (i.e., the likelihood of making a β error)? **This error risk is estimated by the area under the alternative-assumption reference distribution (H_1) that overlaps with the 95% region (to the left of F_{crit}) of the H_0 reference distribution**. In Figure 8.1 this area is identified as representing approximately 20% of the total area under the H_1 curve; however, that proportion is just a visual estimation to illustrate β graphically.

Obviously, experimenters prefer to keep both Type I and Type II errors to a minimum. Unfortunately, there is a tradeoff between these two types of errors. Lowering α (e.g., setting $\alpha = .01$ instead of .05) increases β. This means that reducing the risk of a Type I error (rejecting H_0 when it should not be rejected) increases the risk of making a Type II error (failing to reject H_0 when it should be rejected). Perhaps you can see this interplay as you examine Figure 8.1 by imagining F_{crit} moved to the right, with the resultant shrinkage or decrease in the right-hand tail of the H_0 reference distribution (the α error risk) and the resultant increase in the area in the left-hand tail of the H_1 reference distribution (the β error risk).

For many years β error risk was a relatively unattended concern. Apparently the rationale for this was based on the fact that β could not be specified without having a precise H_1 reference distribution, hence little could be done about it. Over the last few decades effective arguments have been mounted for estimating β and taking it into consideration. Of particular interest is the class of correct decisions involving β, i.e., the probability based on experimental results that H_0 will be rejected when it is actually false ($1-\beta$, or the area under H_1 that is to the right of F_{crit}). This proportion is referred to as "**power**" (sensitivity), and it represents the probability that an experiment will show a significant difference when groups truly differ (obviously an important experimental objective). A high probability of this kind of experimental success is certainly desirable; indeed, if one knew in advance that a planned experiment had a low probability of success (low power), then the experiment would probably be redesigned or abandoned. We will have more to consider about power after the concept of effect size (measured in terms of the proportion of total variance accounted for by the independent variable) and procedures for estimating it are introduced. Once an effect size is estimated and available, it can be applied for a power analysis.

III. Measuring the Size of a Treatment Effect: ω^2

How do you know whether a significant difference between groups reflects a large or a small effect of the independent variable? A first response in addressing this question might be to consider the size of the F ratio: interpreting a large F ratio as an indication of a large treatment effect and an F_{effect} that is only slightly larger than F_{crit} as an indication of a small treatment effect. We live in a society where "big" is stressed and used to try to impress others (big cars, big houses, big …), so it should come as no surprise that even researchers are sometimes swept up with "biggie" sizing. When a researcher informs us in a journal article that the difference between groups was "highly significant," because the F_{effect} was even larger than the F_{crit} required for significance at $p < .001$, the magnitude of the F ratio appears to be used to create an impression about how big the treatment effect is. Technically speaking, we do not have a statistical test for differentiating "significant" from "highly significant," or from "really, really, significant." Although a relationship between effect size and F may seem reasonable, we will see later that a simple translation of "large F means large treatment effect" may be misleading. The difficulty with a straightforward interpretation of F as an index of effect size is that the size of F is sensitive to sample size: for two experiments with the same treatment effect size and different sample sizes (n), F_A will be larger for the experiment with the larger n. There is an obvious problem with using F as an index of effect size when it can be shown that F varies in magnitude even when the proportion of variability accounted for by the independent variable remains constant. Conceptually, the size of a treatment effect should be independent of number of experimental subjects.

At a conceptual level the goal is to understand the cause of variability between groups that experience different treatment conditions as defined by the independent variable. The experimental question concerns whether different treatment conditions produced a systematic difference in performance. Of course, many factors may contribute to variability observed in scores. If the independent variable produced a significant effect but only accounted for a small proportion of total variability, the treatment effect is considered to be small. If the independent variable represented a large proportion of total variability, the treatment effect is considered to be large. "Small" and "large" are vague terms; however, Cohen (1988) provided three reference points for indexing effect size based on the estimated proportion of total variability accounted for by the independent variable using the metric ω^2 (read as "omega squared"). Drawing primarily from an extended literature search, he classified $\omega^2 \geq .15$ as "large," transitioning to a descriptive label of "medium" at $\omega^2 \approx .06$, and "small" at $\omega^2 \approx .01$.

It is becoming more common in the literature to see an estimate of effect size accompanying results of a statistical analysis. That is, in addition to an F_{effect}, df_{effect}, df_{error}, an estimate of error variance (MS_{error}), and α level, ω^2 (or an alternative estimate of effect size) may appear in the text. It has also been argued that a point estimate of effect size bounded by a confidence interval provides an important quantitative summary of results of separate experiments on a common topic (**a meta-analysis**) (e.g., Bird 2002; Faith, Allison, and Gorman 1997; Murphy and Myors 2004). Given that effect size can usually be estimated from the statistical information reported in journal publications, it serves as a convenient metric for evaluating and drawing inferences across a large number of published experiments with similar independent and dependent variables. Although procedures for doing a meta-analysis are beyond the scope of this text, it is still important to be able to estimate the size of a treatment effect. To do this we will use ω^2 exclusively in the context of analysis of variance designs. Alternative measures based on similar principles will appear later in the context of the chi-square (χ^2) statistic (the metric ϕ) and for correlation and regression topics (r^2 or R^2).

For calculating ω^2 we have a relatively simple objective: take an estimate of variance due to the independent variable or treatment factor (an estimate of the population parameter, σ^2_{effect}) and divide it by an estimate of total variance (σ^2_{total}, or $\sigma^2_{effect} + \sigma^2_{error}$); thus $\boldsymbol{\omega^2 = \sigma^2_{effect} \div (\sigma^2_{effect} + \sigma^2_{error})}$. The error term in the analysis of variance ($MS_{S/A}$) provides a direct estimate of random error in the population (σ^2_{error}). However, an estimate of variability due to the population treatment effect is more complicated.

You may recall when the F reference distribution was introduced that the expected value of F under the H_0 assumption was 1.0. The reason given for this expectation was that the numerator estimate in the F ratio (between-group variance) provided an estimate of variance attributed to **both** (1) a potential treatment effect **and** (2) random error. Thus, MS_A is a contaminated estimate of σ^2_{effect} because it not only includes an estimate of variability due to the treatment effect (A), but it also includes an estimate of variability due to random error. Computing ω^2 directly requires a two-step procedure: (1) calculating an estimate for σ^2_{effect} and (2) solving for ω^2 using the estimate for σ^2_{effect} from step 1 and $MS_{S/A}$ as the estimate (**Est**) for σ^2_{error}: $\boldsymbol{\omega^2 = \sigma^2_{effect} \div (\sigma^2_{effect} + MS_{S/A})}$.

The formula below for estimating σ^2_{effect} is presented without formal proof; however, by inspecting the components of the formula, it is readily apparent that a central ingredient involves removing (subtracting) the error variance estimate ($MS_{S/A}$) from the between-groups estimate of variance due to the treatment effect plus error variance (MS_A): Estimated (Est) $\sigma^2_{effect} = (df_A)(\mathbf{MS_A - MS_{S/A}}) \div (a)(n)$; and $\boldsymbol{\omega^2 = \text{Est}\ \sigma^2_{effect} \div (\text{Est}\ \sigma^2_{effect} + \text{Est}\ \sigma^2_{error})}$. Converting this formula for use with sample data, we have:

$$([df_A][MS_A - MS_{S/A}]/[a][n]) \div ([\{df_A\}\{MS_A - MS_{S/A}\} \div \{a\}\{n\}] + MS_{S/A})$$

There is an alternative formula for calculating ω^2, and though "what is going on" in that formula is not readily apparent, it does have two advantages over the first alternative formula: (1) it involves only a single step calculation, and (2) the only components required for the computation are a, n, and F. Since information concerning number of groups (a), number of subjects per group (n), and the F ratio for the effect (F_{effect}) are commonly reported in research publications, effect size can be estimated for research reports even if not provided by the author. This alternative formula is $\boldsymbol{\omega^2 = df_{effect}(F_{effect} - 1) \div (df_{effect}[F_{effect} - 1] + [a][n])}$. It should be noted that a negative effect size is illogical; however, it can occur arithmetically. This problem is avoided by assigning a value of zero to ω^2 whenever $F_{effect} < 1.0$.

These two formulas for directly calculating ω^2 from the results of an analysis of variance are illustrated for the ANOVA summary table that follows. Given the information in this table, we will estimate size of the significant effect for Factor A (ω^2_A) by using the first formula that requires estimating σ_A^2, and then by using the second formula and substituting F, df_A, a, and n appropriately. Regardless of which procedure is used for calculating ω^2, you do not square the value calculated. That is, if $\omega^2 = df_{effect}(F_{effect} - 1) \div (df_{effect}[F_{effect} - 1] + [a][n]) = .20$, **do not square this number** because $\omega^2 = .20$, and $\omega^2 \neq .04$.

ANOVA Summary Table for calculating ω^2

Source	SS	df	MS	F	
A	360	3	120	4.29	$p < .05$
S/A	896	32	28		

1. $\omega^2_{effect} = \text{Est}\ \sigma^2_A \div (\text{Est}\ \sigma^2_A + \text{Est}\ \sigma^2_{error})$; $MS_{S/A}$ provides an estimate of σ^2_{error}; and Est $\sigma^2_A = (df_A)(MS_A - MS_{S/A}) \div (a)(n)$

   The ANOVA summary table provides the following information necessary for estimating σ^2_A and calculating ω^2_A: $df_A = 3$, and $n = 9$ [$df_{S/A} = 32$, and since $df_{S/A} = a(n - 1)$, we have $32 = 4(n - 1)$, or $n = (32 + 4) \div 4 = 9$]; $MS_A = 120$; and $MS_{S/A} = 28$.

   The first step involves estimating σ^2_A:

   Est $\sigma^2_A = (df_A)(MS_A - MS_{S/A}) \div (a)(n) = (3)(120 - 28) \div (4)(9) = (3)(92) \div 36 = \mathbf{7.67}$

   The second step in calculating ω^2 can now be completed:

   $\omega^2_{effect} = \text{Est}\ \sigma^2_{effect} \div (\text{Est}\ \sigma^2_{effect} + \text{Est}\ \sigma^2_{error}) = 7.67 \div (7.67 + 28) = 7.67 \div 35.67 = \mathbf{.215}$

2. The computational alternative for estimating effect size is as follows:

$\omega^2_A = df_A(F_A - 1) \div (df_A[F_A - 1] + [a][n])$ and $df_A = 3$, $a = 4$, $n = 9$, and $F_A = 4.29$.

$\omega^2_A = (3)(4.29 - 1) \div ([3][4.29 - 1] + [4][9]) = (3)(3.29) \div ([3][3.29] + 36) = 9.87 \div 45.87 =$ **.215**

3. Now consider another demonstration relevant to the assertion that the size of the F ratio may be a misleading indicator of effect size. In order to illustrate that F is sensitive to sample size and therefore should not be used as an index of effect size, we will use the formula in number 2 above and work backwards to solve for F holding effect size constant ($\omega^2 = .215$). However, for this demonstration we will assume a larger sample size. The question is, "What would F equal if an effect of exactly the same size had been reported for an experiment with 31, rather than 9 subjects per group?" A little algebra is necessary at this point to see the impact of sample size on F_A. The algebraic task involves making the following substitutions in the formula for ω^2 and solving for F_A: $\omega^2 = .215$, $df_A = 3$, $a = 4$, and $n = 31$.

$\omega^2_A = df_A(F_A - 1) \div (df_A[F_A - 1] + [a][n])$

$.215 = 3(F_A - 1) \div (3[F_A - 1] + [4][31])$, or $.215 = (3F_A - 3) \div (3F_A - 3 + 124)$

Multiplying both sides of the equation by $3F_A - 3 + 124$, or $3F_A + 121$, we have:

$(.215)(3F_A + 121) = 3F_A - 3$, or $.645F_A + 26.015 = 3F_A - 3$

Gathering F_As on the same side of the equation gives us:

$3F_A - .645F_A = 26.015 + 3$, or $2.355(F_A) = 29.015$

$\mathbf{F_A = 29.015 \div 2.355 = 12.32}$: thus, increasing n = 9 to n = 31 nearly tripled the F value even though there was no change in effect size in terms of estimated proportion of variance attributed to Factor A (ω^2).

IV. Power and Sample Size

In terms of probability, the power of an experiment refers to probability that a true difference between groups will be detected (e.g., when H_0 is actually false, an experiment with a high degree of power is likely to produce a significant F ratio). In order to assess power it is essential to have at least a "ballpark" estimate of effect size. Other things equal, power will be greater for detecting large compared to small effect sizes. Power is also determined by sample size, as power will be greater for detecting an effect of a given size for experiments with large n than for experiments with small n. And as was mentioned earlier there is a tradeoff between Type I and Type II errors such that if risk of making a Type I error is decreased (e.g., setting F_{crit} further to the right in the H_0 distribution as would happen if you set α at .01 instead of .05), then the risk of making a Type II (β) error is increased (a larger portion of the H_1 distribution will fall to the left of F_{crit}—i.e., in the "accept H_0" region). Since power = 1 – β, any action that alters β will impact power, and if β is high, power will be low.

In order to estimate the power a completed experiment had for detecting a significant difference, it is necessary to have an estimate of effect size (ω^2), to know the α level used for the H_0 test, and to know the number of groups and the number of subjects in the experiment (both a and n). Questions of power come into play both before and after an experiment. However, the interest and information sought differs in these two situations. For a completed experiment, we obviously know the number of groups, the number of subjects, and the α level. As far as effect size is concerned, this can be estimated from the completed experiment, or the question can be shifted to one about how much power this experiment would have had for detecting some specific, hypothetical effect size. During the planning stage before an experiment is conducted, decisions are made about certain aspects of the design, thus power considerations ought to play a pivotal role in these decisions. Although we may be restricted by α-level convention (i.e., to maintain credibility, researchers normally are not at liberty to set α = .25 instead of .05), we can do something about sample size and ω^2 to get to an acceptable level of power before doing the experiment. Arguably, and for many situations, pre-experimental evaluation is the more important application of power analysis and the one that we will focus on.

There are complicated procedures for computing estimates of power or using power to determine an appropriate sample size. However, it is not necessary to provide details as free software programs are available and relatively easy to access and use. Thomas and Krebs (1997) reviewed 29 power analysis software programs (e.g., Erdfelder, Faul, & Buchner, 1996; Friedman, 1982; Kraemer & Thiemann, 1987). For our purpose, the simplified and abbreviated estimates in Table A-5 in Appendix A will suffice. This table is organized to facilitate determining approximate sample sizes necessary for a limited set of power analyses given α = .05. The reported sample sizes were estimated from materials and formulas presented by Cohen (1988). The table is organized into five blocks, with the top block providing approximate sample sizes for power estimates of .50; the second block for power estimates of .60; the third block for power estimates of .70, the fourth block for power estimates of .80, and the bottom block for power estimates of .90. Each block is organized with rows for 11 estimates of effect size (ω^2) ranging from .01 to .20; and columns for degrees of freedom (ranging from df_A = 1 to df_A = 8). Numbers within the table are the estimated sample sizes (n).

Two experimenter decisions are necessary when using Table A-5 (it is okay for some trial and error before settling on final details): (1) the size of the effect in question, and (2) the power level desired. Before being too idealistic you should realize that the most laudable goals in these respects are likely to be associated with impossibly large sample sizes, so practical and economical factors enter the picture.

To illustrate use of Table A-5 we will start with some reasonable assumptions about effect size and power. Assume you are planning an experiment for an Honor's Thesis, so you and your family have an interest in this experiment being doable within a semester or two. Further, assume that you have little basis for estimating an actual effect size for your independent variable. Of course, you would like to have a high level of power, and even if the actual effect is "small," you would like your experiment to show a significant difference ($\alpha < .05$). You and your advisor agree that three groups are necessary for your experiment. Given this background, you decide that .90 power and ability to detect a small effect (e.g., $\omega^2 = .02$) are noble goals. So off you go to Table A-5, looking at the bottom block of sample sizes; and you enter this section of the table at the intersection of the $\boldsymbol{\omega^2}$ **= .02** row and the $\mathbf{df_A}$ **= 2** column. Assuming you have to test each subject in your experiment individually, you might be a little depressed to see that the approximate number in Table A-5 at this location is **210** [210 subjects per group means you will have to collect data from somewhere around 630 subjects (a x n = 3 x 210 = 630) to achieve these power and effect size goals]. At this point you could decide to postpone your graduation by a few years, or you could opt instead to modify your initial power and/or effect size targets. Since achieving power ≈ .80 is perfectly acceptable, and you would be satisfied with having approximately .80 power for detecting a "medium-to-large" effect at $\alpha \leq .05$; you decide to reevaluate your experimental plan, settling on power ≈ .80 and an estimated effect size of $\omega^2 = .10$. Re-entering Table A-5, this time in the .80 power section (the fourth block in the table) where the $\boldsymbol{\omega^2}$ **= .10** row and $\mathbf{df_A}$ **= 2** column intersect, you see that **n = 30.** If you decide that a total of 90 subjects (a x n = 3 x 30 = 90) is manageable for your experiment, you are ready to go; and you have the satisfaction of knowing that your decision about the number of subjects was made on the basis of relatively sophisticated considerations of power and effect size, rather than by guessing, intuition, or whim.

Within certain limits you may also use Table A-5 to estimate power for detecting an effect of a given size found in a completed experiment. For example, assume an experiment had six groups (a = 6) with 25 subjects per group (n = 25). An appropriate question might be: "What is the approximate power the experiment had for detecting a "medium-size" effect according to Cohen's guidelines ($\omega^2 = .06$)? Table A-5 provides power estimates ranging from .50 to .90, so at least power within this range of values may be estimated. The first step in estimating power is to "scan" the intersections of $df_A = 5$ columns with $\omega^2 = .06$ rows in each of the five sections of Table A-5 to find the value closest to the sample size used in the completed experiment (n = 25). Respective n values given in Table A-5 for these five power levels are n = 45 for power ≈ .90, n = 35 for power ≈ .80, n = 29 for power ≈ .70, n = 24 for power ≈ .60, and n = 19 for power ≈ .50. Thus, for n = 25, it follows that power will be between .60 and .70. For $df_A = 5$ and $\omega^2 = .06$, n = 24 for power ≈ .60 and n = 29 for power ≈ .70. The n = 25 is closer to the n = 24 with power ≈ .60 than it is to the n = 29 with power ≈ .70. Thus, a reasonable estimate of power for $\alpha = .05$, $df_A = 5$, and $\omega^2 = .06$ would be a little above .60 (e.g., **power ≈ .62**). Finally, consider this hypothetical experiment one more time, but for this last illustration assume that the completed experiment had **10** subjects per group (n = 10). Again, scanning the intersections of $df_A = 5$ columns with $\omega^2 = .06$ rows in each of the five sections of Table A-5, the lowest n value occurs with power ≈ .50 (n = 19; well above n = 10 in this example). Thus, using only the information in Table A-5, all we can say is that the power the experiment had for detecting a medium-size effect ($\omega^2 = .06$) at $\alpha \leq .05$ was less than .50. However, this may be sufficient information for a power analysis given that experiments with such low probability of showing a significant difference at $\alpha \leq .05$ given a true effect size of $\omega^2 = .06$ would not be particularly impressive.

V. Homogeneity of Variance and "Normality"

Reference distributions that provide probability information for statistical tests have been developed in accordance with certain assumptions. For example, the reference distributions for the coin-tossing experiments described in Chapter 4 were based on a "fair-coin" assumption—namely, the assumption that "heads" and "tails" were equally probable outcomes. If assumptions underlying the reference distributions we use are not approximated by the data collected, then the reference distribution in question may overestimate or underestimate probabilities associated with statistics calculated from the experimental data. Two important assumptions for the analysis of variance model are (1) that within-group variance estimates of random error are **equal** in the long run (**homogeneity of variance assumption**) and (2) that the experimental data recorded are distributed "normally" around the mean (**normality assumption**), meaning the distribution of scores is roughly symmetrical in shape with a center peak. At one time it was thought that violations of these assumptions did not pose too great a problem because analysis of variance was a robust test and held up well under relatively modest violations (Box, Hunter, and Hunter 1978). However, more recent analyses have questioned earlier claims of robustness (Kline 2004). Monte Carlo simulations (computer simulations of literally thousands of experiments with various violations of assumptions) from data violating these assumptions have revealed a rejection rate higher than 5% with tests conducted at $\alpha = .05$ (Keppel and Wickens 2004). Thus, experiments that do not meet assumptions of homogeneity of variance and/or normality may actually be using an inflated α level without realizing it (i.e., H_0 may be rejected 6% or 7% of the time by chance even though an experimenter assumed his or her operational α level was .05).

Different tests are available for determining whether or not problems exist for underlying assumptions and different remedies are available. For normality, one can always plot data and do a rough eyeball test—that is, assess if the shape of the distribution(s) of scores is roughly normal (a center peak and a symmetrical bell-shaped pattern). For homogeneity of variance, a simple test (not the most highly recommended, but easy to perform) requires calculating an F_{max} index. For this index a separate variance is calculated for each group in the experiment. The **maximum F ratio** is then determined by dividing the largest group variance by the smallest group variance: $F_{max} = s^2_{largest} \div s^2_{smallest}$. A general rule is that you need not worry too much about a violation of the homogeneity of variance assumption unless the largest variance is more than three times larger than the smallest variance ($F_{max} > 3.0$).

So what should experimenters do (at least those who discover these potential violations) besides worry about whether or not the reference distribution is really appropriate for their experiment? Having acknowledged the problem, one may certainly decide just to live with it (after all, we have already seen a sort of "live with it" response when an inflated α_{FW} was recognized for additional meaningful, planned comparisons). Alternatively, a more conservative α (which will result in a larger F_{crit}) may be set for H_0 to compensate for the fact that using $\alpha = .05$ from the H_0 reference distribution table results in a chance rejection rate somewhat higher than 5%. Another alternative is to turn to statistical procedures that are not based on assumptions about homogeneity of variance or normality (distribution-free or non-parametric statistics—for further detail, the interested reader may wish to consult Marascuilo and McSweeney 1977; Siegel and Castellan 1988). However, this latter choice would be made at some cost because non-parametric alternatives are less powerful tests, thus raising the risk of Type II decision errors (an experiment that fails to reject H_0 when in fact H_0 is false).

Several new topics were introduced in this chapter. Given a significant F in an analysis of variance for a multiple group experiment, several exploratory comparisons among pairs of groups and various combinations of groups may be conducted. When this is done, an adjustment should be made lowering α_{PC}, since with multiple comparisons the probability that at least one analytical comparison will be significant by chance alone (α_{FW}) is considerably higher than the α_{PC} level set for each comparison. While we can simply set a more conservative α for each analytical comparison, two formal adjustment procedures were introduced: the Scheffé test and the Tukey test. For both procedures, a new F_{crit} is determined, one associated with a lower α_{PC}. It was noted that researchers may have to compromise and not get too conservative with α_{PC} and α_{FW} levels because it is also important to avoid decision errors that result from failing to reject H_0 for cases in which H_0 is false (β errors). There is a tradeoff between α and β error risks such that a more conservative (lower) setting for α results in a greater risk of a β error; if there is a real effect for the independent variable, experimenters certainly do not want to miss detecting it.

It has been acknowledged that H_0 testing provides rather limited information. A significant F ratio provides an objective justification for regarding an independent variable as effective for the behavior being investigated. Earlier it was noted that reporting a confidence interval for the difference between two means was richer in information than F or t, as it provided the same objective information about whether or not groups differ and additional information about probable lower and upper limits for the difference. In this chapter another index was introduced (ω^2) that provided additional information about relative size of a treatment effect in terms of proportion of total variance attributed to the independent variable. A measure of effect size often accompanies reported F values in the literature.

The ability to estimate the magnitude of a treatment effect (ω^2) clears the way for estimating probability that a given experiment will be successful in rejecting H_0 for the specific estimated effect size (referred to as power). Power analysis plays a particularly important role in planning experiments (see Murphy and Myors 2004). Given an estimated effect size, the number of subjects necessary to achieve a desired level of power can be determined. Thus, experimenters can have some assurance that an experiment has a reasonable chance of successfully detecting an effect (a significant F ratio) at least as large as the effect size originally estimated. Finally, all things considered, one must not lose sight of the fact that certain assumptions were required to establish reference distributions used for test statistics. Two such assumptions were noted: homogeneity of variance and normality. If homogeneity and normality assumptions are not approximated in the experimental data, then probabilities associated with selected cutoff points determined by the reference distribution may be inaccurate (e.g., an F_{crit} point assumed to cutoff the most extreme 5% of a distribution might actually separate the most extreme 7% of a distribution).

The Dunnett (1955) and Newman-Keuls (Keuls 1952; Newman 1939) tests are two additional multiple *post hoc* test procedures that adjust α_{PC} in order to control α_{FW}. The adjustment for the Dunnett test is designed for a limited number of comparisons appropriate for a special class of experiment in which two or more treatments (experimental conditions) are compared separately to a common control condition. Given the limited number of comparisons relevant for contrasting each experimental group with a common control group, it is not necessary to lower α_{PC} to the same degree as is done for the Tukey or Scheffé tests.

The Newman-Keuls test is one that may appeal more to psychologists and social scientists than it does to statisticians. The reason for this is that with Newman-Keuls, α_{FW} is controlled by setting a lower **average** α_{PC} (with variable α_{PC} levels for actual contrasts between pairs of groups—i.e., different α_{PC} criteria for different comparisons). For example, if an experiment involved four groups, there are six possible contrasts between pairs of groups (a_1 vs. a_2, a_1 vs. a_3, a_1 vs. a_4, a_2 vs. a_3, a_2 vs. a_4, and a_3 vs. a_4). If $\alpha_{PC} = .010$ for three between-group contrasts, $\alpha_{PC} = .007$ for two other between-group contrasts, and $\alpha_{PC} = .005$ for the remaining between-group contrast, then the average $\alpha_{PC} = .008$ ($[.010 + .010 + .010 + .007 + .007 + .005] \div 6 = .049 \div 6 = .008$). The family-wise error rate may be estimated by multiplying the number of comparisons (C) times the average α_{PC}; thus, $\alpha_{FW} \approx C(\text{average } \alpha_{PC}) \approx 6(.008) \approx .05$. Instead of setting $\alpha_{PC} = .008$ for each comparison, with the Newman-Keuls procedure, some contrasts would be tested with a more stringent α_{PC} (e.g., $\alpha_{PC} = .005$) and others with a more lenient α_{PC} (e.g., $\alpha_{PC} = .010$).

The procedures described for the Dunnett test are similar to those considered for the Scheffé and Tukey tests. A new adjusted F_{crit} is determined, and each F_{Acomp} calculated from the data for comparisons between an experimental and control group must equal or exceed the adjusted F_{crit} to be significant. In contrast, the Newman-Keuls test involves one major change in approach in comparison to the other multiple comparison adjustment procedures (the Scheffé, Tukey, and Dunnett tests). For Scheffé, Tukey, and Dunnett tests, an F is calculated for each comparison of interest, then a **new** $\mathbf{F_{crit}}$ is calculated to compare to the F_{Acomps} computed from the experimental data. Work required for calculating F ratios for each comparison of interest is not necessary with the Newman-Keuls procedure. The only computations from the data that are necessary involve computing **means** for each group, and then determining a **difference score** (subtracting one group mean from another group mean) between mean scores of specific groups being compared. **Critical mean difference** scores are then determined. For the Newman-Keuls test, a difference score between a given pair of groups must equal or exceed the critical mean difference score for that pair to be significant. Perhaps this approach will become clearer after we work through an example problem.

a. Dunnett Test

The critical comparison F for the Dunnett test (F_D) may be determined with a minimal amount of computational labor: $F_D = q_D^2$. To find the value for q_D, merely go to Table A-6 in Appendix A, and the appropriate q_D is located at the intersection of the row corresponding to the error term degrees of freedom ($df_{S/A}$ or df_{error}) and the column corresponding to the number of groups in the experiment (k = number of treatment means). You should note that the columns in Table A-6, like the columns in Table A-4, are organized in terms of number of groups, and not degrees of freedom between groups. Thus, the only calculation required for finding the critical comparison for a Dunnett test is to square the appropriate number found in Table A-6. We will illustrate determining F_D by referring back to the ANOVA summary table (Table 8.1) in this chapter and $F_{Acomp1} = 6.25$. Assume that this comparison (between Groups a_1 and a_3) represents an experimental-control group contrast. Is this $F_{Acomp1} = 6.25$ significant by a Dunnett test? In this example, a = 5 and $df_{S/A} = 40$. From Table A-6, with α_{FW} set at $p \leq .05$, a = 5 and $df_{S/A} = 40$, we see that $\mathbf{q_D = 2.54}$. Thus, to find the critical F (F_D), you only need to square q_D: $F_D = 2.54^2 = 6.45$; thus, $F_{Acomp1} = 6.25 < F_D$ and is not significant with a Dunnett test.

b. Newman-Keuls Test

For the Newman-Keuls test, α_{FW} is controlled by using "variable" α_{PC} criteria instead of a single, constant α_{PC}. As illustrated in the earlier hypothetical example, if an $\alpha_{PC} \approx .008$ is necessary to hold $\alpha_{FW} = .05$ for the number of comparisons tested, then with Newman-Keuls, some comparisons are tested with a more stringent α_{PC} (e.g., $\alpha_{PC} = .005$) and others are tested with a more lenient α_{PC} (e.g., $\alpha_{PC} = .010$) such that the average α_{PC} for all comparisons tested is $\alpha_{PC} \approx .008$. It was also noted for the Newman-Keuls test that calculating F_{Acomp} values for each comparison tested was not necessary. Rather, the simpler computational task of calculating means for each group and then subtracting one of the means involved in a comparison of two means from the other mean involved in the comparison replaces F_{Acomp} calculations. With the other three test procedures considered (Scheffé, Tukey, and Dunnett) an F_{crit} was calculated, and significance decisions involved comparing each F_{Acomp} to the new F_{crit} (F_S, F_T, or F_D for Scheffé, Tukey, and Dunnett tests, respectively). With the Newman-Keuls procedure, critical differences between means necessary to control α_{FW} (e.g., $\alpha_{FW} = .05$) are calculated (coded as $D_{N\text{-}K}$), and any difference between a pair of group means that exceeds the appropriate critical $D_{N\text{-}K}$ is regarded as significant.

Critical comparison $D_{N\text{-}K}$ values are organized in terms of the magnitude of the differences between respective pairs of group means in the data set. They are set such that the most extreme differences between pairs of group means are tested against the largest $D_{N\text{-}K}$ values (i.e., a more stringent α_{PC} level), and the closest differences between pairs of group means are tested against the smallest $D_{N\text{-}K}$ values (i.e., a more lenient α_{PC} level). Thus, the first step for a Newman-Keuls test is to re-order groups from lowest to highest in terms of the magnitude of their means. The means for the five groups in our example appear below, and as you can see, they happen to be ordered in magnitude the same as they were ordered by their subscript designations.

<u>Experimental Groups</u>	<u>a_1</u>	<u>a_2</u>	<u>a_3</u>	<u>a_4</u>	<u>a_5</u>
Means	3.11	4.78	5.33	6.00	6.78

The key for determining the different $D_{N\text{-}K}$ values is the number of groups that have to be "spanned" in getting to the two groups involved in a comparison. Illustrating what is meant by "number of groups spanned" may be a clearer way of getting the meaning across than trying to provide a verbal definition. In this example, the comparison between Group a_1 vs. Group a_5 spans all five groups. Comparisons between a_1 vs. a_4 and a_2 vs. a_5 span four groups. Comparisons between a_1 vs. a_3, a_2 vs. a_4, and a_3 vs. a_5 span three groups. All of the remaining comparisons span two groups: a_1 vs. a_2, a_2 vs. a_3, a_3 vs. a_4, and a_4 vs. a_5.

There are two general computational tasks with the Newman-Keuls test: First, difference scores between all pairs of group means must be determined, and second, four critical $D_{N\text{-}K}$ values must be determined for (1) comparison spanning five group means, (2) comparisons spanning four group means, (3) comparisons spanning three group means, and (4) comparisons spanning two group means. The formula for $D_{N\text{-}K}$ is $D_{N\text{-}K} = q_K\sqrt{(MS_{S/A} \div n)}$, where appropriate q_K values are found in the same table that was used for the Tukey test (Table A-4), only now with columns corresponding to the number of means spanned and rows corresponding to $df_{S/A}$ ($df_{S/A} = 40$). The 10 mean differences and the appropriate comparison $D_{N\text{-}K}$ values are given in the table that appears following the computations for the critical $D_{N\text{-}K}$ for the a_1 vs. a_5 comparison (spanning five groups), and for the critical $D_{N\text{-}K}$ for the a_1 vs. a_4 and the a_2 vs. a_5 comparisons that span four groups. The $D_{N\text{-}K}$ values for spanning three groups and two groups are **2.17** and **1.80**, respectively (you may wish to confirm these calculations by referring to Table A.4).

a_1 vs. a_5; $n = 9$; $q_K = 4.04$; $MS_{S/A} = 3.55$; and $D_{N\text{-}K} = (4.04)\sqrt{(3.55 \div 9)} = (4.04)(.63) = \mathbf{2.55}$

a_1 vs. a_4; $n = 9$; $q_K = 3.79$; $MS_{S/A} = 3.55$; and $D_{N\text{-}K} = (3.79)\sqrt{(3.55 \div 9)} = (3.79)(.63) = \mathbf{2.39}$

a_2 vs. a_5; $n = 9$; $q_K = 3.79$; $MS_{S/A} = 3.55$; and $D_{N\text{-}K} = (3.79)\sqrt{(3.55 \div 9)} = (3.79)(.63) = \mathbf{2.39}$

Results of Newman-Keuls Significance Tests of 10 Pair-wise Comparisons

Comparison	$\bar{Y}_{Ai}-\bar{Y}_{Aj}$	# Spanned	q_K	$D_{N\text{-}K}$	Significant
a_1 vs. a_5	6.78 – 3.11 = **3.67**	5	4.04	**2.55**	YES
a_1 vs. a_4	**2.89**	4	3.79	**2.39**	YES
a_2 vs. a_5	**2.00**	4	3.79	**2.39**	NO
a_1 vs. a_3	**2.22**	3	3.44	**2.17**	YES
*a_2 vs. a_4	1.22	3	3.44	**2.17**	NO
*a_3 vs. a_5	1.45	3	3.44	**2.17**	NO
a_1 vs. a_2	**1.67**	2	2.86	**1.80**	NO
*a_2 vs. a_3	.55	2	2.86	**1.80**	NO
*a_3 vs. a_4	.67	2	2.86	**1.80**	NO
*a_4 vs. a_5	.78	2	2.86	**1.80**	NO

Of the 10 possible pair-wise contrasts, the difference between the two group means that spanned all five groups (Group a_1 vs. Group a_5) is significant. One of the two pair-wise contrasts that spanned four groups (Group a_1 vs. Group a_4) is significant and one of the contrasts that spanned three groups (Group a_1 vs. Group a_3) is significant. The remaining seven contrasts between pairs of group means are not significant. You can see from this table that if the contrast between Group a_1 vs. Group a_3 had been tested at either of the more stringent α_{PC} levels (setting the critical $D_{N\text{-}K}$ value at 2.55 or 2.39) it would not have been judged to be significant. Finally, there is a reasonable and minor labor-saving ground rule to consider. We may conclude, without a formal test, that a difference between any pair of groups included within a non-significant span is also not significant (e.g., Keppel and Wickens 2004). In our example, the pair-wise comparison that spanned five groups (a_1 vs. a_5) was significant, thus we proceeded to test the two comparisons that spanned four groups (a_1 vs. a_4 and a_2 vs. a_5). The point emphasized here is that since the Group a_2 vs. Group a_5 comparison was not significant, pair-wise contrasts within this span are automatically judged to be non-significant without a formal test (the set of comparisons between groups that are marked with an asterisk in the table: a_2 vs. a_3; a_2 vs. a_4; a_3 vs. a_4; a_3 vs. a_5; and a_4 vs. a_5).

8.1: Model Problem 1. In a small *pilot* study, 18 students were randomly assigned to 3 groups (a = 3), with 6 students in each group (n = 6). The experimenter was interested in how "level of processing" affected memory. Students were shown a list of 25 common words, with each word exposed for 5 seconds. As each word was shown, Group C subjects were asked to write down the number of different **consonants** in the word, Group R subjects were asked to estimate number of rhyming words, and Group S subjects were asked to estimate number of **synonyms** for each word. The experimenter's thesis was that elaborative study (focusing on sound, and more importantly, meaning) would be beneficial, hence she predicted that both Groups R and S would recall more words than Group C. Ten minutes after study, students were asked for recall. The mean number of words correctly recalled for each group was Group C = 8.17; Group R = 11.00; and Group S = 14.17. The analysis of variance summary table appears below.

Source	SS	df	MS	F	(F_{crit})
A (study method)	108.11	2	54.06	**3.98**	3.68
S/A	203.67	15	13.58		

1. Provide an estimate of the size of the effect for Factor A.

For this question, we will calculate ω^2_A in two ways. First, we will use the two-step formula [Step 1: **Estimated** $\boldsymbol{\sigma^2_{effect} = (df_A)(MS_A - MS_{S/A}) \div (a)(n)}$; and Step 2: $\boldsymbol{\omega^2_A = \text{Est } \sigma^2_{effect} \div (\text{Est } \sigma^2_{effect} + \text{Est } \sigma^2_{error})}$]. Second, we will calculate ω^2_A using the single-step formula: $\boldsymbol{\omega^2_A = df_A(F_A - 1) \div (df_A[F_A - 1] + [a][n])}$.

a. Step 1: Est $\sigma^2_{effect} = (df_A)(MS_A - MS_{S/A}) \div (a)(n) = (2)(54.06 - 13.58) \div (3)(6)$

Est $\sigma^2_{effect} = (2)(40.48) \div 18 = 4.50$; and Est $\sigma^2_{error} = MS_{S/A} = 13.58$

Step 2: $\omega^2_A = \text{Est } \sigma^2_{effect} \div (\text{Est } \sigma^2_{effect} + \text{Est } \sigma^2_{error})$

$\boldsymbol{\omega^2_A = 4.50 \div (4.50 + 13.58) = .249}$

b. $\omega^2_A = df_A(F_A - 1) \div (df_A[F_A - 1] + [a][n]) = (2)(3.98 - 1) \div ([2][3.98 - 1] + [3][6])$

$\boldsymbol{\omega^2_A = 5.96 \div 23.96 = .249}$

2. As you can see, by Cohen's guidelines this ω^2 represents a large effect. The experimenter, encouraged by this effect size, planned to repeat the study. Given these results she felt comfortable setting a goal of being able to detect a large effect ($\omega^2 = .15$) at $p \le .05$, and with power ≈ .90. Approximately how large a sample size will she need in order to achieve her goal?

a. To obtain the estimate for a sample size for an experiment with a = 3, $\omega^2 = .15$, $\alpha = .05$, and power ≈ .90, simply refer to the bottom section of Table A-5 in Appendix A. The intersection of the $\omega^2 = .15$ row and the "$df_A = 2$" column has **n = 25.**

8.2: Model Problem 2. For this problem we will consider an experiment with six levels of Factor A and three subjects in each group (a = 6, n = 3). The respective group means are $\bar{Y}_{A1} = 2.33$, $\bar{Y}_{A2} = 1.67$, $\bar{Y}_{A3} = 5.00$, $\bar{Y}_{A4} = 3.33$, $\bar{Y}_{A5} = 6.33$, $\bar{Y}_{A6} = 8.00$. The ANOVA summary table is given below:

ANOVA Summary Table

Source	SS	df	MS	F	(F_{crit})
Factor A	89.78	5	17.96	**5.05**	3.11
S/A	42.67	12	3.56		

1. Using the Tukey test, determine the critical F_T for analytical comparisons at $\alpha_{FW} = .05$.

This question does not ask you to test any specific comparison among groups or combinations of groups; it only asks for F_T. The formula for F_T is $F_T = q_T^2 \div 2$ and q_T is obtained from Table A-4 in Appendix A. In this table we want column 6 (a = 6 groups) and row 12 ($df_{S/A} = 12$) and at $\alpha_{FW} = .05$, $q_T = 4.75$. Therefore, $\mathbf{F_T = q_T^2 \div 2 = 4.75^2 \div 2 = 22.56 \div 2 = 11.28}$.

2. Now repeat #1 above for the Scheffé test: $F_S = df_A(F_{crit})\{df_A, df_{S/A}\}$.

For this formula, we need df_A and $F_{crit}(5, 12)$; remember, $(F_{crit})\{df_A, df_{S/A}\} = F_{crit}(5, 12) = 3.11$ (see Table A-1 in Appendix A). Therefore, $\mathbf{F_S = (5)(3.11) = 15.55}$.

3. We will complete this illustration by computing one specific analytical comparison to determine if an experimenter doing a Tukey test would have decided that the difference between the groups involved was significant at $\alpha_{FW} \leq .05$ and whether or not the significance decision would have been different had a Scheffé test been used. (1) Is the difference between Group a_1 and Group a_6 significant with a Tukey test? (2) Is the difference between Group a_1 and Group a_6 significant with a Scheffé test?

   a. Step 1: Calculate the SS_{Acomp} for this comparison. Let $c_1 = 1$ and $c_6 = -1$.

$SS_{Acomp} = (n)(\Sigma c_i \bar{Y}_{Ai})^2 \div \Sigma c_i^2$; and n = 3, $\bar{Y}_{A1} = 2.33$ and $\bar{Y}_{A6} = 8.00$

$SS_{Acomp} = (3)([1][2.33] + [-1][8.00])^2 \div (1^2 + [-1]^2)$

$SS_{Acomp} = (3)(-5.67)^2 \div 2 = 96.45 \div 2 = 48.22$

$df_{Acomp} = 1$, therefore, $MS_{Acomp} = 48.22$

b. Step 2: Calculate $\mathbf{F_{Acomp} = MS_{Acomp} \div MS_{S/A} = 48.22 \div 3.56 = 13.54}$

c. From Part 1, $F_T = 11.28$, and $F_{Acomp} = 13.54$ is significant with a Tukey test.

d. From Part 2, $F_S = 15.55$, and $F_{Acomp} = 13.54$ is not significant with a Scheffé test.

8.3: Practice Problem 1. Estimate the magnitude of the effect of Factor A in Practice Problem 2, Chapter 6 (page 111).

8.4: Practice Problem 2. For a 4-group experiment, approximately how many subjects per group are necessary to detect a medium-sized effect ($\omega^2 = .06$) at $\alpha \leq .05$ with power $\approx .80$?

8.5: Practice Problem 3. Data from a 3-group experiment were presented in Practice Problem 2, Chapter 7 (page 129). Three analytical comparisons between pairs of groups are possible (a_1 vs. a_2; a_1 vs. a_3; a_2 vs. a_3). Determine the critical comparison F values (F_T and F_S) that must be equaled or exceeded for significance (for $\alpha_{FW} \leq .05$) with Tukey and Scheffé tests. Are any comparisons significant with these more stringent tests?

8.6: Practice Problem 4. An ANOVA summary table (Table 6.2) is presented on page 97. For these fictitious data, the difference between groups was not significant [F (1, 38) = 1.94]. Based on the effect size estimated for these data, (a) provide an approximate power estimate this experiment had for finding a significant F at p = .05 for a 'large' effect ($\omega^2 = .15$), and (b) determine how many subjects (n) the experimenter would need to have in each group in order to be able to detect an effect of this magnitude that would be significant at the $\alpha = .05$ level with power equal to .80.

8.7: Practice Problem 5. Conduct an F_{max} test for the data in the following table. Given the results from this test, would you advise the experimenter to proceed with an analysis of variance on these data, or is a more cautious approach in order, such as using a more conservative α level or considering further analyses with non-parametric statistics?

		Experimental Groups					
a_1	a_2	a_3	a_4	a_5	a_6	a_7	a_8
0	7	3	3	6	18	21	23
4	8	9	6	9	10	19	31
7	7	5	9	12	9	24	40
3	2	6	9	8	11	14	16
5	5	10	17	13	22	18	27
9	4	8	7	4	20	30	19

8.8: Practice Problem 6. The following tables are copied from Practice Problem 4, Chapter 7 (page 130).

1. For the first part of this problem, assume that Group a_1 is a control group and do a Dunnett test with $\alpha_{FW} = .05$ to compare each experimental group (a_2, a_3, and a_4) with the a_1 control group.
2. For the second part of this problem, compare each group with every other group using the Newman-Keuls procedure to adjust α_{PC} levels as necessary to hold α_{FW} at the .05 level. Which pair-wise comparisons, if any, are significant?

Data Matrix: Number of Missed Shots

	a_1	a_2	a_3	a_4
	8	9	16	20
	11	13	20	15
	12	18	22	25
	5	8	15	17
$\bar{Y}$	**9.00**	**12.00**	**18.25**	**19.25**

ANOVA SUMMARY TABLE

Source	SS	df	MS	F
A	292.25	3	97.42	6.44
S/A	181.50	12	15.12	

IX. Practice Test 2: Chapters 6–8

(answers given in Appendix B)

I. Circle the alternative that provides the best answer for each question.

1. To increase the power an experiment has for detecting a significant difference, you should
   a. increase the sample size
   b. decrease the sample size
   c. use confidence intervals
   d. do an F_{max} test

2. A β error refers to
   a. probability you will do the wrong experiment
   b. probability you will conclude that there is a difference among groups when there is not
   c. probability you will conclude that there is no difference among groups when there is
   d. probability there is heterogeneity of variance among groups when there is not

3. If n = 10, $\bar{Y}$ = 8.5, and s = 4.28, what is the standard error of the mean (s_E)?
   a. 1.15
   b. 1.35
   c. 1.55
   d. 1.75

4. The pairs of statements below describe two types of errors that are possible when drawing inferences from the results of the statistical analysis of experimental data, identified as Type I (α) and Type II (β) errors, respectively. Which pair of statements is correct?
   a. A Type I error occurs if the experimenter claims that the null hypothesis is true when it is really true
      A Type II error occurs if the experimenter claims that the null hypothesis is false when it is really false
   b. A Type I error occurs if the experimenter claims that the null hypothesis is false when it is really false
      A Type II error occurs if the experimenter claims that the null hypothesis is true when it is really true
   c. A Type I error occurs if the experimenter claims that the null hypothesis is true when it is really false
      A Type II error occurs if the experimenter claims that the null hypothesis is false when it is really true
   d. A Type I error occurs if the experimenter claims that the null hypothesis is false when it is really true
      A Type II error occurs if the experimenter claims that the null hypothesis is true when it is really false

5. For the data in the table below and to the right, SS_A = ____?

a. .5

b. 1.6

c. 2.9

d. 18.8

$\underline{a_1}$	$\underline{a_2}$	$\underline{a_3}$
3	4	2
5	7	0
4	4	9

6. An experimenter plans to do an experiment having one independent variable (Factor A) with 5 levels (a = 5). A reasonable effect size estimate for Factor A is $\omega^2 = .12$. In order to achieve a power of .90, what should n (number of subjects per group) equal?
   a. 19
   b. 24
   c. 30
   d. 50

7. An experiment involves 5 groups. After completing an ANOVA, the experimenter decided to do 8 analytical comparisons. If he sets the "per comparison" error rate at .05 ($\alpha_{PC} = .05$), what is the **estimated** "family-wise" error rate (α_{FW})?
   a. .05
   b. .15
   c. .30
   d. .40

8. Which of the following procedures for controlling $\alpha_{FW} \approx .05$ is the most conservative (i.e., the procedure that sets α_{PC} at the smallest value)?
   a. Newman-Keuls
   b. Dunnett
   c. Scheffé
   d. Tukey

9. If an experiment involves a single control group and several experimental groups, and the experimenter is only concerned with follow-up comparisons between each experimental group and the control group, which procedure for controlling $\alpha_{FW} \approx .05$ is most appropriate?
   a. Newman-Keuls
   b. Dunnett
   c. Scheffé
   d. Tukey

10. Which of the following changes will result in an increase in the power of an experiment?
   a. An increase in the risk of making a Type I (α) error
   b. An increase in sample size (n)
   c. An increase in the size of the treatment effect
   d. All of the above

II. Do an Analysis of Variance on the data that follow and provide the necessary information to complete the Analysis of Variance Summary Table that appears below the data matrix, indicating whether or not Factor A is significant at $p \leq .05$:

DATA MATRIX:

Groups

	a_1	a_2	a_3	a_4	a_5	a_6
	5	6	3	7	7	9
	5	7	5	8	7	10
	2	4	6	9	9	12
	4	3	7	8	10	15
ΣY	**16**	**20**	**21**	**32**	**33**	**46**
ΣY_i^2	**70**	**110**	**119**	**258**	**279**	**550**

ANALYSIS OF VARIANCE SUMMARY TABLE

Source	SS	df	MS	F
A				
s/A				

III. Using the data in Problem II, test to see if an F ratio for Linear Trend is significant at the .05 level.

IV. A data matrix and ANOVA summary table are presented below ($a = 3$, $n = 5$):

DATA MATRIX

	a_1	a_2	a_3
	3	9	9
	1	7	7
	5	5	12
	7	6	8
	4	3	9
ΣY_i	20	30	45
$\bar{Y}$	4.0	6.0	9.0

ANOVA SUMMARY TABLE

Source	SS	df	MS	F
A	63.33	2	31.66	7.04
S/A	54.00	12	4.50	

A. Do an analytical comparison to determine if there is a significant difference between Group A_1 and Group A_3.

B. Indicate the critical F_S (Scheffé F) that the F_{comp} calculated in Part A above would have to equal or exceed to be significant at $\alpha_{FW} \leq .05$ (i.e., $\mathbf{F_S = ?}$).

V. Determine the 95% confidence interval for the difference between the means for Group a_1 and Group a_3 in Part III (the data for these two groups follow).

DATA MATRIX

	$\underline{a_1}$	$\underline{a_3}$
	3	9
	1	7
	5	12
	7	8
	$\underline{4}$	$\underline{9}$
ΣY_i	20	45
$\bar{Y}$	4.0	9.0
s^2	5.0	3.5

Chapter 9

The Analysis of Variance with Two Independent Variables

A judicious man uses statistics, not to get knowledge, but to save himself from having ignorance foisted upon him.

—Carlye

I. The Factorial Design
II. Evaluating Three Effects in a Single Experiment
  a. Computational Components
  b. An Illustration: Revisiting a Previous Problem
III. The Two-Factor ANOVA with More than Two Levels for Each Factor
IV. A Computer Software Application with SPSS
V. Summary
VI. Practice Problems

I. The Factorial Design

Taking stock of our progress to this point, you should now be able to use the analysis of variance to test the viability of the null hypothesis (all group means are effectively equal) for comparing any number of groups. Given that a significant difference is observed in an experiment involving more than two groups, you can do follow-up (analytical) comparisons between any two groups (original or re-combined), calculating a between-group sum of squares restricted to the data of interest and conducting the appropriate F_{Acomp} test. The analysis of variance was introduced in the simplest context with contrived data sets that had a single independent variable (**single-factor designs**), and with different groups of subjects randomly assigned to the various treatment conditions. Using reward magnitude as an example of an independent variable (Factor A), groups of college students could be differentiated in terms of hourly pay for serving in a "sensory deprivation" experiment. Pay would vary across groups (e.g., a_1 = \$2 per hr.; a_2 = \$3 per hr.; a_3 = \$5 per hr.; a_4 = \$8 per hr.; and a_5 = \$12 per hr.). Participants have absolutely nothing to do but lie on a bed in a sound-deadened chamber with minimal visual, tactile, etc., stimulation. The dependent variable would be the length of time each volunteer subject was able to stay in the chamber. While this may sound like a "dream" job, in actual experiments of this sort, volunteers usually chose to leave the chamber after a relatively short stay.

In this chapter we will expand on the single-factor experiment by considering a second independent variable so that two factors are simultaneously manipulated within the same experiment. An experiment with two independent variables (Factor A and Factor B) is commonly referred to as a **factorial experiment**, given that every combination of the different levels of Factors A and B is represented; it is symbolized as an A × B (read as "A by B") design. For example, consider adding a second factor to a sensory deprivation experiment. Factor A is hourly pay rate ($a = 5$ as described in the preceding paragraph), and we will couple that independent variable with a second independent variable, *viz.*, occasional *time-outs* or *no time-outs* (Factor B, with $b = 2$). Half of the participants in each pay condition will have a 30-minute time-out from sensory deprivation once every three hours, during which time they will be able to listen to a statistics lecture. The remaining subjects in each group will not have a time-out break. The number of groups needed for a factorial experiment is determined by multiplying number of levels of Factor A by number of levels of Factor B ($5 \times 2 = 10$). If b_1 is a lecture time-out condition and b_2 is a no-time-out condition, then 10 groups are necessary for the experiment: a_1b_1 ($2 per hr. and lecture time-outs), a_1b_2 ($2 per hr. and no time-outs), a_2b_1 ($3 per hr. and lecture time-outs), etc., up to the final a_5b_2 combination ($12 per hr. and no time-outs).

It is probably fair to say that the majority of experiments published today involve two or more independent variables, thus it is important to examine both the logic and analysis of the two-factor design. Including a second independent variable has some obvious advantages, such as enabling a researcher to evaluate not one, but two independent variables within a single experiment. You may recall that an essential requirement for causal reasoning from an experiment is that groups are equivalent in all respects except one, namely the different condition or level of the independent variable experienced. Can it be argued that introducing Factor B violates this objective? If we only compared the a_1b_1 group ($2 per hr. and lecture time-outs) with the a_2b_2 group ($3 per hr. and no time-outs), we would certainly have a problem, because a performance difference between these two groups could result from the fact that a_1 produces a different effect from a_2, or that b_1 produces a different effect from b_2, or both. However, for this very reason, the a_1b_1 vs. a_2b_2 comparison is not of interest. Rather, to evaluate Factor A (does $a_1 = a_2 = a_3 = a_4 = a_5$?), each level of Factor A involves equal combinations of the two Factor B levels (i.e., $a_1b_1 + a_1b_2 = \mathbf{a_1}$, $a_2b_1 + a_2b_2 = \mathbf{a_2}$, etc.). Although Factor B is not held constant when evaluating Factor A, it is equally represented across the different levels of Factor A (e.g., half of the participants in the $2 per hr. group [$a_1$] have the 30 minute time-out lecture [b_1], while remaining participants in this group do not have a time-out period [b_2]). With equal numbers of subjects in all groups, the question of whether students earning $12 per hr. will remain in the sensory deprivation chamber longer than students earning $2 per hr. is not contaminated by whether or not students get to hear a 30-minute statistics lecture as a time-out from sensory deprivation, as half of the students in each reward condition hear the lecture. If the time-out lectures drive volunteers out of the chamber sooner, the lectures will impact an equal number of subjects in each reward condition.

Being able to evaluate two (or more) independent variables within a single experiment (rather than having to conduct two or more experiments) certainly offers advantages in terms of economy and efficiency. In addition, one gains information about whether or not the effects of one independent variable generalize across the levels of another independent variable that were included in the experiment. For the contrived sensory deprivation example, we might find that time spent in a sensory deprivation chamber is directly related to hourly pay, and further, that this effect generalizes (is basically the same) across both the time-out and no time-out conditions. However, and more importantly, we may observe that the effect does not generalize. This might occur if we observed that the direct relation between hourly pay and length of stay was present for subjects in the time-out groups (b_1), but not for subjects who did not get time-outs (b_2). This more complex relationship is called an **interaction** (or A × B interaction) and can be statistically tested.

Three F ratios are calculated in the two-factor ANOVA: a *main effect* for Factor A (e.g., to evaluate if the five reward conditions differ), a *main effect* for Factor B (e.g., do the two time-out conditions differ?), and the A × B *interaction*. The interaction may indicate that one independent variable does not have the same effect at each level of the other independent variable (i.e., the main effect of Factor A for b_1 groups is not the same as the main effect of Factor A for b_2 groups). Arithmetically, the presence of an interaction in a 2 × 2 design (two levels of Factor A and two levels of Factor B) translates to $\bar{Y}_{a1b1} - \bar{Y}_{a1b2} \neq \bar{Y}_{a2b1} - \bar{Y}_{a2b2}$, or $\bar{Y}_{a1b1} - \bar{Y}_{a1b2} - \bar{Y}_{a2b1} + \bar{Y}_{a2b2} \neq 0$.

II. Evaluating Three Effects in a Single Experiment

Three effects can be assessed in a two-factor A × B design. Two effects are "main effects," and they basically enable us to evaluate each independent variable (Factor A and Factor B) just as if two single-factor experiments had been conducted. The third effect evaluated is the A × B interaction, and this effect indicates whether or not the descriptions of main effects have to be qualified. If there were no interaction, then a direct, unqualified description of a significant main effect is possible (e.g., a main effect for Factor B may be found if scores are higher for Group a_1b_1 than for Group a_1b_2, scores are higher for Group a_2b_1 compared to Group a_2b_2, scores are higher for Group a_3b_1 compared to Group a_3b_2, etc.). A significant interaction forces one to qualify the previous summary statement (e.g., scores are **higher** for Group a_1b_1 than for Group a_1b_2; however, the opposite is true or scores are equal for other b_1-b_2 contrasts, such as **lower** scores for Group a_2b_1 compared to Group a_2b_2).

a. Computational Components

In performing the necessary calculations for a single-factor analysis of variance we began by calculating three basic ratios: [A], [T], and [Y]. From these three calculations we computed two sums of squares: $SS_A = [A] - [T]$ and $SS_{S/A} = [Y] - [A]$. From here it was a simple matter to calculate the two corresponding variance estimates: $MS_A = SS_A \div df_A$ and $MS_{S/A} = SS_{S/A} \div df_{S/A}$; and the ANOVA test statistic: $F = MS_A \div MS_{S/A}$. In moving to the computations required for the two-factor ANOVA some of the same calculations are required, and some new calculations are required. A minor additional complication that may create confusion occurs because the addition of a second factor in the experiment (Factor B) has implications for the letter-coding scheme.

As far as some of the necessary computations are concerned, it does not matter whether you have a single-factor design or a two-factor design, you still have to calculate [Y] and [T]. That is, it is still necessary to square each individual score and sum these squared scores ($[Y] = \Sigma Y_i^2$), and it is still necessary to sum all of the individual scores, square this sum, and divide it by the total number of scores or N, where N = (a)(n) in the single-factor design and N = (a)(b)(n) in the two-factor design $\{[T] = (\Sigma Y_i)^2 \div (a)(b)(n)\}$. These two mathematical operations were part of the three basic ratio calculations (bracketed terms) that had to be done for the single-factor analysis of variance. The third basic ratio for the **between group** sum of squares (SS_A) is also computed in the same way for the two-factor design; however, it is coded as [**AB**] instead of [A] because groups are now differentiated by both the A and the B treatment they received. To calculate [AB], each group sum ($AB_{i,j}$) is squared, summed, and then divided by the number of subjects in each group $\{[AB] = (\Sigma AB_{i,j})^2 \div (n)\}$.

The organization of data in a two-factor design permits meaningful combinations of subgroups. That is, different original groups that received the a_1 treatment can be combined into a new single a_1 group, and similarly for a_2, a_3, etc. These new combinations provide the basis for calculating [A] for the two-factor design. Here we come face to face with a source for confusion because [A] for the single-factor design lines up with [AB] for the two-factor design, as both are central to calculating the Sum of Squares between groups (SS_{BG}) SS_{BG} in the respective designs. For the two-factor design, basic ratios for SS_A and for SS_B are new and require combining original groups before calculating [A] and [B]. For the respective calculations, we have: $[A] = (\Sigma A_i)^2 \div (b)(n)$ and $[B] = (\Sigma B_j)^2 \div (a)(n)$. The two denominators ("b times n" for [A] and "a times n" for [B]) merely "count" the number of individual scores that had to be summed to determine the values of (ΣA_i) and (ΣB_j). A small data set of 12 scores is given in the following table to illustrate similarities and differences between basic-ratio calculations for single-factor and two-factor designs:

Single-Factor Design

	a_1	a_2	a_3	a_4
	3	7	5	2
	6	2	0	2
	4	1	3	2
ΣY_i	**13**	**10**	**8**	**6**

$[A] = (13^2 + 10^2 + 8^2 + 6^2) \div 3 = 369 \div 3$

[A] = **123.00**

$[T] = 37^2 \div 12 = 1{,}369 \div 12 =$ **114.08**

$[Y] = 3^2 + 6^2 + \ldots + 2^2 =$ **161.00**

Two-Factor Design

a_1b_1	a_1b_2	a_2b_1	a_2b_2
3	7	5	2
6	2	0	2
4	1	3	2
13	**10**	**8**	**6**

$[AB] = (13^2 + 10^2 + 8^2 + 6^2) \div 3 = 369 \div 3$

[AB] = **123.00**

$[T] = 37^2 \div 12 = 1{,}369 \div 12 =$ **114.08**

$[Y] = 3^2 + 6^2 + \ldots + 2^2 =$ **161.00**

For the two-factor design (but not for the single-factor design) two new meaningful data matrices can be created from the original data matrix. It should be apparent that these new data matrices involve nothing more than appropriately converting two columns of three scores (e.g., a_1b_1 and a_1b_2) to single columns of six scores (e.g., a_1): one set appropriate for calculating [A] (a_1 vs. a_2) and one set appropriate for calculating [B] (b_1 vs. b_2).

For Factor A:

	a_1	a_2
	3	5
	6	0
	4	3
	7	2
	2	2
	1	2
ΣY_i	**23**	**14**

For Factor B:

	b_1	b_2
	3	7
	6	2
	4	1
	5	2
	0	2
	3	2
ΣY_i	**21**	**16**

$[A] = (23^2 + 14^2) \div (b)(n) = 725 \div 6 = \mathbf{120.83}$

$[B] = (21^2 + 16^2) \div (a)(n) = 697 \div 6 = \mathbf{116.17}$

Four sums of squares may be calculated for the two-factor design, whereas only two sums of squares were calculated for the single-factor design. Obviously since actual scores for the four A groups in this illustration have not been altered by changing to AB coding labels, variability within and between the four groups will not be changed (i.e., given that we are using the same scores for each of the four groups; between-group variance, within-group variance, and total variance will not be affected by whether groups are labeled a_1, a_2, a_3, and a_4 or a_1b_1, a_1b_2, a_2b_1, and a_2b_2). Given this common data set, the sum of squares for within-group variability ($SS_{S/A}$) for a single-factor analysis ([Y] – [A]) is the same as $SS_{S/AB}$ for a two-factor analysis ([Y] – [AB]). Also, the **single-factor between-group sum of squares** (SS_{BG} = [A] – [T]) is the same as the **two-factor between-group sum of squares** (SS_{BG} = [AB] – [T]).

At this point you should be able to anticipate the basic analytical change in the analysis of variance from the single-factor to the two-factor design. **The SS_{BG} in the two-factor design can be partitioned into three components**: part of SS_{BG} comes from variability due to Factor A (SS_A), part from variability due to Factor B (SS_B), and the remainder, or residual, is from the A × B interaction ($SS_{A \times B}$). The respective sums of squares for the two-factor design are SS_A = [A] – [T], SS_B = [B] – [T] and $SS_{A \times B}$ = [AB] – [A] – [B] + [T]. This latter formula for $SS_{A \times B}$ follows from the fact that $SS_{BG} = SS_A + SS_B + SS_{A \times B}$. The basic ratio for the between-group sum of squares in the two-factor design is SS_{BG} = [AB] – [T]. Since SS_A = [A] – [T] and SS_B = [B] – [T], it follows that $SS_{A \times B}$ = [AB] – [A] – [B] + [T]. You may check this by replacing the sums of squares in the formula above ($SS_{BG} = SS_A + SS_B + SS_{A \times B}$) with their respective basic ratios.

The addition of Factor B justifies an analysis of the between-group variance into component parts. In the single-factor design, $a = 4$, $n = 3$, $df_A = a - 1 = 3$, and $df_{S/A} = a(n - 1) = 8$. In the two-factor design, $a = 2$, $b = 2$, $n = 3$, $df_A = a - 1 = 1$, $df_B = b - 1 = 1$, $df_{A \times B} = (a - 1)(b - 1) = 1$, and $df_{S/AB} = (a)(b)(n - 1) = 8$. For comparison purposes, single-factor and two-factor ANOVA summary tables are presented in the following tables to highlight the similarities and differences.

Single-Factor ANOVA Summary Table

Source	df	SS	MS	F
A	a – 1	[A] – [T]	$SS_A \div df_A$	$MS_A \div MS_{S/A}$
S/A	a(n – 1)	[Y] – [A]	$SS_{S/A} \div df_{S/A}$	
Total	(a)(n) – 1	[Y] – [T]		

Two-Factor ANOVA Summary Table

Source	df	SS	MS	F
A	a – 1	[A] – [T]	$SS_A \div df_A$	$MS_A \div MS_{S/AB}$
B	b – 1	[B] – [T]	$SS_B \div df_B$	$MS_B \div MS_{S/AB}$
A × B	(a – 1)(b – 1)	[AB] – [A] – [B] + [T]	$SS_{AB} \div df_{AB}$	$MS_{AB} \div MS_{S/AB}$
S/AB	(a)(b)(n – 1)	[Y] – [AB]	$SS_{S/AB} \div df_{S/AB}$	
Total	(a)(b)(n) – 1	[Y] – [T]		

b. An Illustration: Revisiting a Previous Problem

Is testing the significance for three effects within a single experiment something new? You may recall the trend analysis procedures from Chapter 7 where we followed up a single-factor experiment involving four groups by testing the significance of a linear trend, a quadratic trend, and a cubic trend. Also, in Model Problem 7.1 (page 128) three analytical comparisons were illustrated, along with a warning that we would re-visit this set of comparisons in a future chapter. That time has come because the specific three complex comparisons in that illustration are actually equivalent to the two main effects and interaction had those data come from a two-factor (2 × 2) experiment. We will re-work Model Problem 7.1 with the same number set, but a slightly different made-up story about where the numbers came from. For this scenario, assume that there are two levels of Factor A (a = 2) and two levels of Factor B (b = 2), so all that has changed from Model Problem 7.1 in the data set are the column headings.

For these data we need to change the storyline as (1) there are only two levels of sleep deprivation, 15 and 30 hours, and (2) there is a second variable (Factor B) with groups having either difficult (Diff) or easy reasoning problems to solve. However, none of the numbers have changed, so the calculations that were done when the data were analyzed as a single-factor, four-group experiment will not change. For [Y], each individual score had to be squared and then summed; for [T], all the scores were first summed, with this total then squared and divided by the total number of scores. There are still the original 40 scores in the data set; however, with the single-factor example in Chapter 7 that number was arrived at by multiplying a = 4 times n = 10. With the new A × B configuration, a = 2, b = 2, and n = 10, and the total number of scores is determined by (a)(b)(n) = (2)(2)(10) = 40. There are still four groups, and SS_{BG} is determined by squaring each group total and dividing by n. The re-coded data are presented in Table 9.1.

Table 9.1: Data from Model Problem 7.1 Re-Coded as a Two-Factor Experiment

Hours of Sleep Deprivation and Performance on Difficult or Easy Reasoning Tests

	15 hrs-Diff (a_1b_1)	15 hrs-Easy (a_1b_2)	30 hrs-Diff (a_2b_1)	30 hrs-Easy (a_2b_2)
	7.0	9.5	8.3	10.0
	6.5	9.0	7.5	9.6
	5.9	8.4	7.5	9.5
	5.5	8.0	7.1	9.4
	5.0	7.6	6.2	8.0
	4.7	7.5	6.2	7.8
	4.0	6.0	5.0	7.8
	3.8	5.9	4.8	7.2
	3.1	4.0	4.0	7.2
	2.7	3.8	3.9	7.1
ΣY_i	**48.2**	**69.7**	**60.5**	**83.6**
ΣY_i^2	**250.74**	**521.07**	**387.93**	**710.54**
$\bar{Y}$	**4.82**	**6.97**	**6.05**	**8.36**

Notice what was compared in the three analytical comparisons in Model Problem 7.1. The group means, coefficients, and sums of squares for that problem appear below, and columns are re-coded in terms of both Factors A and B. In addition, sums of squares for the three comparisons illustrated in Model Problem 7.1 are reproduced this table.

Group Means and Weighting Coefficients for Three Complex Comparisons

Group Means:	4.82(a_1b_1)	6.97(a_1b_2)	6.05(a_2b_1)	8.36(a_2b_2)
A_{comp1}	+1	+1	-1	-1
A_{comp2}	+1	-1	+1	-1
A_{comp3}	+1	-1	-1	+1

Comp 1: SS_{Acomp1} = **17.16**
Comp 2: SS_{Acomp2} = **49.73**
Comp 3: SS_{Acomp3} = **.08**

The plus and minus signs identify the two sides of a comparison; it is clear that A_{comp1} compared columns 1 and 2 (a_1b_1 and a_1b_2) with columns 3 and 4 (a_2b_1 and a_2b_2). That is, the two a_1 conditions (with positive signed coefficients) were compared to the two a_2 conditions (with negative signed coefficients). Hence, the effect for Factor A (a_1 vs. a_2) will be the same as A_{comp1}! A_{comp2} compared columns 1 and 3 (a_1b_1 and a_2b_1) with columns 2 and 4 (a_1b_2 and a_2b_2). That is, the two b_1 conditions (with positive-signed coefficients) were compared to the two b_2 conditions (with negative-signed coefficients). The effect for Factor B (b_1 vs. b_2) will be the same as A_{comp2}! The interaction comparison is A_{comp3}, however, seeing this relation may not be as obvious. Recall that H_0 for an interaction means that the difference between a_1 and a_2 will be the same for both b_1 and b_2 conditions. That is, according to H_0, $a_1b_1 - a_1b_2 = a_2b_1 - a_2b_2$, or $a_1b_1 - a_1b_2 - a_2b_1 + a_2b_2 = 0$. The positive- and negative-signed coefficients used for calculating SS_{Acomp3} by the coefficient method are consistent with this interaction definition (+1, -1, -1, and +1 for c_1, c_2, c_3, and c_4, respectively).

Although in this simple 2 × 2 example problem the calculations could be done by the coefficient method, we will complete the illustration by using the more general two-factor analysis of variance procedures, beginning by calculating sums of squares using the basic ratio components of the SS formulas. The three component parts of SS_{BG} will be the same (within rounding error) as reported for the three analytical comparisons in Model Problem 7.1: $SS_{Acomp1} = SS_A$; $SS_{Acomp2} = SS_B$; and $SS_{Acomp3} = SS_{A\times B}$. The basic ratios for these calculations are $[A] = (A_1^2 + A_2^2) \div (b)(n)$, $[B] = (B_1^2 + B_2^2) \div (a)(n)$, and $[AB] = ([A_1B_1]^2 + [A_1B_2]^2 + [A_2B_1]^2 + [A_2B_2]^2) \div n$, where upper case letters denote summed scores (e.g., A_1 is the sum of all scores for subjects in a_1 groups).

$[A] = ([48.2 + 69.7]^2 + [60.5 + 83.6]^2) \div (2)(10) = (13{,}900.41 + 20{,}764.81) \div 20 = 34{,}665.22 \div 20 = \mathbf{1{,}733.26}$

$[B] = ([48.2 + 60.5]^2 + [69.7 + 83.6]^2) \div (2)(10) = (11{,}815.69 + 23{,}500.89) \div 20 = 35{,}316.58 \div 20 = \mathbf{1{,}765.83}$

$[AB] = (48.2^2 + 60.5^2 + 69.7^2 + 83.6^2) \div 10 = (2{,}323.24 + 3{,}660.25 + 4{,}858.09 + 6{,}988.96) \div 10 = 17{,}830.54 \div 10 = \mathbf{1{,}783.05}$

$[T] = (7.0 + 6.5 + \ldots + 7.1)^2 \div 40 = 262^2 \div 40 = 68{,}644 \div 40 = \mathbf{1{,}716.10}$

$[Y] = (7.0^2 + 6.5^2 + \ldots + 7.1^2) = \mathbf{1{,}870.28}$

$SS_A = [A] - [T] = 1{,}733.26 - 1{,}716.10 = \mathbf{17.16}$ (Same as SS_{Acomp1})

$SS_B = [B] - [T] = 1{,}765.83 - 1{,}716.10 = \mathbf{49.73}$ (Same as SS_{Acomp2})

$SS_{A\times B} = [AB] - [A] - [B] + [T] = 1{,}783.05 - 1{,}733.26 - 1{,}765.83 + 1{,}716.10 = \mathbf{.06}$ (Within rounding error, the same as SS_{Acomp3})

$SS_{S/AB} = [Y] - [AB] = 1{,}870.28 - 1{,}783.05 = \mathbf{87.23}$ (Same as the SS_{error} from the single-factor design, however, it is now coded as $SS_{S/AB}$ instead of $SS_{S/A}$)

Since df_A, df_B, and $df_{A\times B}$ all equal 1.0, mean squares and corresponding sums of squares will be the same: $MS_A = 17.16$, $MS_B = 49.73$, and $MS_{A\times B} = .06$.

ANOVA Summary Table:

Source	SS	df	MS	F	F_{crit} (using df_{error} = 30)
A	17.16	1	17.16	7.09*	4.17
B	49.73	1	49.73	20.55*	4.17
A × B	.06	1	.06	< 1.0	4.17
S/AB	87.23	36	2.42		

* $p \leq .05$

III. The Two-Factor ANOVA with More than Two Levels for Each Factor

In the preceding problem, each of the three effects (Factor A, Factor B, and the A × B interaction) was evaluated using the same reference distribution with F_{crit} set at 4.17 for $p \leq .05$. If Factor A and Factor B have different degrees of freedom, the three F_{crit} values will not be identical. We will illustrate this by going through an ANOVA for a more complex design involving three levels of Factor A and four levels of Factor B (a 3 × 4 design). Since 3 × 4 = 12, this between-group design calls for 12 different groups of subjects. For this example problem, let a = 3, b = 4, and n = 5; the data matrix in Table 9.2 is organized with three columns for A conditions, four rows for B conditions, and five individual scores in each a_ib_j cell.

Table 9.2: Fictitious Data from a 3 × 4 Experimental Design

		Factor A						
		a_1	$\Sigma\Sigma A_1B_j$	a_2	$\Sigma\Sigma A_2B_j$	a_3	$\Sigma\Sigma A_3B_j$	ΣB_j
	b_1	3, 6, 4, 2, 4	**19**	9, 7, 9, 5, 8	**38**	7, 2, 3, 8, 5	**25**	**82**
Factor B	b_2	4, 4, 8, 4, 5	**25**	6, 7, 8, 4, 6	**31**	6, 7, 6, 9, 7	**35**	**91**
	b_3	2, 6, 5, 7, 5	**25**	5, 7, 1, 5, 3	**21**	9, 9, 9, 7, 8	**42**	**88**
	b_4	4, 2, 5, 5, 3	**19**	8, 4, 6, 6, 5	**29**	7, 6, 5, 8, 9	**35**	**83**
		ΣA_i	**88**		**119**		**137**	**T = 344**

Table 9.2 provides the (a)(b)(n) = 60 individual scores, plus summary information (indicated in boldface type) in terms of each A_iB_j sum, each A_i sum, each B_j sum, and the sum of all 60 scores (T). So we will begin with the hardest part: calculating the five basic ratios, and building up to presentation of the results of the analysis in an ANOVA summary table (we will see that with this fictitious data set, that Factor B is not significant, however both Factor A and the A × B interaction are significant at $p \leq .05$).

Step 1: Calculate the five Basic Ratios

$[A] = \Sigma A_i^2 \div (b)(n) = (88^2 + 119^2 + 137^2) \div (4)(5) = 40{,}674 \div 20 = \mathbf{2{,}033.70}$

$[B] = \Sigma B_j^2 \div (a)(n) = (82^2 + 91^2 + 88^2 + 83^2) \div (3)(5) = 29{,}638 \div 15 = \mathbf{1{,}975.87}$

$[AB] = \Sigma(A_iB_j)^2 \div n = (19^2 + 25^2 + \ldots + 35^2) \div 5 = 10{,}498 \div 5 = \mathbf{2{,}099.60}$

$[T] = (\Sigma Y_{i,j})^2 \div (a)(b)(n) = 344^2 \div (3)(4)(5) = 118{,}336 \div 60 = \mathbf{1{,}972.27}$

$[Y] = \Sigma Y_{i,j}^2 = 3^2 + 6^2 + \ldots + 9^2 = \mathbf{2{,}236.00}$

Step 2: Calculate the four Sums of Squares (**Note: SS values still must be positive numbers!**)

$SS_A = [A] - [T] = 2{,}033.70 - 1{,}972.27 = \mathbf{61.43}$

$SS_B = [B] - [T] = 1{,}975.87 - 1{,}972.27 = \mathbf{3.60}$

$SS_{A\times B} = [AB] - [A] - [B] + [T] = 2{,}099.60 - 2{,}033.70 - 1{,}975.87 + 1{,}972.27 = \mathbf{62.30}$

$SS_{S/AB} = [Y] - [AB] = 2{,}236.00 - 2{,}099.60 = \mathbf{136.40}$

Step 3: Determine the respective Degrees of Freedom

$df_A = a - 1 = 3 - 1 = 2$

$df_B = b - 1 = 4 - 1 = 3$

$df_{A\times B} = (a - 1)(b - 1) = (2)(3) = 6$

$df_{S/AB} = (a)(b)(n - 1) = (3)(4)(5 - 1) = 48$

Step 4: Calculate the four Mean Squares

$MS_A = SS_A \div df_A = 61.43 \div 2 = \mathbf{30.72}$

$MS_B = SS_B \div df_B = 3.60 \div 3 = \mathbf{1.20}$

$MS_{A\times B} = SS_{AB} \div df_{AB} = 62.30 \div 6 = \mathbf{10.38}$

$MS_{S/AB} = SS_{S/AB} \div df_{S/AB} = 136.40 \div 48 = \mathbf{2.84}$

Step 5: Calculate the three F ratios

$$F_A = MS_A \div MS_{S/AB} = 30.72 \div 2.84 = \mathbf{10.82}$$

$$F_B = MS_B \div MS_{S/AB} = 1.20 \div 2.84 = \mathbf{.42}$$

$$F_{A \times B} = MS_{A \times B} \div MS_{S/AB} = 10.83 \div 2.84 = \mathbf{3.65}$$

Step 6: Present the Results in an ANOVA Summary Table

ANOVA Summary Table

Source	SS	df	MS	F	F_{crit}
A	61.43	2	30.72	**10.82***	3.18
B	3.60	3	1.20	**.42**	2.79
AxB	62.30	6	10.38	**3.65***	2.29
S/AB	136.40	48	2.84		

IV. A Computer Software Application with SPSS

The two-factor ANOVA problem just completed had n = 5 subjects per group and a total of 60 scores. This sample size (n = 5) was deliberately small to make hand calculations manageable; much larger sample sizes are more realistic for actual experiments (e.g., n = 20 or more is quite common). Since total number of subjects (and scores to analyze) is determined by multiplying a times b times n, you can see that for n = 20 there would have been 3 × 4 × 20 = 240 scores in the data matrix. The most tedious of the basic ratio calculations is [Y], which requires squaring each individual score and summing the squared scores. Thus, it is important to take advantage of computer software, which normally is readily available. Getting a computer to do this work is not particularly difficult, although instruction for doing this is easier while at a computer rather than reading a book description. Nonetheless, this section represents an attempt to write a simplified description of the necessary computer inputs and outputs in the context of the data from the 3 × 4 design described in the previous section. For this illustration we will refer to a popular software package in the behavioral sciences, SPSS.

Once you open SPSS by checking the "type in data" box, a spreadsheet appears on the screen, with numbered rows and "*VAR*" at the top of each column. Two options appear in the lower left-hand corner of the screen: "Data View" and "Variable View." The "Data View" screen is open, so click on "Variable View." This screen enables you to name columns, so one thing we could do is type in "FactorA" for the first column, "FactorB" for the second column, and "Scores" for the third column. Now go back to the spreadsheet (click on "Data View" in the lower left-hand corner of the screen) to input appropriate information in the first three columns, now labeled "FactorA," "FactorB," and "Scores." What appears on the screen will be similar to Table 9.3:

Table 9.3: Data View Screen for SPSS

	FactorA	FactorB	Scores
1			
2			
3			
4			
5			
6			
7 (continuing through 60 rows to match the number of scores in this example data set).			

Although the computer will do the computations, the task of entering data still remains. The computer works with numbers, so group distinctions within Factors A and B have to be converted to number codes. We have several options here, we just must be extremely careful to make sure that information in the spreadsheet is organized correctly. There are three different conditions within Factor A, so one option is to code these appropriately with the numbers 1, 2, and 3. Similarly, with four different conditions within Factor B, they can be distinguished with respective number codes of 1, 2, 3, and 4. We will enter the actual scores in the third column, making sure that a row with "1" in the Factor A column and "1" in the Factor B column has a score from a participant in Group a_1b_1. The completed data spreadsheet for SPSS is given in Table 9.4.

Table 9.4 Data View Screen for SPSS

	FactorA	FactorB	Scores
1	1.00	1.00	3.00
2	1.00	1.00	6.00
3	1.00	1.00	4.00
4	1.00	1.00	2.00
5	1.00	1.00	4.00
6	1.00	2.00	4.00
7	1.00	2.00	4.00
8	1.00	2.00	8.00

9	1.00	2.00	4.00
10	1.00	2.00	5.00
11	1.00	3.00	2.00
12	1.00	3.00	6.00
13	1.00	3.00	5.00
14	1.00	3.00	7.00
15	1.00	3.00	5.00
16	1.00	4.00	4.00
17	1.00	4.00	2.00
18	1.00	4.00	5.00
19	1.00	4.00	5.00
20	1.00	4.00	3.00
21	2.00	1.00	9.00
22	2.00	1.00	7.00
23	2.00	1.00	9.00
24	2.00	1.00	5.00
25	2.00	1.00	8.00
26	2.00	2.00	6.00
27	2.00	2.00	7.00
28	2.00	2.00	8.00
29	2.00	2.00	4.00
30	2.00	2.00	6.00
31	2.00	3.00	5.00
32	2.00	3.00	7.00
33	2.00	3.00	1.00
34	2.00	3.00	5.00
35	2.00	3.00	3.00
36	2.00	4.00	8.00
37	2.00	4.00	4.00
38	2.00	4.00	6.00
39	2.00	4.00	6.00
40	2.00	4.00	5.00
41	3.00	1.00	7.00
42	3.00	1.00	2.00
43	3.00	1.00	3.00

44	3.00	1.00	8.00
45	3.00	1.00	5.00
46	3.00	2.00	6.00
47	3.00	2.00	7.00
48	3.00	2.00	6.00
49	3.00	2.00	9.00
50	3.00	2.00	7.00
51	3.00	3.00	9.00
52	3.00	3.00	9.00
53	3.00	3.00	9.00
54	3.00	3.00	7.00
55	3.00	3.00	8.00
56	3.00	4.00	7.00
57	3.00	4.00	6.00
58	3.00	4.00	5.00
59	3.00	4.00	8.00
60	3.00	4.00	9.00

Given this number of data entries, it is obviously important to proof the spreadsheet to make sure that all entries are correct. After the data have been entered and checked, all that remains to be done is to provide the computer with appropriate computational instructions. For this next step, move the cursor to "Analyze" on the tool bar at the top of the screen (located between *Transform* and *Graphs* on our version of SPSS). The pull-down menu for "Analyze" has several options. If we were doing a t test or a single-factor analysis of variance, we would move the cursor down to the "Compare Means" option, and the box to the right of this option provides choices for t and for a One-Way ANOVA. However, since we are doing a two-factor analysis of variance, we need to go further down the options list to the one labeled "**General Linear Model**." The box that opens to the right of this option also provides choices. The statistical designs we are concerned with involve a single dependent variable, and the appropriate option for this class of design is the one labeled "Univariate." Move your cursor to *Univariate* to open a dialog box that should look similar to the one that appears in Table 9.5.

Table 9.5: SPSS Dialog Box for Two-Factor ANOVA

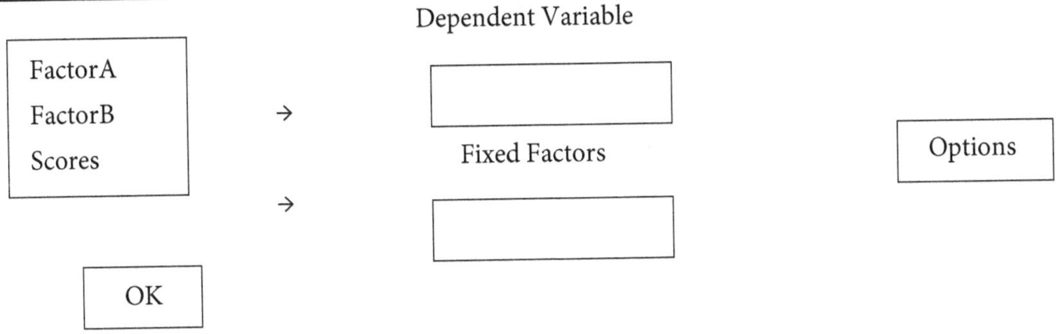

Other information appears in the SPSS dialog box, but this is all we need for doing a two-factor ANOVA. The "scores" are the dependent variable, so we need to highlight "Scores" and click on the arrow to the left of the dependent variable box. After you do this, "Scores" will now appear in the dependent variable box and disappear from the box on the left. Factor A and Factor B are the independent variables, referred to as "Fixed Factors." Highlight "FactorA" and click on the arrow to the left of the fixed factors box, and this will move "FactorA" from the left-hand box to the "Fixed Factors" box. Do the same for "FactorB." Your dialog box should now resemble the one in Table 9.6:

Table 9.6: Completed SPSS Dialog Box for Two-Factor ANOVA

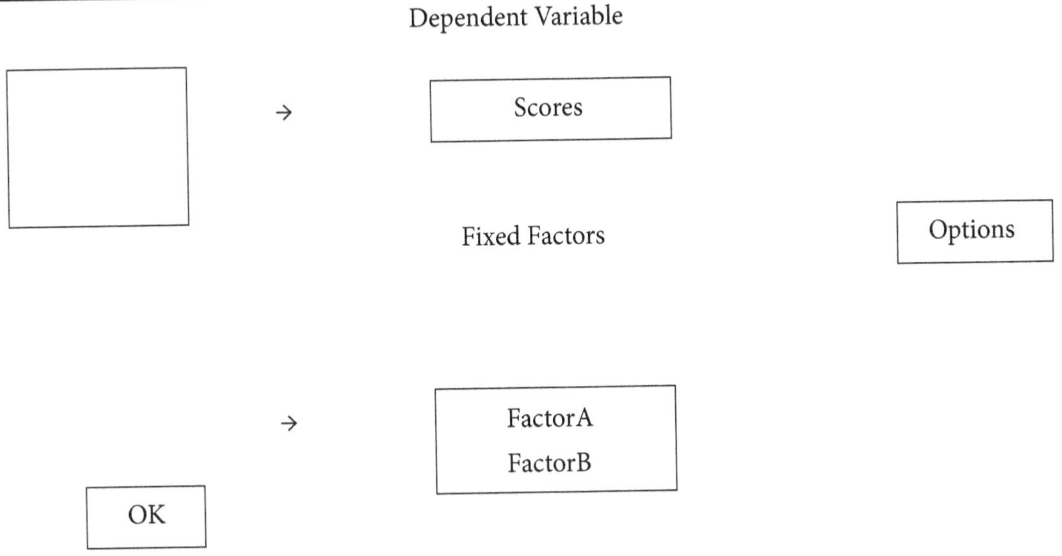

If you now click on the "OK" box, in just a few seconds SPSS will display a completed ANOVA summary table. Clicking on the "Options" box before clicking on "OK" allows you to instruct the computer to provide additional information, such as means and standard deviations for each of the 12 groups in the experiment. The SPSS ANOVA table is similar to what has been presented in this text, although it includes things not needed for our present purpose. It is copied for these data in Table 9.7, with the information relevant for our hand calculations typed in bold. Although SPSS takes the results to three-decimal place accuracy, you can see that it confirms the hand calculations reported earlier.

Table 9.7: SPSS ANOVA Summary Table

Tests of Between-Subjects Effects

Source	Type III Sum of Squares	df	Mean Square	F	Sig
Corrected Model	127.333	11	11.576	4.074	.000
Intercept	1972.267	1	1972.267	694.053	.000
FactorA	**61.433**	**2**	**30.717**	**10.809**	**.000**
FactorB	**3.600**	**3**	**1.200**	**.422**	**.738**
FactorA*FactorB	**62.300**	**6**	**10.383**	**3.654**	**.005**
Error	**136.400**	**48**	**2.842**		
Total	2236.000	60			
Corrected Total	263.733	59			

V. Summary

The factorial design was introduced in this chapter. Although more than two independent variables (factors) can certainly be included within a single experiment, we will not go beyond consideration of two-factor designs. Inclusion of a second factor in a single experiment brought an important new concept into the analysis: namely, evaluation of the *interaction* between two factors. A significant interaction indicates that effects of one variable (e.g., Factor A) are not the same at all levels of the second independent variable—that is, Factor A may affect performance one way when examined under one level of Factor B (at b_1) and in an entirely different way when examined under a different level of Factor B (at b_2).

Computational procedures for the two-factor ANOVA were detailed, noting the coding changes that are required when each group is designated jointly by its specific Factor A treatment (a_i) and its specific Factor B treatment (b_j). Given the meaningful organization of groups in an A × B design, it is possible to sub-divide the variance **between** all groups into three component parts: Factor A, Factor B, and the Factor A × Factor B interaction. We began the computational illustrations by re-visiting a four-group experiment (a_1 vs. a_2 vs. a_3 vs. a_4) and changing the storyline (but not the scores) to make it appropriate for a 2 × 2 factorial design. With this A × B configuration, it was shown how the three analytical comparisons in Model Problem 7.1 represented Factor A, Factor B, and the A × B interaction, respectively. We went on to demonstrate that the results of the 2 × 2 analysis of variance and the results of the analytical comparisons calculated in Chapter 7 were the same: $SS_A = SS_{Acomp1}$, $SS_B = SS_{Acomp2}$, and

$SS_{A \times B} = SS_{Acomp3}$. A second example problem for a more complex two-factor design (3 × 4) completed this section on "how to do it" with a hand-held calculator.

Even with relatively small data sets, doing ANOVAs with a hand calculator may put a bit of a strain on one's patience. Thus, this seemed to be an appropriate time to call upon the assistance of a computer software program. Although entering data into the computer can be a time-consuming process, much as it is for entering data into a hand-held calculator, once entered into the computer, ANOVA computations are done swiftly and accurately (of course, only within the accuracy limits of the data-entering steps). The final "computer appreciation" section of this chapter involved a description of the SPSS application in terms of data input and computational output. The last column in the SPSS ANOVA summary table on page 178 indicated the exact probability (p values carried out to three decimal place accuracy) for finding an F this large or larger from the appropriate reference distribution for F. For Factor A, $p = .000$, meaning that the probability of $F \geq 30.717$ is less than .0005 (e.g., $p = .0006$ rounded off to three decimal place accuracy would have appeared as .001 in the computer summary table). For the A × B interaction, $p = .005$, and the probability level associated with the non-significant B effect was $p = .738$ (i.e., nearly 74% of the time $F \geq .422$ would be expected to occur by chance alone).

9.1: Model Problem. This ANOVA problem requires a little creative thinking in order to get the data ready for analysis. Means and standard deviations in the table below are actually taken from a published study (Ballardini, Yamashita, and Wallace 2008); however, we will take some liberties and treat the data as if they involved a 2 × 3 "between-group" design. Assume there were 12 different subjects in each group (n = 12). Subjects had a brief study exposure of 18 different 16-word lists. Words on each list were presented successively, and Factor A (a = 3) was exposure duration. Each word was flashed on a screen for approximately 20 ms for participants in the a_1 group, 50 ms for participants in the a_2 group, and 250 ms for participants in the a_3 group. These are very brief study exposures, with the longest being only one-quarter of a second! Words on each list were related either semantically (b_1) or phonetically (b_2). After lists were presented, subjects had a recognition memory test. The data in the table are proportions of false recognitions (incorrect reports that a word had been presented on a study list when it had not been presented) to critical test words that were not presented during study, but were related to studied words (e.g., *awake*, *dream*, *snore*, *blanket*, *slumber*, *pillow*, etc. were included on one semantic study list, and the critical test word **sleep** was not among the study words).

Data Table: Mean Proportions of False Recognition (Standard Deviations in Parentheses)

	Presentation Duration (Factor A)		
Inter-Item Similarity (Factor B)	20 ms (a_1)	50 ms (a_2)	250 ms (a_3)
Semantic (b_1)	.11 (.107)	.12 (.149)	.25 (.208)
Phonetic (b_2)	.44 (.128)	.32 (.170)	.31 (.166)

Based on this descriptive information, report the results for a two-factor ANOVA.

Moving directly to the "Step 1" basic ratio calculations, we soon discover that we have two problems. We cannot simply use the numbers in this table to compute basic ratios necessary to get us started on ANOVA calculations. **Mean scores** are presented in the table, and we have been calculating basic ratios using **individual scores** for [Y] and various combinations of **sums of individual scores** for [A], [B], [AB], and [T]. There are alternative formulas using means; however, you are not expected to be able to develop these on your own. Can the data in the table be used to obtain (1) individual scores and (2) various group totals? The answer to Part 2 of this question is "yes." Multiplying each group mean by the number of subjects in each group (n = 12) provides the sums necessary to calculate [A], [B], [AB], and [T]. However, the problem in Part 1 is not as easily solved. Scores for each individual subject necessary to calculate [Y] cannot be recovered from this table of means and standard deviations. Since $SS_{S/AB}$ = [Y] – [AB], and $MS_{S/AB} = SS_{S/AB} \div df_{S/AB}$, and since [Y] cannot be calculated without knowing individual scores, we have to look to an alternative method for calculating $MS_{S/AB}$ in order to complete the analysis of variance (i.e., you have to think "outside the box" at this point). To solve this problem, you may recall that the within-group variance in the ANOVA is simply the average of the variances for each separate group: $\mathbf{MS_{S/AB} = (s^2_{1,1} + s^2_{2,1} + s^2_{3,1} + s^2_{1,2} + s^2_{2,2} + s^2_{3,2}) \div (a)(b)}$. Thus, an analysis of variance can be completed from a table of group sums and variances, so the initial step for doing an ANOVA with these data requires converting means to sums by multiplying each group mean by n = 12, and converting standard deviations to variances by squaring each standard deviation. $MS_{S/AB}$ can then be determined by taking the average of the six group variances.

Converted Table: Sum of Proportions of False Recognitions (Variances in Parentheses)

	Presentation Duration (Factor A)			
Inter-Item Similarity (Factor B)	20 ms (a_1)	50 ms (a_2)	250 ms (a_3)	ΣB_j
Semantic (b_1)	1.32 (.011)	1.44 (.022)	3.00 (.043)	**5.76**
Phonetic (b_2)	5.28 (.016)	3.84 (.029)	3.72 (.028)	**12.84**
ΣA_i	**6.60**	**5.28**	**6.72**	**T = 18.60**

Computational steps:

Step 1: Basic ratios that can be calculated

$[A] = (6.60^2 + 5.28^2 + 6.72^2) \div (b)(n) = 116.60 \div 24 = \mathbf{4.86}$

$[B] = (5.76^2 + 12.84^2) \div (a)(n) = 198.04 \div 36 = \mathbf{5.50}$

$[AB] = (1.32^2 + 1.44^2 + 3.00^2 + 5.28^2 + 3.84^2 + 3.72^2) \div n = 69.28 \div 12 = \mathbf{5.77}$

$[T] = 18.60^2 \div (a)(b)(n) = 345.96 \div 72 = \mathbf{4.80}$

Step 2: Calculate the three between-group sums of squares

$SS_A = [A] - [T] = 4.86 - 4.80 = \mathbf{.06}$

$SS_B = [B] - [T] = 5.50 - 4.80 = \mathbf{.70}$

$SS_{A\times B} = [AB] - [A] - [B] + [T] = 5.77 - 4.86 - 5.50 + 4.80 = \mathbf{.21}$

Step 3: Calculate the respective mean squares

$MS_A = SS_A \div df_A = .06 \div 2 = \mathbf{.030}$

$MS_B = SS_B \div df_B = .70 \div 1 = \mathbf{.700}$

$MS_{A\times B} = SS_{A\times B} \div df_{A\times B} = .21 \div 2 = \mathbf{.105}$

Step 4: Determine $MS_{S/AB}$ by taking the average of the six group variances

$MS_{S/AB} = \Sigma s^2_{i,j} \div (a)(b) = (.011 + .022 + .043 + .016 + .029 + .028) \div 6 = \mathbf{.025}$

Step 5: The ANOVA Summary Table

ANOVA Summary Table

Source	SS	df	MS	F	F_{crit}
A	.06	2	.030	**1.20**	3.15
B	.70	1	.700	**28.00***	4.00
AxB	.21	2	.105	**4.20***	3.15
S/AB		66	.025	* $p \leq .05$	

9.2: Practice Problem 1: For the data that appear in the following table,

(a) Calculate [Y], [T], [A], [B], and [AB].

(b) Report values for SS_A, SS^B, SS_{AxB}, and $SS_{S/AB}$.

(c) Present the corresponding Mean Squares and F ratios in an ANOVA Summary Table, indicating which effects are significant at the .05 level.

Data Table:

Groups

a_1b_1	a_1b_2	a_2b_1	a_2b_2
7	9	4	4
9	10	6	5
5	8	6	4
5	6	3	2
6	7	7	6
4	8	5	3

9.3: Practice Problem 2. Assume the fictitious data for this problem involve reading speed scores measured in terms of average number of words read per minute. Children at four grade levels (Factor A) were tested. Half of the children at each grade level learned to read using a textbook that used an "alphabet-training" approach, and half of the children at each grade level learned to read using a textbook that used a "phonics-training" approach (Factor B). Determine which (if any) effects are significant at $p \leq .05$.

Data Set: Reading Speed Test (Words per Minute) for Children at Grade Levels 1st through 4th Following Either Alphabet (A) or Phonics (P) Training

Training and Grade Level

1st A	2nd A	3rd A	4th A	1st P	2nd P	3rd P	4th P
90	100	140	180	90	80	100	160
110	100	120	160	120	120	130	150
80	90	100	120	70	110	150	180
90	110	110	160	80	100	120	190
60	120	120	150	100	130	150	200
70	100	120	190	90	90	140	160
100	110	150	200	80	100	130	160
70	100	140	180	80	110	120	150

9.4: Practice Problem 3. Repeat the analysis of variance for Practice Problem 2 using SPSS (or alternative statistical software). This exercise will provide a check on your hand calculations in Practice Problem 2.

9.5: Practice Problem 4. Do an ANOVA based on the information provided about **Group Means** (in Bold) and **Standard Deviations** (in Parentheses) that appear in the following table (n = 15 subjects in each a_ib_j group):

Table of Group Means and Standard Deviations (in Parentheses)

		Factor A		
		a_1	a_2	a_3
Factor B	b_1	**4.50** (1.60)	**6.00** (2.40)	**7.00** (2.50)
	b_2	**5.00** (2.30)	**5.70** (2.25)	**7.80** (2.55)
	b_3	**5.10** (2.10)	**5.00** (1.75)	**8.50** (2.90)

9.6: Practice Problem 5. For this problem, a = 3, b = 4, and n = 2. Scores for the 24 subjects in this experiment are given in the following table. Determine if the main effects and interaction are significant at the .05 significance level. After you complete this problem using a hand calculator, confirm your results using SPSS (or alternative statistical software).

Data Matrix:

a_1b_1	a_1b_2	a_1b_3	a_1b_4	a_2b_1	a_2b_2	a_2b_3	a_2b_4	a_3b_1	a_3b_2	a_3b_3	a_3b_4
3	3	1	5	3	7	4	6	9	8	4	5
4	1	2	3	6	3	5	8	12	10	3	2

Chapter 10

Analytical Comparisons for the Two-Factor Design

Statistical thinking will one day be as necessary for efficient citizenship as the ability to read and write.

—Wells

Even though three different effects were evaluated in the two-factor analysis of variance, still more detailed analytical comparisons may be carried out, as was the case for the single-factor design. Researchers should always try to obtain as much meaningful information as possible from the data they have collected, so meaningful, interpretable, analytical comparisons should at least be considered. What is meant here by "interpretable" may be illustrated by a counter-example from the last chapter. In that chapter it was noted that a specific comparison between an a_1b_1 group and an a_2b_2 group from a two-factor experiment is not interpretable because the a_1b_1 and a_2b_2 groups differ in two ways: Group a_1b_1 experiences level 1 of Factor A compared to level 2 of Factor A for Group a_2b_2; and Group a_1b_1 experiences level 1 of Factor B compared to level 2 of Factor B for Group a_2b_2. Thus, neither different treatment levels in Factor A nor different treatment levels in Factor B could provide an unambiguous account for why performance differences may be observed between Group a_1b_1 and Group a_2b_2.

Analytical comparisons for a two-factor design are more complicated than for the single-factor design. First, there are three results to consider: main effects for Factor A and Factor B and the A × B interaction. (More detailed analyses of each effect may be possible.) Second, there are common-sense guidelines governing decisions about which, if any, of the three effects from the analysis of variance are eligible for further analytical comparisons. As was true for the single-factor design, we should agree that more detailed analytical comparisons will only be tested for significant effects. That is, if Factor A was not significant in the ANOVA, then Factor A would not be considered for further analytical comparisons. Also, if $df_{effect} = 1$, then at least for main effects, a_1 vs. a_2 (a main effect for Factor A) or b_1 vs. b_2 (a main effect for Factor B), the ANOVA provides comparisons that are as detailed as we can get. A third guideline to follow imposes some order on the process, indicating that one should consider analytical comparisons of main effects *if and only if* the A × B interaction is **not** significant. If the A × B interaction is significant, then the exclusive focus of attention will be on further detailed analyses of this significant interaction, and any main effects would not be analyzed further.

I. A General Strategy for Deciding on Follow-Up Questions

The statements in the preceding paragraph actually provide an outline for how to proceed with analytical comparisons following an analysis of variance with the two-factor design. We will repeat these in a more organized step-wise format before moving on to descriptions and computational illustrations. The recommended strategy to follow contains a number of contingencies or choice points in considering analytical comparisons. This maze of steps and contingencies may seem a bit confusing at first, so perhaps the summary outline in Table 10.1 will be helpful.

Table 10.1 Contingencies for Analytical Comparisons to Consider for the Two-Factor Design

A × B interaction is not significant				A × B interaction is significant		
Factor A ($df_A \geq 2$)		Factor B ($df_B \geq 2$)		**Simple Effects**		**Interaction Contrasts**
↓	↓	↓	↓	↓	↓	
Not sig	Sig	Not Sig	Sig	Not sig	Sig	
↓	↓	↓	↓	↓	↓	
No tests	**Main comps**	No tests	**Main comps**	No tests	**Simple comps**	

Upon completion of the analysis of variance, the first contingency checkpoint involves the A × B interaction. We will begin with an illustration *as if* the **interaction is not significant**, allowing us to drop down to the second contingency checkpoint and consider analytical comparisons (tests of **main comparisons**) given that Factor A and/or Factor B are significant (with $df \geq 2$). After illustrating procedures for doing main comparisons, we will go back to the first contingency checkpoint, the **A x B interaction is significant**, and consider doing follow-up analyses (tests of **simple effects** or **interaction contrasts**) that will provide more detailed information about the interaction.

The absence of an interaction means that interpretations of main effects do not have to be qualified. The second contingency checkpoint along the non-significant interaction path involves the question of whether either or both main effects are significant. We will not consider doing analytical comparisons for main effects that are not significant in the overall analysis of variance. However, for significant main effects with $df_A \geq 2$ and/or $df_B \geq 2$, follow-up analyses on Factor A and/or Factor B, may be considered. Notice that the language here is "permissive," meaning that conducting follow-up statistical analyses are not obligatory. However, further analyses are relevant for many research designs. Analytical comparisons for main effects are referred to as *main comparisons*, and we will see later that they are, for the most part, carried out just as they were for the single-factor design (e.g., using coefficients to weight and define contrasts of interest).

Returning now to the significant A × B interaction path, one may choose to ask either of two types of follow-up questions: one involving *simple-effects analyses* and the other involving *interaction contrasts.* A simple-effects analysis tests the significance of one factor (e.g., Factor A) successively at each level of the other factor (e.g., Factor B). For example, if a = 3 and b = 4, four separate simple-effects tests may be conducted subsequent to a significant A × B interaction, with one test for each of the following four questions: (1) Is Factor A significant when only b_1 data are considered (is Factor A significant at b_1)? (2) Is Factor A significant at b_2? (3) Is Factor A significant at b_3? (4) Is Factor A significant at b_4? Of course, if interests dictate, the questions tested statistically could revolve around Factor B, which would require statistical tests for three follow-up questions (i.e., Is Factor B significant at a_1? Is Factor B significant at a_2? Is Factor B significant at a_3?). If necessary, additional follow-up analyses subsequent to simple-effects analyses can be done for any significant simple effect for which the degrees of freedom are greater than 1.0. For example, if Factor A is significant at b_1, since $df_{A\ at\ b1} = 3$, one might wish to follow this simple-effects analysis with a test of whether the a_1b_1 group is significantly different from the a_2b_1 group (this analytical comparison is "a_1 vs. a_2 at b_1"). This level of analytical comparison is referred to as a *simple comparison.*

The interest with follow-up analyses involving the second alternative, *interaction contrasts*, concerns whether the A × B interaction holds up when the analysis is limited to a particularly meaningful subset of data. If the original design was a 3 × 4 factorial, and if within this set of 12 groups there were two levels of Factor A that were of particular interest (e.g., a_1 and a_2) and two levels of Factor B that were of particular interest (e.g., b_2 and b_3), then you could proceed to test this smaller scale interaction contrast to determine if the A × B interaction is significant using only a_1b_2, a_1b_3, a_2b_2, and a_2b_3 groups (literally a new 2 × 2 factorial design for an $A_{comparison} \times B_{comparison}$ interaction).

II. Main Comparisons Given a Non-Significant Interaction

The contingencies for considering main comparisons are (1) the A × B interaction was not significant, (2) the main effect in question was significant, and (3) the main effect in question has two or more degrees of freedom. Given these contingencies are met, one may consider doing more detailed statistical analyses directed at experimental questions of interest. In a single-factor experiment, the effect for Factor A in the analysis of variance is a main effect, thus procedures for analytical comparisons of interest following a significant main effect have already been described. Illustrations were provided in Chapter 7 comparing intact pairs of groups (e.g., a_1 vs. a_2), comparing complex rearrangements of groups (e.g., treating Group a_1 and Group a_2 as if they were a single group to compare with Group a_3), and testing for different functional relations in the data (e.g., linear, quadratic, and cubic trends). The same comparisons may be made in the same way (i.e., using coefficients to define and appropriately weight groups involved in the comparisons of interest) for main comparisons in the two-factor design as was done for the one-treatment factor in the single-factor design. However, there is a "catch." There is one additional operation required for main comparisons in the two-factor design. The data have to be **re-grouped or appropriately defined by repeating coefficients** to reflect, in effect, a single-factor organization. For example, consider the fictitious scores in Table 10.2 from a two-factor design with a = 3, b = 2 and n = 5:

Table 10.2: Fictitious scores for a 3 × 2 factorial design

a_1b_1	a_2b_1	a_3b_1	a_1b_2	a_2b_2	a_3b_2
3	5	4	6	6	8
1	3	7	2	4	5
5	5	6	4	5	7
2	6	5	3	3	4
0	4	8	1	5	5

It turns out for this data set that neither the A × B interaction nor the main effect for Factor B is significant (you can practice your ANOVA skills by confirming this claim). However, the main effect for Factor A is significant, $F(2, 24) = 10.24$, $MS_{S/AB} = 2.53$, $p < .05$. Thus, we could consider doing one or more main comparisons on Factor A. For example, we might want to compare Group a_1 with Group a_2 ($A_{\text{comparison 1}}$). For this contrast we would need to calculate a mean for each group and create appropriate coefficients. Either the coefficients assigned to the three a_i groups may be repeated for the three a_2 groups, or the set of three weighting coefficients may be assigned to three new a_i' groups created by combining respective b_1 and b_2 scores within each a_i group. The two options are illustrated in Table 10.3: Option A shows the original six groups with repeated coefficients (+1, -1, and 0 for the a_ib_1 groups are repeated for the a_ib_2 groups), and Option B shows the three weighting coefficients with a recombined set of three groups, where each new a_i' group includes the respective a_ib_1 and a_ib_2 scores.

The formula for the sum of squares for an analytical comparison, using weighting coefficients, is $SS_{Acomp} = n(\Sigma c_i \bar{Y}_{Ai})^2 \div \Sigma c_i^2$. The required calculations for SS_{Acomp} are done two consecutive times for this example problem: (1) the first time for the six-group arrangement and (2) the second time for the recombined set of three new a_i' groups. In both cases, all of the a_1 conditions (+1 coefficients) are contrasted with all of the a_2 conditions (-1 coefficients). However, you should note that when groups are recombined across the two levels of Factor B in the single a_1', a_2', and a_3' groups, then each new combined group has 10 subjects in it [n' = (b)(n) = (2)(5) = 10]. For the six-group arrangement n = 5, and we have:

A. $SS_{Acomp1} = n(\Sigma c_i \bar{Y}_{Ai})^2 \div \Sigma c_i^2$
$SS_{Acomp1} = 5([+1][2.2] + [-1][4.6] + [+1][4.0] + [-1][4.6])^2 \div ([1^2 + [-1]^2 + 1^2 + [-1]^2)$
$SS_{Acomp1} = 5(-3)^2 \div 4 = 45 \div 4 =$ **11.25**

For the recombined three-group arrangement n' = 10, and we have:

Table 10.3 Getting Data Ready for a Main Comparison of Factor A

I. Option A:

	a_1b_1	a_2b_1	a_3b_1	a_1b_2	a_2b_2	a_3b_2
	3	5	4	6	6	8
	1	3	7	2	4	5
	5	5	6	4	5	7
	2	6	5	3	3	4
	0	4	8	1	5	5
$\bar{Y}_i$	2.2	4.6	6.0	4.0	4.6	5.8
c_i	+1	-1	0	+1	-1	0

II. Option B:

	a_1'	a_2'	a_3'
	3	5	4
	1	3	7
	5	5	6
	2	6	5
	0	4	8
	6	6	8
	2	4	5
	4	5	7
	3	3	4
	1	5	5
$\bar{Y}_i$	3.1	4.6	5.9
c_i	+1	-1	0

B. $SS_{Acomp1} = (b)(n)(\Sigma c_i \bar{Y}_{Ai})^2 \div \Sigma c_i^2$
$SS_{Acomp1} = 10([+1][3.1] + [-1][4.6])^2 \div (1^2 + [-1]^2)$
$SS_{Acomp1} = 10(-1.5)^2 \div 2 = 22.5 \div 2 =$ **11.25**

As you can see, the SS_{Acomp1} equals 11.25 regardless of which way the calculations are carried out. Since MS_{error} was given ($MS_{S/AB} = 2.53$) for the original analysis of variance, we can complete the problem: $df_{Acomp1} = 1$, $MS_{Acomp1} = 11.25$, and $F_{Acomp1}(1, 24) = 11.25 \div 2.53 =$ **4.45**, which is significant at $p < .05$. Make sure you note that the second set of computations had 10 subjects ($b \times n$) in each of the new a_i' groups (e.g., five a_1b_1 and five a_1b_2 scores appear in the a_1' column). It is recommended that you use the organizational procedure you find less confusing.

For this same data set we will work through one more example problem; however, this time we will test a complex contrast comparing Group a_1 with a combination of Group a_2 and Group a_3 ($A_{Comparison\ 2}$). The illustration is completed using the second procedure that involved combining b_1 and b_2 scores for each respective a_i' column (you may wish to confirm the results reported for this example by using the first procedure that involved setting appropriate coefficients for each of the six original groups). Means and coefficients for this contrast follow:

	a_1'	a_2'	a_3'
$\bar{Y}_i$	3.1	4.6	5.9
c_i	+2	-1	-1

$SS_{Acomp2} = 10([2][3.1] + [-1][4.6] + [-1][5.9])^2 \div ([+2]^2 + [-1]^2 + [-1]^2)$

$SS_{Acomp2} = 10(6.2 - 4.6 - 5.9)^2 \div 6 = 10(-4.3)^2 \div 6$

$SS_{Acomp2} = 10(18.49) \div 6 = \mathbf{30.82}$; and $MS_{Acomp2} = SS_{Acomp2} \div 1 = \mathbf{30.82}$

$F_{Acomp2}(1, 24) = MS_{Acomp2} \div MS_{error} = 30.42 \div 2.53 = \mathbf{12.18}$, $p < .05$

III. Simple Effects Given a Significant Interaction and Simple Comparisons Given a Significant Simple Effect

A significant A × B interaction means that the effect on performance that one independent variable has (e.g., Factor A) is not the same at all levels of the other independent variable (e.g., Factor B). For example, scores may increase across the different levels of Factor A for the b_1 groups ($\bar{Y}_{a1b1} < \bar{Y}_{a2b1} < \bar{Y}_{a3b1}$); however, for the b_2 groups scores may not change or may decrease across the different levels of Factor A ($\bar{Y}_{a1b2} \geq \bar{Y}_{a2b2} \geq \bar{Y}_{a3b2}$). Simple-effects analyses refer to statistical test procedures for determining if one independent variable is significant at each level of the second independent variable. Simple-effects analyses for a 3 × 3 experiment would require three separate F tests. Given that the interest is in assessing the simple effects of Factor A at each level of Factor B (although one could just as readily assess the simple effects of Factor B at each level of Factor A), a complete simple-effects analysis requires computing an $F_{A\ at\ b1}$ to address the question of whether Factor A is significant when only the b_1 data are considered, an $F_{A\ at\ b2}$, and an $F_{A\ at\ b3}$. A significant A x B interaction implies that conclusions drawn from the three simple effects will not be the same.

The steps involved in doing simple-effects analyses are illustrated by considering the number set in Table 10.4. The numbers are organized for an A × B design, with a = 3, b = 3, and n = 4. The ANOVA results for this number set are given in Table 10.5 (again, you have an opportunity for more ANOVA practice by doing the calculations to confirm the SS, MS, and F values in Table 10.5).

Table 10.4: Sample Data for 3 × 3 ANOVA and Simple-Effects Analysis

	Factor A						
Factor B	a_1	Σ	a_2	Σ	a_3	Σ	ΣY_{Bj}
b_1	3, 6, 2, 4	**15**	9, 7, 5, 9	**30**	7, 2, 3, 8	**20**	**65**
b_2	4, 4, 8, 4	**20**	6, 7, 8, 4	**25**	6, 7, 6, 9	**28**	**73**
b_3	2, 6, 5, 7	**20**	5, 7, 1, 5	**18**	10, 9, 13, 8	**40**	**78**
	ΣY_{Ai}	**55**		**73**		**88**	**T = 216**

Table 10.5 ANOVA Summary Table for Data in Table 10.4

Source	SS	df	MS	F	
Factor A	45.50	2	22.75	5.14	$p < .05$
Factor B	7.17	2	3.58	.81	
A × B Interaction	65.83	4	16.46	3.72	$p < .05$
Error (S/AB)	119.50	27	4.43		

Since the A × B interaction is significant, we will ignore the main effect for Factor A and proceed directly to a simple-effects analysis. For this analysis, let's test to see if Factor A is significant at each level of Factor B. Given that b = 3, three F ratios must be calculated to complete this simple-effects analysis: $F_{A\ at\ b1}$, $F_{A\ at\ b2}$, and $F_{A\ at\ b3}$. For each test, all we need to do is compute the corresponding $MS_{A\ at\ bj}$ because $F_{A\ at\ bj} = MS_{A\ at\ bj} \div MS_{S/AB}$. From the overall analysis of variance, we know that $MS_{S/AB} = 4.43$. A simple-effects question asks about a limited set of data; consequently, only groups identified by the question are relevant for calculating respective sums of squares. We will begin by extracting b_1 data from Table 10.4 (the first row of numbers), and calculating $[A_{at\ b1}]$, $[T_{at\ b1}]$, and $SS_{A\ at\ b1}$ from this restricted data set.

1. Is Factor A significant at b_1? The relevant b_1 data are given in the following table:

	a_1	a_2	a_3
	3	9	7
	6	7	2
	4	5	3
	2	9	8
Σ	15	30	20

$[A_{at\ b1}] = (15^2 + 30^2 + 20^2) \div 4 = 1{,}525 \div 4 = \mathbf{381.25}$

$[T_{at\ b1}] = 65^2 \div 12 = 4{,}225 \div 12 = \mathbf{352.08}$

$SS_{A\ at\ b1} = [A_{\ at\ b1}] - [T_{\ at\ b1}] = 381.25 - 352.08 = \mathbf{29.17}$

Since $df_{A\ at\ b1} = 2$ (Groups a_1, a_2, and a_3 are involved in this analysis), $MS_{A\ at\ b1} = 29.17 \div 2 =$ **14.58**

$\mathbf{F_{A\ at\ b1}(2, 27) = 14.58 \div 4.43 = 3.29}$, which is not significant since $F_{A\ at\ b1} < F_{crit}(2, 27) = 3.37$.

2. Is Factor A significant at b_2? The relevant b_2 data are given in the following table:

	a_1	a_2	a_3
	4	6	6
	4	7	7
	8	8	6
	4	4	9
Σ	20	25	28

$[A_{\ at\ b2}] = (20^2 + 25^2 + 28^2) \div 4 = 1{,}809 \div 4 = \mathbf{452.25}$

$[T_{\ at\ b2}] = 73^2 \div 12 = 5{,}329 \div 12 = \mathbf{444.08}$

$SS_{A\ at\ b2} = [A_{\ at\ b2}] - [T_{\ at\ b2}] = 452.25 - 444.08 = \mathbf{8.17}$

Since $df_{A\ at\ b2} = 2$, $MS_{A\ at\ b2} = 8.17 \div 2 = \mathbf{4.08}$

$\mathbf{F_{A\ at\ b2}(2, 27) = 4.08 \div 4.43 = .92}$, which also is not significant.

3. Is Factor A significant at b_3? The relevant b_3 data are given in the following table:

	a_1	a_2	a_3
	2	5	10
	6	7	9
	5	1	13
	7	5	8
Σ	20	18	40

$[A_{\ at\ b3}] = (20^2 + 18^2 + 40^2) \div 4 = 2{,}324 \div 4 = \mathbf{581.00}$

$[T_{\ at\ b3}] = 78^2 \div 12 = 6{,}084 \div 12 = \mathbf{507.00}$

$SS_{A\ at\ b3} = [A_{\ at\ b3}] - [T_{\ at\ b3}] = 581.00 - 507.00 = \mathbf{74.00}$

Since $df_{A\ at\ b3} = 2$, $MS_{A\ at\ b3} = 74.00 \div 2 = \mathbf{37.00}$

$\mathbf{F_{A\ at\ b3}(2, 27) = 37.00 \div 4.43 = 8.35}$, which is significant at $p \leq .05$ since $F_{A\ at\ b3}(2, 27) > F_{crit}(2, 27) = 3.37$.

For this set of simple-effects questions we can conclude that there is a significant difference among the a_1, a_2, and a_3 groups only when restricted to the b_3 condition. By the way, simple effects merely take the sum of squares for Factor A and the sum of squares for the A × B interaction and partition them somewhat differently. That is, each simple effect of Factor A has a component of the overall A effect and a component of the A × B interaction; within rounding error, we will find that $SS_A + SS_{A \times B}$ (from the ANOVA) = $SS_{A\ at\ b1} + SS_{A\ at\ b2} + SS_{A\ at\ b3}$ (i.e., 45.50 + 65.83 = 29.17 + 8.17 + 74.00; or 111.33 ≈ 111.34).

In the preceding example, Factor A was significant at level 3 of Factor B. Since Factor A still has two degrees of freedom when restricted to the b_3 subset of data, the A at b_3 effect could be subjected to additional analytical comparisons (referred to as *simple comparisons* since they follow from the simple-effects analysis). Factor A was not significant at either b_1 or b_2, thus, A at b_1 and A at b_2 will not be considered for further analytical comparisons. Any further analytical comparisons on Factor A that are restricted to the b_3 data are defined as simple comparisons, whether the comparison is between two intact groups (e.g., a_1 vs. a_2 at b_3), a new combination of groups as illustrated in Model Problem I.C. at the end of this chapter (e.g., $a_1 + a_2$ vs. a_3 at b_3), or the following illustration to test a simple comparison involving a functional relation or trend (e.g., A_{linear} at b_3).

Computational procedures for a simple comparison are basically the same as analytical comparisons for the single-factor design and for main comparisons considered earlier in this chapter. You only need to extract the data relevant to the simple effect that will be analyzed further. Factor A was only significant with the b_3 data, so any additional analyses will be restricted to the means for the a_1b_3, a_2b_3, and a_3b_3 groups. As with other analytical comparisons, from here we "let the coefficients do the talking" in terms of defining the comparisons we want to make. We will illustrate how this works by doing a trend analysis to determine if either a linear or quadratic trend relating a_1, a_2, and a_3 groups in the b_3 treatment condition is significant. The respective group means and coefficients (the trend coefficients from Table A-3 in Appendix A) are listed below. From this information we will need to calculate $SS_{Alinear}$ and $SS_{Aquadratic}$.

Groups:	a_1 at b_3	a_2 at b_3	a_3 at b_3
$\bar{Y}$	5.00	4.50	10.00
c_{linear}	-1	0	+1
$c_{quadratic}$	+1	-2	+1

A. Is there a significant Linear Trend?

Using the c_{linear} coefficients, we have: $SS_{Alin\ at\ b3} = n(\Sigma c_i \bar{Y}_{Ai})^2 \div \Sigma c_i^2 = 4([-1][5.00] + [0][4.50] + [+1][10.00])^2 \div ([-1]^2 + [0]^2 + [+1]^2)$; $SS_{Alin\ at\ b3} = 4(5)^2 \div 2 = 100 \div 2 = \mathbf{50.00}$; and with $df_{Alin\ at\ b3} = 1$, $MS_{Alin\ at\ b3} = \mathbf{50.00}$; $F_{Alin\ at\ b3}(1, 27) = MS_{Alin\ at\ b3} \div MS_{S/AB} = 50.00 \div 4.43$; $\mathbf{F_{Alin\ at\ b3} = 11.29, p < .05}$

B. Is there a significant Quadratic Trend?

Using the $c_{quadratic}$ coefficients, we have: $SS_{Aquad\ at\ b3} = n(\Sigma c_i \bar{Y}_{Ai})^2 \div \Sigma c_i^2 = 4([1][5.00] + [-2][4.50] + [1][10.00])^2 \div ([1]^2 + [-2]^2 + [1]^2)$; $SS_{Aquad\ at\ b3} = 4(6)^2 \div 6 = 144 \div 6 = \mathbf{24.00}$

(You may recall from Chapter 7 that for $df_A = 2$, $SS_{Alin} + SS_{Aquad} = SS_A$. It is also the case that $SS_{A\ at\ bj} = SS_{Alin\ at\ bj} + SS_{Aquad\ at\ bj}$. Therefore, we could have found $SS_{Aquad\ at\ b3}$ by substituting $SS_{A\ at\ b3} = 74.00$ and $SS_{Alin\ at\ b3} = 50.00$ in this formula and solved for $SS_{Aquad\ at\ b3}$: $SS_{Aquad\ at\ b3} = 74.00 - 50.00 = 24.00$.) With $df_{Aquad\ at\ b3} = 1$, $MS_{Aquad\ at\ b3} = \mathbf{24.00}$; $F_{Aquad\ at\ b3}(1, 27) = MS_{Aquad\ at\ b3} \div MS_{S/AB} = 24.00 \div 4.43$; $\mathbf{F_{Aquad\ at\ b3} = 5.42, p < .05}$

IV. Controlling α_{FW}

Two formal correction procedures were introduced in Chapter 8: the Scheffé test and the Tukey test. Nothing new is required for testing main comparisons or simple comparisons with the more conservative α_{PC} levels used with these tests (the same is true for the Dunnett and Newman-Keuls tests introduced in Chapter 8, Appendix VIII.A). Remember, only one step was added to the analytical comparison procedures when the more stringent Scheffé or Tukey criteria was used instead of the standard .05 or .01 α_{PC} cutoffs. These test procedures did not change how the F_{comp} was computed; rather, the procedures merely adjusted F_{crit} (the critical comparison F). For example, one could ask if these simple comparisons for linear and quadratic trends are significant for $\alpha_{FW} \leq .05$ with a Scheffé test (generally, a trend analysis would be a "planned" analysis, thus this extension is included for illustration purposes only). The formula for the critical F with the Scheffé test is $F_S = (a - 1)F\{df_A, df_{error}\}$. For the simple trend comparisons at b_3, $df_{A\ at\ b3} = 2$ and from the overall ANOVA, $df_{S/AB} = 27$. We will continue using $df_{S/AB} = 26$ from the table of critical F values (Table A-1 in Appendix A) since this table does not have a listing for $df_{denom} = 27$: $F_{crit}(2, 27) = 3.37$. Thus, the critical Scheffé $F_S = (3 - 1)(3.37) = \mathbf{6.74}$. Had the simple trend comparisons been done with an α_{PC} Scheffé adjustment, we would have concluded that there was a significant linear trend ($F_{Alin\ at\ b3} = 11.29$) for Factor A at b_3; however, the quadratic trend for Factor A at b_3 would not have been significant ($F_{Aquad\ at\ b3} = 5.42$).

V. Interaction Contrasts

An interaction contrast involves analyzing a smaller scale A' × B' interaction than the original A × B interaction. This approach to a more detailed analysis of an interaction might be considered when only certain levels of one or both factors are of special interest. For example, consider the data in Table 10.4 from a 3 × 3 factorial design. The interaction was significant, $F(4, 27) = 3.72$, $MS_{S/AB} = 4.43$, $p < .05$. This is a complex interaction with $df_{A\times B} = 4$ and it is not clear from this whether the interaction is present with a subset of groups. If the a_1b_1, a_1b_2, a_2b_1, and a_2b_2 groups were of particular interest, we could proceed to test this smaller scale (2 × 2) interaction directly to see if it was significant. Note that selected groups must include the same levels of Factor A and the same levels of Factor B (e.g., an interaction contrast using the a_1b_1, a_1b_2, a_3b_1, and a_2b_3 groups is not interpretable). To do an interaction contrast, extract the data from the groups of interest and calculate $SS_{A'\times B'}$ (where primes identify the relevant subsets of the original data). It is not necessary to do a complete ANOVA because the only interest is in the component interaction. Since it is appropriate to use $SS_{S/AB}$ from the 3 × 3 ANOVA as the error term, then $SS_{A'\times B'}$ is the only sum of squares that must be calculated. Unfortunately, of the five basic ratios that have to be calculated for a complete two-factor analysis of variance, four must be computed to determine the sum of squares for the A × B interaction (the exception is the basic ratio for Y). Table 10.6 lists only the relevant data of interest from the four-group subset of the original nine groups from Table 10.4: Group a_1b_1, Group a_1b_2, Group a_2b_1, and Group a_2b_2. Only the data from these four groups will be used in calculating the sum of squares for the follow-up 2 × 2 interaction contrast (i.e., for calculating $SS_{A'\times B'}$ = [A'B'] – [A'] – [B'] + [T'], where A' contrasts a_1 vs. a_2 and B' contrasts b_1 vs. b_2).

Table 10.6: The a_1b_1, a_1b_2, a_2b_1, and a_2b_2 Groups from Table 10.4

Groups:	a_1b_1	a_1b_2	a_2b_1	a_2b_2
	3	4	9	6
	6	4	7	7
	2	8	5	8
	4	4	9	4
$\Sigma Y_{i,j}$	**15**	**20**	**30**	**25**

$[A'] = (35^2 + 55^2) \div 8 = 4{,}250 \div 8 = \mathbf{531.25}$

$[B'] = (45^2 + 45^2) \div 8 = 4{,}050 \div 8 = \mathbf{506.25}$

$[T'] = 90^2 \div 16 = 8{,}100 \div 16 = \mathbf{506.25}$

$[A'B'] = (15^2 + 20^2 + 30^2 + 25^2) \div 4 = 2{,}150 \div 4 = \mathbf{537.50}$

$SS_{A'\times B'} = 537.50 - 531.25 - 506.25 + 506.25 = \mathbf{6.25}$

$df_{A'\times B'} = 1$, therefore $MS_{A'\times B'} = 6.25$ and $\mathbf{F_{A'\times B'}(1, 27) = 6.25 \div 4.43 = 1.41}$, which is not significant, as $F_{crit}(1, 27) = 4.23$.

VI. Summary

A two-factor analysis of variance provides information about three general effects: (1) whether Factor A is significant, (2) whether Factor B is significant, and (3) whether the A × B interaction is significant. Given the contingency that the interaction is not significant and one or both main effects are significant, then analytical comparisons of the significant main effects (*main comparisons*) may be considered just as they were with the single-factor design (including, if necessary, consideration of α_{PC} adjustment procedures to control α_{FW}). Of course, additional analytical comparisons need only be considered if $df_{effect} \geq 2$, because for $df_{effect} = 1$, the main effect from the overall ANOVA would have already provided a statistical test for the only two groups that could be compared. Procedures for testing selected main comparisons are identical to procedures for testing analytical comparisons for the single-factor design; however, one must be careful when collapsing data over the factor not being tested, since this operation results in a new sample size [e.g., n' = (b)(n)]. For example, a main comparison contrasting the a_1 vs. a_2 conditions in a 3 × 3 factorial design involves comparing a new recombined group composed of the original a_1b_1, a_1b_2, and a_1b_3 groups for Group a_1' with a second recombined group composed of the original a_2b_1, a_2b_2, and a_2b_3 groups for Group a_2'.

New procedures were introduced in this chapter for analytical comparisons that would be appropriate given the contingency that the A × B interaction in the analysis of variance was significant. A researcher may examine the interaction in more detail through a series of *simple-effects* analyses, or, alternatively, by testing the significance of a smaller scale interaction (an *interaction contrast* using a subset of the original groups). For both procedures, only the data relevant to the comparison of interest are entered into the analysis: in order to calculate the F value, it is only necessary to compute the basic ratios and sum of squares to determine a variance estimate for the simple effect (e.g., $MS_{A\ at\ b1}$) or for the interaction contrast ($MS_{Acomp \times Bcomp}$), as the denominator for F is the error term from the original ANOVA ($MS_{S/AB}$). The process may be further refined for any significant simple effect or interaction contrast. An illustration was provided for testing a *simple comparison* following a significant simple effect. Again, the procedures followed from analytical comparison procedures for single-factor designs once the data relevant for the comparison had been identified and selected (e.g., if $F_{A\ at\ b1}$ is significant, then one may do a simple comparison contrasting a_1 vs. a_2 at b_1, and for this contrast, only the data from the a_1b_1 and a_2b_1 groups are evaluated).

10.1: Model Problem. Several procedures will be illustrated in this Model Problem using the data in Table 10.4 (a 3 × 3 design). We will begin by (A) doing a selected main comparison for Factor A (of course, given that the A × B interaction is significant for this data set, a main comparison would normally not be considered, so it is being done here only for illustration purposes). We will complete the example problem by breaking down the significant interaction by doing (B) some simple-effects analyses, (C) a selected simple comparison for the one significant simple effect (B at a_3), and (D) an interaction contrast as an alternative to the simple-effects approach.

A. Do a main comparison for Factor A to test the difference between a_1 vs. $a_2 + a_3$. To calculate SS_{Acomp} for this contrast we need to know the group means and an appropriate set of weighting coefficients. Remember, each group for this main comparison will be a composite of the three levels of Factor B (e.g., the a_1' group includes all those subjects who received the a_1b_1, a_1b_2, and a_1b_3 treatments), hence the number of subjects for each of these three new a_i' groups is n' = (b)(n).

Groups:	a_1'	a_2'	a_3'
$\bar{Y}_{Ai}$	4.58	6.08	7.33
c_i	+2	-1	-1

$$SS_{Acomp} = (b)(n)(\Sigma c_i \bar{Y}_{Ai})^2 \div \Sigma c_i^2$$

$$SS_{Acomp} = (3)(4)([+2][4.58] + [-1][6.08] + [-1][7.33])^2 \div ([+2]^2 + [-1]^2 + [-1]^2)$$

$$SS_{Acomp} = 12(-4.25)^2 \div 6 = 216.75 \div 6 = 36.12, \text{ and since } df_{Acomp} = 1,$$

$$MS_{Acomp} = 36.12 \text{ and } \mathbf{F_{Acomp}(1, 27) = 36.12 \div 4.43 = 8.15, p < .05}$$

B. Even though Factor B was not significant, the presence of an A × B interaction may mean that Factor B affected performance in opposite ways at different levels of Factor A, thus washing out an overall B effect. To see if there is merit to this possibility, we will do simple-effects analyses of Factor B at each level of Factor A. A complete simple-effects analysis for Factor B at the different levels of Factor A requires the calculation of three sums of squares: $SS_{B\ at\ a1}$, $SS_{B\ at\ a2}$, and $SS_{B\ at\ a3}$.

1. For B at a_1,

$$[B_{at\ a1}] = (15^2 + 20^2 + 20^2) \div 4 = 1{,}025 \div 4 = \mathbf{256.25}$$

$$[T_{at\ a1}] = 55^2 \div 12 = 3{,}025 \div 12 = \mathbf{252.08}$$

$SS_{B\ at\ a1} = 256.25 - 252.08 = \mathbf{4.17}$; $df_{B\ at\ a1} = 2$; and $MS_{B\ at\ a1} = 4.17 \div 2 = \mathbf{2.08}$

$\mathbf{F_{B\ at\ a1}(2, 27) = 2.08 \div 4.43 = .47}$ **(not significant)**

2. For B at a_2,

   $[B_{at\ a2}] = (30^2 + 25^2 + 18^2) \div 4 = 1{,}849 \div 4 = \mathbf{462.25}$

   $[T_{at\ a2}] = 73^2 \div 12 = 5{,}329 \div 12 = \mathbf{444.08}$

   $SS_{B\ at\ a2} = 462.25 - 444.08 = \mathbf{18.17}$; $df_{B\ at\ a2} = 2$; $MS_{B\ at\ a2} = 18.17 \div 2 = \mathbf{9.08}$

   $\mathbf{F_{B\ at\ a2}(2, 27) = 9.08 \div 4.43 = 2.05}$ **(not significant)**

3. For B at a_3,

   $[B_{at\ a3}] = (20^2 + 28^2 + 40^2) \div 4 = 2{,}784 \div 4 = \mathbf{696.00}$

   $[T_{at\ a3}] = 88^2 \div 12 = 7{,}744 \div 12 = \mathbf{645.33}$

   $SS_{B\ at\ a3} = 696.00 - 645.33 = \mathbf{50.67}$; $df_{B\ at\ a3} = 2$; $MS_{B\ at\ a3} = 50.67 \div 2 = \mathbf{25.33}$

   $\mathbf{F_{B\ at\ a3}(2, 27) = 25.33 \div 4.43 = 5.72, p < .05}$

C. Since B at a_3 is significant, simple comparisons for B at a_3 are appropriate. We will do one simple comparison: $b_1 + b_2$ vs. b_3 at a_3. To do this simple comparison analysis, we merely need to isolate the a_3b_1, a_3b_2, and a_3b_3 data and determine the appropriate means and coefficients.

Groups:	b_1 at a_3	b_2 at a_3	b_3 at a_3
Y_{bj}	5.00	7.00	10.00
c_j	+1	+1	-2

$SS_{Bcomp\ at\ a3} = (n)(\Sigma c_j \bar{Y}_{Bj})^2 \div \Sigma c_j^2$

$SS_{Bcomp\ at\ a3} = (4)([+1][5.00] + [+1][7.00] + [-2][10.00])^2 \div ([+1]^2 + [+1]^2 + [-2]^2)$

$SS_{Bcomp\ at\ a3} = 4(-8.00)^2 \div 6 = 256.00 \div 6 = 42.67$

Since $df_{Bcomp\ at\ a3} = 1$, $MS_{Bcomp\ at\ a3} = 42.67$

$\mathbf{F_{Bcomp\ at\ a3}(1, 27) = 42.67 \div 4.43 = 9.63, p < .05}$

D. Interaction Contrasts: Assume the researcher opts to do an interaction contrast instead of a simple-effects analysis. For this exercise, the interest is in testing the 2 × 2 interaction involving a_1 and a_2 conditions for Factor A and b_1 and b_3 conditions for Factor B. The relevant A × B matrix of group sums for these four groups has been isolated; we only need to calculate $MS_{Acomp \times Bcomp}$ for these data since we can use the error term from the overall analysis of variance ($MS_{S/AB} = 4.43$) for $F_{Acomp \times Bcomp}$. The primes (e.g., A') in the basic ratios for A, B, T, and AB again signify that these calculations are done on a subset of the original A × B data matrix.

Matrix of Group Sums for the Interaction Contrast

	a_1	a_2	ΣB_j
b_1	15	30	**45**
b_2	20	18	**38**
ΣA_i	**35**	**48**	

$[A'] = (35^2 + 48^2) \div 8 = 3{,}529 \div 8 = \mathbf{441.12}$

$[B'] = (45^2 + 38^2) \div 8 = 3{,}469 \div 8 = \mathbf{433.62}$

$[T'] = 83^2 \div 16 = 6{,}889 \div 16 = \mathbf{430.56}$

$[A'B'] = (15^2 + 20^2 + 30^2 + 18^2) \div 4 = 1{,}849 \div 4 = \mathbf{462.25}$

$SS_{A' \times B'} = 462.25 - 441.12 - 433.62 + 430.56 = \mathbf{18.07}$

Since $df_{A' \times B'} = 1$, $MS_{A' \times B'} = \mathbf{18.07}$

$\mathbf{F_{A' \times B'}(1, 27) = 18.07 \div 4.43 = 4.08}$, which is not significant at $p \leq .05$ since $F_{crit}(1, 27) = 4.23$

10.2: Practice Problem 1. The data matrix for this problem lists group sums for an experiment with a = 4, b = 3, and n = 3. The only significant effect in the analysis of variance was the main effect for Factor A: $F_A(3, 24) = 5.78$, $\mathbf{MS_{S/AB} = 6.28}$, $p < .05$. Assume the four levels of Factor A represent equally spaced intervals along a quantitative dimension and do a **main comparison trend analysis** (i.e., test the significance of the linear, quadratic, and cubic trends for Factor A).

Matrix of Group Sums

	a_1	a_2	a_3	a_4
b_1	18	26	30	47
b_2	27	30	26	32
b_3	25	32	36	35

10.3: Practice Problem 2. For the AB matrix of group sums and the ANOVA summary table for this problem, you can see that a = 3, b = 4, and n = 2. There are three parts to this problem to provide practice for more detailed analyses of the significant A × B interaction. Part A calls for a simple-effects analysis to test the significance of Factor A at each level of Factor B. Part B calls for a simple comparison as a follow-up for each significant simple effect found in Part A. For Part C you are asked to test the A' × B' interaction contrast involving the a_2b_3, a_2b_4, a_3b_3, and a_3b_4 groups.

Matrix of Group Sums

	a_1	a_2	a_3
b_1	9	13	26
b_2	10	10	14
b_3	10	20	32
b_4	6	27	26

ANOVA Summary Table

Source	SS	df	MS	F
A	249.08	2	124.54	27.68 sig.
B	80.46	3	26.82	5.96 sig.
A × B	96.92	6	16.15	3.59 sig.
S/AB	54.00	12	4.50	

A. Do **simple-effects analyses** and indicate whether $F_{A\ at\ b1}$, $F_{A\ at\ b2}$, $F_{A\ at\ b3}$, and $F_{A\ at\ b4}$ are significant at $p \leq .05$.

B. For any significant simple effect, do a follow-up test to determine if the **simple comparison** that contrasts the combined a_1 and a_2 groups with the a_3 group ($a_1 + a_2$ vs. a_3) is significant.

C. Do the appropriate statistical test to determine whether the **A x B interaction contrast** is significant based on the a_2 and a_3 groups for Factor A and the b_3 and b_4 groups for Factor B.

D. Repeat the analysis in Part C above using SPSS or an alternative software program. In all likelihood, the $F_{Acomp \times Bcomp}$ in the computer analysis will not be the same as the $F_{Acomp \times Bcomp}$ calculated in Part C. Can you explain why the two interaction F ratios (one done on a hand-held calculator and one done on a computer) are not identical? If you are having trouble explaining this discrepancy, you should direct your attention to the estimates of MS_{error} ($MS_{S/AB}$) in the two analyses.

10.4: Practice Problem 3. For the data in Model Problem 9.1 (page 180), do a **simple-effects analysis** to determine if the proportion of false recognitions as a function of semantic vs. phonological similarity differ at each of the presentation speeds (20, 50, and 250 msec). That is, evaluate if Factor B (semantic vs. phonological similarity) is significant at a_1 (20 msec), at a_2 (50 msec), and at a_3 (250 msec).

10.5: Practice Problem 4. Using the data from Model Problem 9.1, compute the F ratio for an interaction contrast comparing semantic vs. phonetic similarity just for the 20 msec and 250 msec groups.

10.6: Practice Problem 5. The purpose of this problem is to test your skills and understanding for applying both the Tukey and the Scheffé test procedures for evaluating the significance of a **main comparison**. We'll use the data from Model Problem 10.1 and test a specific main comparison for Factor A comparing Group a_1 vs. Group a_2 + Group a_3. Don't worry about whether one or the other α_{FW} control procedure is more appropriate; just test whether this particular comparison is significant at $\alpha_{FW} = .05$ with (1) a Tukey adjustment for α_{PC}, and with (2) a Scheffé adjustment for α_{PC}, and report the critical F_T and F_S values.

Chapter 11

Data Analysis with the Same Subjects in All Conditions

All those who drink this remedy recover in a short time, except those whom it does not help, who die. Therefore, it is obvious that it fails only in incurable cases.

—Galen

I. The Single-Factor Within-Subject Design (A × S)

Thus far we have considered analysis of variance procedures appropriate for experimental designs for which different levels of the independent variable(s) are represented by different groups of subjects. For example, consider an experiment designed to compare performance on a word-guessing task among groups who hear only the first 25 ms, the first 50 ms, or the first 100 ms of recorded words that take from 900–1,000 ms to pronounce completely. If we had 30 volunteers available for this experiment, we would proceed by creating three groups with 10 individuals randomly assigned to each group (a_1 = 25 ms, a_2 = 50 ms, and a_3 = 100 ms exposures). This design is referred to as an *independent samples, random groups*, or *between-subject* design. Another way of designing the experiment (that has both advantages and disadvantages that we will consider later) would be to have each volunteer tested at all three spoken word durations, a procedure referred to as a *dependent samples, repeated measures,* or *within-subject* design. That is, each subject would hear the first 25 ms for one-third of the recorded words, the first 50 ms for another one-third of the words, and the first 100 ms for remaining one-third of the words. One difference you can see immediately is that if you wanted to have 10 subjects in each condition of the experiment, the between-subject design requires a total of 30 participants, while the within-subject design is more economical in this regard as it would only require a total of 10 participants, since each participant would serve in all three partial exposure duration conditions (a_1, a_2, and a_3).

The single-factor, within-subject design is coded as an A × S design, as if there were a Factor A and a Factor S (the "S" is the coding letter for the subjects in the experiment). We will see shortly that the analysis-of-variance operations for this design are similar to the A × B two-factor design. Because sums of squares for the single-factor within-subject design are calculated for Factor A, for the subject Factor S, and for the A × S interaction, it is introduced here after the concept of interaction was introduced with the two-factor between-subject design rather than immediately after consideration of the single-factor between-subject design.

Is the choice of whether to use a between-subject or a within-subject design merely a matter of personal preference or convenience? Probably in some cases it is. Nonetheless, there are important differences to consider. We have already alluded to the fact that the within-subject design is more economical in terms of number of participants required to complete the experiment. However, the major virtue of the within-subject design lies with its greater *power* compared to the between-subject design. Recall that the denominator for the F ratio in the analysis of variance is an estimate of random error ($MS_{S/A}$ in the single-factor between-subject design and $MS_{S/AB}$ in the two-factor design). Certainly an important contributor to random error is variation among individual participants (individual differences) randomly assigned to different groups with the between-subject design. What happens to individual variability with the within-subject design? In the between-subject design, the first subject in Group a_1 is a different person from the first subject in Group a_2 and the same for Group a_3. In the within-subject design, the first subject in Group a_1, Group a_2, and Group a_3 is the same person. Would you expect more or less subject variability (a contributor to random error) as a function of whether different individuals are randomly assigned to each experimental group or the same individuals participate in all experimental conditions? Of course, having each subject participate in all experimental conditions should reduce individual difference variability, and hence reduce the overall error variance. A reduction in error variance will result in an increase in power because a smaller denominator in the F ratio means that there is, in effect, less "noise" when assessing variance due to the independent variable.

Given that the repeated measures design is more economical and more powerful, why shouldn't this always be the design of choice? The answer is that the experimental situation and task requirements must be carefully scrutinized before deciding on whether a within-subject design is appropriate because there are potential problems when the same subject participates in more than one experimental condition. Potential problems stem from the fact that the experience of being in one experimental condition (e.g., receiving a large reward for correct responses) may impact how a subject performs in a second condition (e.g., receiving a small reward for correct responses). That is, after experiencing one treatment condition an individual may be "older and wiser" (thereby affecting performance in a future treatment condition), or more practiced or more fatigued later, or possibly changed by events and experiences that occur between successive testing occasions. *Counterbalancing* conditions (methodological procedures for arranging sequences of different treatment conditions defined by the independent variable) across subjects such that each condition occurs equally often as 1st, 2nd, 3rd, etc. in order of presentation is normally relied upon to try to minimize damage done by extraneous factors that could change an individual from one testing session to another. However, if suspected extraneous factors that could change subjects are not bi-directional ("differential carry-over effects," see Keppel et al. 1992), then the fundamental principle in experimental design, *viz.*, that all conditions are equivalent with the exception of differences due to the independent variable, could be compromised. A between-subject design may be the only viable option if differential carry-over effects threaten the integrity of a within-subject design.

Hopefully, an intuitive or common-sense analysis of a human learning paradigm that was popular many years ago will better illustrate this problem. Studies of "incidental learning" appeared to some researchers to be relevant to issues of motivation and attention in learning. Researchers who were curious about intention to learn and the extent to which learning is reduced when formal instructions to learn list materials are not given to subjects, employed an incidental learning paradigm (e.g., Mechanic 1964; Postman 1964). A common experimental procedure in studying incidental learning is to compare how much is recalled after presenting a list of words or nonsense syllables (e.g., "COZ," "DAQ," "WUB") when subjects in one treatment condition are told in advance that a recall test will follow a stimulus presentation, compared to a second treatment condition for which there is no mention of an impending recall test. Experimenters have been well aware that subjects (often college students) are pretty smart, and most would not be clueless about the possibility of a future memory test if they were shown a list of words, even if formal instructions to try to remember were not given. Thus, a plausible cover story or orienting task is included in this paradigm to minimize the likelihood that subjects will suspect that they are participating in a memory experiment.

The paradigm requires three groups: An incidental-learning group that performs an orienting task such as rating words along some dimension or characteristic (e.g., "pleasantness"), an intentional-learning group given instructions to learn along with the requirement to perform the orienting task, and an unencumbered, intentional-learning group given instructions to learn but not required to perform the orienting task. Including both an orienting task and formal instructions to learn is important in order to assess the influence, if any, of the orienting task on performance.

At this point we are not concerned about the details for evaluating instructions to learn by comparisons among these three conditions. The purpose for mentioning the incidental learning paradigm is to illustrate a likely problem that could well be present if a within-subject design were used with this paradigm. While list items and conditions can be counterbalanced (i.e., different lists of nonsense syllables used equally often for each instructional condition, and equal numbers of subjects having the three instructional conditions presented equally often as the 1st, 2nd, and 3rd task to perform), order effects will likely not be bi-directional. That is, it would not be surprising to find differential transfer or carryover effects based on the order in which the experimental conditions are presented. Imagine yourself and a friend as participants in a within-subject version of such an experiment. If you happen to have the incidental learning condition first, it is most likely that your focus will be on the assigned task (e.g., rating items along some dimension) and not on preparing for a future memory test. So for you, the announcement of a memory test after the rating task is completed would likely come as a surprise. If the incidental learning condition is the 2nd or 3rd task for your friend, then prior experiences with a memory test or two following this type of list presentation format might create a more wary individual who would be suspicious about the possibility of a future memory test, even though formal instructions to try to remember the items were not given. Hence, your friend might, in effect, provide his or her own "self instruction" to try to remember items from the list.

a. The ANOVA for a = 2 Designs

The procedures for conducting an analysis of variance for the A × S single-factor within-subject design bear a close resemblance to ANOVA procedures for the A × B two-factor between-subject design described in Chapter 9. The calculations that can and cannot be done may be best illustrated by considering a minimal example problem with a = 2 and **n or s** = 2 (two experimental conditions and two subjects). For this example, we will assume the total data set is as follows:

	a_1	a_2	ΣS_j
s_1	9	5	**14**
s_2	4	3	**7**
ΣA_i	**13**	**8**	

You should have no trouble calculating basic ratios for A, S, T, and A × S (nothing is really different from the A × B design at this point except that rows are labeled s_1, s_2 instead of b_1, b_2); thus, what you would have done to calculate [B] for the two-factor A × B design, you now do to calculate [S] for the single-factor A × S design: $[A] = (13^2 + 8^2) \div 2 = 116.5$, $[S] = (14^2 + 7^2) \div 2 = 122.5$, $[T] = 21^2 \div 4 = 110.25$, and $[AS] = (9^2 + 4^2 + 5^2 + 3^2) \div 1 = 131.0$. However, we needed to calculate one additional basic ratio for the A × B design in order to calculate $SS_{S/AB}$ (to determine variability among the subjects within groups), as $SS_{S/AB} = [Y] - [AB]$. So what does [Y] look like for the A × S design? To find [Y] you need to square each individual score and sum these squared scores: $[Y] = 9^2 + 4^2 + 5^2 + 3^2 = 131$. However, this is precisely what we did to find [AS]—that is, [Y] and [AS] are exactly the same. Thus, if you tried to calculate a sum of squares for subjects "within the AS conditions," you would find that there is only one score in each unique combination of a level of Factor A and a level of Factor S; $SS_{S/AS} = [Y] - [AS] = 0$. Given that there is only one subject within each $a_i s_j$ condition, there is no variability for subjects within $a_i s_j$ conditions to measure in the A × S design. A minimum of two scores is necessary in order to engage the concept of variability.

In comparing calculations required for A × B and A × S designs, we have one fewer basic ratio and one fewer sum of squares to compute for the A × S design. However, if we cannot calculate $SS_{S/AS}$, what is the appropriate error term (denominator for F) for evaluating the effect of Factor A in the A × S design? It turns out that both MS_A and $MS_{A\times S}$ have the same components influencing them (random error and a potential A × S interaction). However, variability due to Factor A (if any) influences MS_A but not $MS_{A\times S}$. For the analysis of variance model, $MS_{A\times S}$ is an appropriate error term for calculating F_A ($F_A = MS_A \div MS_{A\times S}$, and if H_0 is true, then the Factor A effect in the numerator of the F ratio is 0 and F_A has an expected value of 1.0 since the remaining contributors of variability in both the numerator and denominator are random error and an A × S interaction). A meaningful F ratio for the Subject Factor

(F_S) cannot be calculated because there is not an appropriate error term for S, so the only F ratio that can be evaluated for the A × S design is F_A.

We will illustrate procedures for doing a within-subject ANOVA for a design with a = 2 (two experimental conditions) and n = 5 (number of subjects in each condition may be denoted by either "s" or "n," and since we have coded sample size with a lower-case "n" in the past, we will continue using this code). With a two-condition contrast, the data can also be tested with a t test (referred to as a *paired-sample t* when the same subjects serve in both conditions) or by determining corresponding confidence intervals. For each illustration we will use the same data set. Assume these data are from a small-scale experiment in which five children (n = 5) worked on mathematical reasoning problems on two occasions (a = 2). On one occasion (a_1) the five children received positive (Pos) comments while they worked on the problems (e.g., "that's very good," "you are doing fine") and on the other occasion (a_2) no evaluative comments were made while they worked on problems. Fictitious data (number of correct solutions) are given in Table 11.1.

Table 11.1: Number of Problems Correctly Solved

	a_1(Pos)	a_2(Silence)	ΣS_j
s_1	10	10	**20**
s_2	7	4	**11**
s_3	6	3	**9**
s_4	9	8	**17**
s_5	13	10	**23**
ΣA_i	**45**	**35**	

Step 1: Calculate Basic Ratios

$[A] = (45^2 + 35^2) \div n = 3{,}250 \div 5 = \mathbf{650.00}$

$[S] = (20^2 + 11^2 + 9^2 + 17^2 + 23^2) \div a = 1{,}420 \div 2 = \mathbf{710.00}$

$[T] = 80^2 \div (a)(n) = 6{,}400 \div 10 = \mathbf{640.00}$

$[Y] = [AS] = 10^2 + 7^2 + \ldots + 10^2 = \mathbf{724.00}$

Step 2: Calculate Sums of Squares

$SS_A = [A] - [T] = 650.00 - 640.00 = \mathbf{10.00}$

$SS_S = [S] - [T] = 710.00 - 640.00 = \mathbf{70.00}$

$SS_{A \times S} = [Y] - [A] - [S] + [T] = 724.00 - 650.00 - 710.00 + 640.00 = \mathbf{4.00}$

Step 3: Determine Degrees of Freedom

$df_A = a - 1 = 2 - 1 = \mathbf{1}$

$df_S = n - 1 = 5 - 1 = \mathbf{4}$

$df_{A \times S} = (a - 1)(n - 1) = (1)(4) = \mathbf{4}$

Step 4: Calculate Mean Squares

$MS_A = SS_A \div df_A = 10.00 \div 1 = \mathbf{10.00}$

$MS_S = SS_S \div df_S = 70.00 \div 4 = \mathbf{17.50}$

$MS_{A \times S} = SS_{A \times S} \div df_{A \times S} = 4.00 \div 4 = \mathbf{1.00}$

Step 5: Present F_A and Results in an ANOVA Summary Table (you will note from this table that an F_S cannot be calculated because there is no appropriate error term [denominator] for Factor S)

ANOVA Summary Table

Source:	SS	df	MS	F	F_{crit}
A	10.00	1	10.00	10.00 $p < .05$	7.71
S	70.00	4	17.50	*cannot be determined*	
A × S (error term)	4.00	4	1.00		

b. Paired-Sample t Test

As was true for a between-subject design with a = 2, a t test for a within-subject design would produce the same significance decision that was arrived at using an analysis of variance. In this section the analysis of these data is repeated using a t test for dependent samples (referred to as a paired-sample t test), which is the appropriate version of t when the same subject serves in both conditions of the experiment. Since $F = t^2$, we know within rounding error that $t = \sqrt{10} = 3.16$. With dependent samples, the numerator in the formula for t is the same as it is with independent samples; however, the estimate of random error (s_{diff}, which is the standard error of the difference that appears in the denominator for t) is impacted by the relationship between paired scores from the same subject. For this estimate we will begin by (1) creating a difference score (D) for each subject by subtracting the respective a_2 score from the a_1 score, (2) computing a sum of squares for D scores (SS_D), (3) determining the mean square (MS_D) by dividing SS_D by the degrees of freedom ($n - 1 = 4$), and finally, (4) determining the standard error of the difference by taking the square root of MS_D divided by n. In formula terms, $t = (\bar{Y}_{A1} - \bar{Y}_{A2}) \div \sqrt{(MS_D}$

$\div$ n). The $a_1 - a_2$ scores for the five subjects in Table 11.1 are $D_1 = 0$, $D_2 = 3$, $D_3 = 3$, $D_4 = 1$, and $D_5 = 3$. Given $n = 5$ and the necessary computations for the two group means and MS_D, we have:

$\bar{Y}_{A1} = 9.00$, $\bar{Y}_{A2} = 7.00$, and $\bar{Y}_{A1} - \bar{Y}_{A2} = 9.00 - 7.00 = \mathbf{2.00}$

$SS_D = [D] - [T] = (0^2 + 3^2 + 3^2 + 1^2 + 3^2) - (10^2 \div 5) = 28.00 - 20.00 = \mathbf{8.00}$

$MS_D = SS_D \div (n - 1) = 8.00 \div 4 = \mathbf{2.00}$

Substituting these values in the formula for t: $t = 2.00 \div \sqrt{(2.00 \div 5)} = 2.00 \div \sqrt{.40}$

$t = 2.00 \div .6325 = \mathbf{3.16}$ (which equals $\sqrt{F}$)

c. Confidence Intervals for the Difference Between Group Means

The formula for a confidence interval with repeated measures is similar to the formula given in Chapter 6, with the exception that for dependent samples the standard error of the difference is given by the formula $s_{diff} = (MS_D \div n)$ and replaces the standard error of the mean in the formula for CI:

$$CI_{.95}: (\bar{Y}_{A1} - \bar{Y}_{A2}) - (t_{.05})\sqrt{(MS_D \div n)} \le \mu_1 - \mu_2 \le (\bar{Y}_{A1} - \bar{Y}_{A2}) + (t_{.05})\sqrt{(MS_D \div n)}$$

The critical $t_{.05}$ with four degrees of freedom is found in the statistical tables at the back of the book (from Table A-2, $t_{.05} = 2.78$); and we have already determined that $(\bar{Y}_{A1} - \bar{Y}_{A2}) = 2.00$ and $MS_D = 2.00$. Therefore, $CI_{.95}: 2.00 - (2.78)\sqrt{(2.00 \div 5)} \le \mu_1 - \mu_2 \le (2.00) + (2.78)\sqrt{(2.00 \div 5)}$

$CI_{.95}: 2.00 - (2.78)\sqrt{.40} \le \mu_1 - \mu_2 \le 2.00 + (2.78)\sqrt{.40}$

$CI_{.95}: 2.00 - (2.78)(.6325) \le \mu_1 - \mu_2 \le 2.00 + (2.78)(.6325)$

$CI_{.95}: 2.00 - 1.76 \le \mu_1 - \mu_2 \le 2.00 + 1.76$, or $\mathbf{CI_{.95}: .24 \le \mu_1 - \mu_2 \le 3.76}$ (and we can see that $CI_{.95}$ does not include 0; thus the significant F and t values at $p \le .05$ are consistent with this 95% confidence interval).

II. The A × S ANOVA for Experiments with a > 2

Nothing new is added for calculating sums of squares by increasing number of conditions and number of subjects in an A × S design. However, with a > 2, analytical comparisons may be considered after the ANOVA is completed; and at this point new procedures will be involved. Thus, it is useful to work through one more example problem. The previous ANOVA data are repeated here with two new conditions added to the original two conditions. For this problem assume there is a 3rd condition (a_3) with negative comments (Neg) during the mathematical reasoning tests (e.g., "that's not very good," "you need to try harder") and a 4th condition (a_4) with both positive and negative comments during the problem-solving session. Fictitious data for all four conditions are presented in Table 11.2.

Table 11.2: Number of Problems Correctly Solved

	a_1(Pos)	a_2(Silence)	a_3(Neg)	a_4(Pos & Neg)	ΣS_j
s_1	10	10	5	12	**37**
s_2	7	4	1	5	**17**
s_3	6	3	3	6	**18**
s_4	9	8	4	8	**29**
s_5	13	10	7	9	**39**
ΣA_i	**45**	**35**	**20**	**40**	

Step 1: Calculate Basic Ratios

$[A] = (45^2 + 35^2 + 20^2 + 40^2) \div n = 5{,}250 \div 5 = \mathbf{1{,}050.00}$

$[S] = (37^2 + 17^2 + 18^2 + 29^2 + 39^2) \div a = 4{,}344 \div 4 = \mathbf{1{,}086.00}$

$[T] = 140^2 \div (a)(n) = 19{,}600 \div 20 = \mathbf{980.00}$

$[Y] = [AS] = 10^2 + 7^2 + \ldots + 5^2 = \mathbf{1{,}174.00}$

Step 2: Calculate Sums of Squares

$SS_A = [A] - [T] = 1{,}050.00 - 980.00 = \mathbf{70.00}$

$SS_S = [S] - [T] = 1{,}086.00 - 980.00 = \mathbf{106.00}$

$SS_{A \times S} = [Y] - [A] - [S] + [T] = 1{,}174.00 - 1{,}050.00 - 1{,}086.00 + 980.00 = \mathbf{18.00}$

Step 3: Determine Degrees of Freedom

$df_A = a - 1 = 4 - 1 = \mathbf{3}$; $df_S = n - 1 = 5 - 1 = \mathbf{4}$; $df_{A \times S} = (a - 1)(n - 1) = (3)(4) = \mathbf{12}$

Step 4: Calculate Mean Squares

$MS_A = SS_A \div df_A = 70.00 \div 3 = \mathbf{23.33}$

$MS_S = SS_S \div df_S = 106.00 \div 4 = \mathbf{26.50}$

$MS_{A \times S} = SS_{A \times S} \div df_{A \times S} = 18.00 \div 12 = \mathbf{1.50}$

Step 5: Present F_A and Results in an ANOVA Summary Table

ANOVA Summary Table

Source:	SS	df	MS	F	F_{crit}
A	70.00	3	23.33	15.55 $p < .05$	3.49
S	106.00	4	26.50		
A × S	18.00	12	1.50		

Based on the results of this analysis of variance, the conclusion is that the difference among these four conditions is significant at $p < .05$. Before considering more detailed analyses of the significant A effect (analytical comparisons), we will illustrate procedures for doing an A × S ANOVA using SPSS.

III. Using SPSS for the Within-Subject ANOVA

Entering data in an SPSS spreadsheet for a within-subject design is more direct and logical than entering data for the between-subject design. In the latter case a "dummy" code had to be created to represent different levels of the independent variable(s). For the within-subject design, each column in the spreadsheet corresponds to a different condition, thus for each of the first four columns in the SPSS spreadsheet, simply enter the data directly. The five scores for the a_1 condition are entered in the 1st column, the five scores for the a_2 condition are entered in the 2nd column, the five scores for the a_3 condition are entered in the 3rd column, and the five scores for the a_4 condition are entered in the 4th column. It is essential when entering these data into the SPSS spreadsheet that each row corresponds to four scores for a given subject (e.g., data in row 1, columns 1–4 are the a_1s_1, a_2s_1, a_3s_1, and a_4s_1 scores, respectively). Table 11.3 lists scores from this example problem for each of the five subjects as they would appear in an SPSS spreadsheet.

Table 11.3: Data View Screen for SPSS

Subjects	a_1	a_2	a_3	a_4
1	10.00	10.00	5.00	12.00
2	7.00	4.00	1.00	5.00
3	6.00	3.00	3.00	6.00
4	9.00	8.00	4.00	8.00
5	13.00	10.00	7.00	9.00

In order to analyze data from a within-subject (repeated measures) design using SPSS, after entering the data on the spreadsheet, you proceed to "Analyze" on the tool bar at the top of the computer screen. The option you want to follow from the draw-down menu again is the "General Linear Model." However, instead of selecting "Univariate" as was done for the two-factor between-subject design, you need to click on "Repeated Measures" for this within-subject design. Clicking on "repeated measures" will open a dialog box similar to the one that appears in Table 11.4. Other information appears in the SPSS dialog box, but the parts displayed are all that we need. "Factor 1" appears in the upper box, and since a = 4, we need to type the number "4" (which appears in italics) in the box below "number of levels." After you have indicated the number of levels, click on "Add" and "factor 1(4)" (shown in italics) will appear in the box below number of levels. This first Dialog Box simply identifies the number of within-subject factors in the experiment (which is one in our example problem) and the number of conditions (levels) for each within-subject variable identified. From here we need to move to a second Dialog Box to complete the instructions for the computer. Open the second Dialog Box by clicking on "Define" at the bottom of the screen. This action opens a screen that appears similar to the one sketched in Table 11.5).

Table 11.4: SPSS Dialog Box for Repeated Measures ANOVA

factor 1

Number of Levels

4

Add

factor1(4)

Define

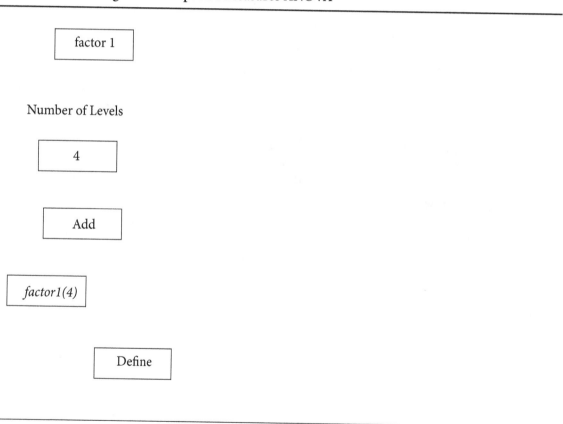

Table 11.5: Completed SPSS Dialog Box for Repeated Measures ANOVA

Within-subject variables

a1
a2
a3
a4

→

Between-subject variables

Options

→

OK

Covariates

You complete the "define" operations by highlighting "a1" and clicking on the arrow to the left of the "within-subject variables" box to move "a1" to that box. Do the same for "a2" through "a4." Those of you who are experienced with cursor functions will discover that this highlighting and moving operation for all conditions can be accomplished in a single step. After all your "a_i" conditions have been moved to the "within-subject variable" box, position the cursor on "OK" and click to produce a completed ANOVA summary table (a task that only takes the computer a second or two to complete). If you click the "Options" box before clicking on "OK" you can instruct the computer to provide additional information, such as means and standard deviations for each of the four conditions in the experiment. The SPSS ANOVA table is similar to what has been presented in this text, although it includes things we do not need. All we need for this problem is the first rows ("sphericity assumed") in each section of the table labeled "tests of within-subject effects" (factor 1 and error). It is copied for these data in Table 11.6. These SPSS results confirm the hand calculations reported earlier.

Table 11.6: SPSS ANOVA Summary Table

Tests of Within-Subject Effects

Dependent Variable: Scores

Source		Type III Sum of Squares	df	Mean Square	F	Sig
Factor 1	Sphericity Assumed	70.000	3	23.333	15.556	.000
Error	Sphericity Assumed	18.000	12	1.500		

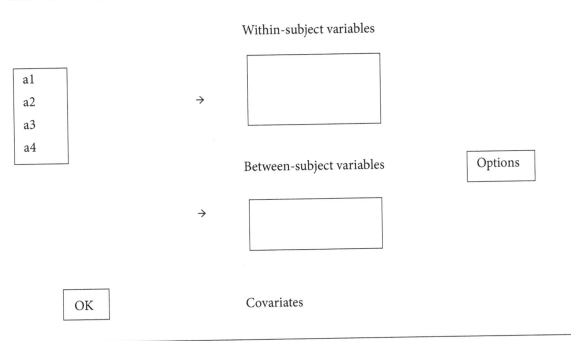

As was true for between-subject designs with $df_A \geq 2$, additional analytical comparisons may be tested following a within-subject analysis of variance. Further, calculating SS_{Acomp} is done in the same way for both types of designs. There is one basic change, however, which increases the workload required when analytical comparisons are done on a hand-held calculator for within-subject designs. With the between-subject design, once SS_{Acomp} was calculated, the computational work was virtually completed because we were able to use $MS_{S/A}$ (or $MS_{S/AB}$) from the overall analysis of variance as the error term (denominator in the F ratio) for F_{Acomp}. For the within-subject design the A × S interaction serves as the error term, and the fact that this term is an interaction implies that using all levels of Factor A in determining $MS_{A\times S}$ is only appropriate as the denominator in the F ratio when all levels of Factor A are considered in the numerator. For a weighted subset of Factor A conditions the appropriate error term needs to be a similarly weighted A × S interaction. This explanation is the long-winded version of stating the requirement for evaluating analytical comparisons in a within-subject design, *viz.*, that a new error term must be calculated: $F_{Acomp} = MS_{Acomp} \div \mathbf{MS_{Acomp \times S}}$.

In the remainder of this section we will work through two analytical comparison example problems to illustrate procedures for doing follow-up comparisons in the single-factor, within-subject design, making use of the fictitious data from the example problem completed earlier in the chapter (a = 4 and n = 5). For the first illustration we will contrast the positive verbal feedback condition (a_1) with the silence or no feedback condition (a_2). Determining F_{Acomp1} for the contrast between a_1 and a_2 at this stage can be completed very quickly—so quickly, in fact, that it can be reported that $F_{Acomp1}(1, 4) = 10.00$, $MS_{Acomp \times S} = 1.00$, $p < .05$. The reason we know F_{Acomp1} for the a_1 vs. a_2 contrast is that the example problem at the beginning of this chapter for the single-factor, within-subject design for a = 2 involved calculating MS_A and $MS_{A\times S}$ for this two-group data set. This particular contrast also serves to emphasize an important procedural point: *viz.*, the computations necessary for MS_A and $MS_{A\times S}$ for contrasting these two conditions involve nothing more than doing a complete single-factor within-subject analysis of variance using only the two groups involved in the comparison (a_1 and a_2).

The second illustration involves a more complex comparison, so we will need to use weighting coefficients not only to calculate SS_{Acomp2}, but also to calculate $SS_{Acomp2 \times S}$. Of course, this could have been done as well for the first illustration, and the same F_{Acomp1} value would have been produced (you may want to confirm this assertion after this demonstration). You should recall that the formula for calculating a sum of squares for a comparison of interest involved (1) multiplying the mean for each experimental condition by its respective weighting coefficient, (2) summing these products, (3) squaring the summed products, (4) multiplying the sum of the squared products by n, and (5) dividing this result by the sum of the squared weighting coefficients. These operational instructions are expressed more succinctly by the formula we used: $SS_{Acomp} = (n)(\Sigma c_i \bar{Y}_{Ai})^2 \div \Sigma c_i^2$.

In order to calculate $SS_{Acomp \times S}$, each individual score in the original data matrix must be **transformed** into a new score by multiplying the original score by the corresponding weighting coefficient. This is the first step in getting the data ready for calculating $SS_{Acomp \times S}$. After this step is completed we still have a data matrix with four columns of numbers. An analytical comparison is a contrast between two conditions, so the second step involves converting the four columns of numbers to the two that are being compared. Remember, positive-signed weighting coefficients identify one side of the comparison, and negative-signed weighting coefficients identify the other side. So by summing all like-signed scores, a new set of scores in two columns (positive signed scores vs. negative signed scores) is created that define the contrast of interest. The data are now ready for performing a single-factor, within-subject ANOVA with a = 2 (note **positive and negative signs are only used to identify the two conditions being compared**—that is, the children provided only positive scores, and the actual scores that are compared between the two conditions are all positive numbers).

For this second analytical comparison we will contrast conditions in which the children received at least some positive verbal feedback ($a_1 + a_4$) with the silence or no feedback condition (a_2). Coefficients that may be used for this A_{comp2} contrast are $c_1 = +1$, $c_2 = -2$, $c_3 = 0$, and $c_4 = +1$. Multiplying each individual score in Table 11.2 by its respective weighting coefficient yields the data matrix of weighted scores shown in Table 11.7. Combining scores that have the same positive signs and combining scores that have the same negative signs gives the two-condition data matrix in Table 11.8 that will be used for the ANOVA calculations. As noted earlier, you can see that all scores in Table 11.8 are positive numbers. Positive and negative signs are used only to identify the two conditions involved in the contrast (the column headings).

Table 11.7: Weighted Data Matrix for Number of Problems Correctly Solved

	a_1(Pos)	a_2(Silence)	a_3(Neg)	a_4(Pos & Neg)
s_1	+10	-20	0	+12
s_2	+7	- 8	0	+5
s_3	+6	-6	0	+6
s_4	+9	-16	0	+8
s_5	+13	-20	0	+9

Table 11.8: Data Matrix for ANOVA Calculations Contrasting + vs. – Signed Conditions

	+ Signed	– Signed	ΣS_i
s_1	22	20	**42**
s_2	12	8	**20**
s_3	12	6	**18**
s_4	17	16	**33**
s_5	22	20	**42**
ΣA_j	**85**	**70**	

The data in Table 11.8 are in the proper form for doing an analysis of variance: the F computed for this data set is F_{Acomp2}, which contrasts the two conditions with positive comments ($a_1 + a_4$) with the silent condition (a_2). You may agree that getting data ready for the analysis of variance is probably the most difficult part of evaluating a complex analytical comparison.

Step 1: Calculate Basic Ratios

$[A] = (85^2 + 70^2) \div n = 12{,}125 \div 5 = \mathbf{2{,}425.00}$

$[S] = (42^2 + 20^2 + 18^2 + 33^2 + 42^2) \div a = 5{,}341 \div 2 = \mathbf{2{,}670.50}$

$[T] = 155^2 \div (a)(n) = 24{,}025 \div 10 = \mathbf{2{,}402.50}$

$[Y] = [AS] = 22^2 + 12^2 + \ldots + 20^2 = \mathbf{2{,}701.00}$

Step 2: Calculate Sums of Squares

$SS_{Acomp2} = [A] - [T] = 2{,}425.00 - 2{,}402.50 = \mathbf{22.50}$

$SS_S = [S] - [T] = 2{,}670.50 - 2{,}402.50 = \mathbf{268.00}$

$SS_{Acomp2 \times S} = [Y] - [A] - [S] + [T] = 2{,}701.00 - 2{,}425.00 - 2{,}670.50 + 2{,}402.50 = \mathbf{8.00}$

Step 3: Determine Degrees of Freedom

$df_{Acomp2} = a - 1 = 2 - 1 = \mathbf{1}$

$df_S = n - 1 = 5 - 1 = \mathbf{4}$

$df_{Acomp2 \times S} = (a - 1)(n - 1) = (1)(4) = \mathbf{4}$

Step 4: Calculate Mean Squares

$MS_{Acomp2} = SS_{Acomp2} \div df_{Acomp2} = 22.50 \div 1 = \mathbf{22.50}$

$MS_S = SS_S \div df_S = 268.00 \div 4 = \mathbf{67.00}$

$MS_{Acomp2 \times S} = SS_{Acomp2 \times S} \div df_{Acomp2 \times S} = 8.00 \div 4 = \mathbf{2.00}$

Step 5: Present F_{Acomp2} and Results in an ANOVA Summary Table

ANOVA Summary Table

Source	SS	df	MS	F	F_{crit}
A	22.50	1	22.50	11.25*	7.71
S	268.00	4	67.00		
A × S	8.00	4	2.00		

There is one footnote to add to this analysis. The transformation of the scores in the original data impacted variance estimates for both Factor A and the A × S interaction. Since the transformations are the same in both cases, the $F_{Acomp2} = 11.25$ is correct. However, if it is necessary to indicate the actual error variance estimate ($MS_{Acomp2 \times S}$), as might be the case for a published journal article, then you need to know that the value in this ANOVA summary table for $MS_{Acomp2 \times S}$ is inflated and must be corrected (i.e., the transformation operation must be "undone"). Dividing MS_{Acomp2} and $MS_{Acomp2 \times S}$ by the appropriate correction factor ($\Sigma c_i^2 \div 2$ for this example) will take care of the necessary corrections. For this problem, $\Sigma c_i^2 \div 2 = (1^2 + [-2]^2 + 0^2 + 1^2) \div 2 = 6 \div 2 = 3$. Thus, the corrected variance estimate for MS_{Acomp2} is $22.50 \div 3 = 7.50$, and the corrected variance estimate for $MS_{Acomp2 \times S}$ is $2.00 \div 3 = .67$ (and $7.50 \div .67 = 11.19$, a value within rounding error of the F_{Acomp2} reported in Step 5). The correction factor for the first comparison is $\Sigma c_i^2 \div 2 = (1^2 + [-1]^2) \div 2 = 2 \div 2 = 1$, hence this comparison of the two existing conditions from the original set of four conditions does not require a correction of the variance estimates (i.e., MS_{Acomp1} and $MS_{Acomp1 \times S}$ would be "corrected" by dividing each by $\Sigma c_i^2 \div 2 = 1$, leaving them unchanged).

V. Summary

A second type of experimental design was introduced in this chapter. Up to this point we considered experiments with individuals randomly assigned to different groups (between-subject designs). Each group received a different treatment as defined by the corresponding level of the independent variable(s). An alternative approach involves having each subject participate in the experiment multiple times, thus serving in every experimental condition as defined by the corresponding level of the independent variable (within-subject designs). The within-subject design does not require as many volunteers to achieve a given sample size; however, more importantly, this type of design has an advantage in terms of power. The reason that it is normally a more powerful design is that individual differences that contribute to the estimate of error variance ought to be attenuated in the within-subject design compared to the between-subject design. In theory, using the same subject in all conditions of an experiment should result in variance due to individual differences being largely removed from the error variance estimate. A smaller error variance means a smaller denominator for the test statistic (F), and consequently, in most cases, a more sensitive design for detecting differences produced by the independent variable. We say "in most cases" because you may have noticed that there is also an accompanying reduction in the degrees of freedom for the error term—it was noted earlier that power is also directly related to sample size. For example, given a = 3 and n = 10, the critical comparison F (2, 27) at the .05 level for a between-subject design is 3.39. For the same values for 'a' and 'n' for a within-subject design, the critical comparison F (2, 18) is 3.55.

A potential problem with the within-subject design is that participating in an experiment may change how an individual will perform on subsequent occasions. This situational reactivity is particularly serious if the effects on behavior of being in Condition a_1 first and Condition a_2 second are not the same as being in Condition a_2 first and Condition a_1 second (differential transfer or carryover effects).

The single-factor, within-subject design has one feature in common with the two-factor, between-subject design. Both designs involve an interaction: A × B in the between-subject design and A × S in the within-subject design. For computational purposes, this difference in coding labels for the second factor does not impact the calculation of basic ratios ([B] or [S]), sums of squares (SS_B or SS_S), and mean squares (MS_B or MS_S). The major computational difference for between- and within-subject ANOVAs is that variation among subjects within each a_ib_j group can be estimated for the A × B design. For the single-factor A × S design, there is only one score for each subject in a given a_is_j condition, thus the notion of variability for a single subject score within a specific a_is_j condition is meaningless. The estimate of random error is $MS_{S/AB}$ for the A × B design. With only a single score in each a_is_j cell, a comparable $MS_{S/AS}$ for the A × S design is undefined, as [Y] – [AS] will be 0 for all data sets. For the A × B design, three F ratios could be calculated: F_A, F_B, and $F_{A \times B}$. For the A × S design, F_A can be calculated, but a meaningful F_S and $F_{A \times S}$ cannot be computed because we do not have an appropriate error term for evaluating the subject factor. $MS_{A \times S}$ does serve as the appropriate error term for computing an F ratio for Factor A: $F_A = MS_A \div MS_{A \times S}$.

As was done earlier for the single-factor between-subject design, computational procedures for a = 2 contrasts were illustrated using F, t for dependent samples, and confidence intervals for the difference between means. A second example problem with a = 4 was presented, followed by a description for using SPSS and examples for doing analytical comparisons. The A × S interaction provides the error term for the within-subject ANOVA, and given that an A × S interaction implies that subject variation is dependent on specific levels of Factor A, it follows that an estimate of error variance based on all levels of Factor A may not be an appropriate error term for comparing a limited set of A conditions. Thus, analytical comparisons for the within-subject design require calculating both SS_{Acomp} and $SS_{Acomp \times S}$ (and MS_{Acomp} and $MS_{Acomp \times S}$) from a data matrix derived from the original data set. The new data matrix was derived in two steps: (1) multiplying each score in the original data set by the coefficients defining the contrast of interest (e.g., if c_1 = -2, then each score in the a_1 condition has to be multiplied by -2), and (2) creating a new matrix consisting of two scores for each subject, with one being the sum of his or her positive signed scores and the other being the sum of his or her negative signed scores (the **positive and negative signs are used only as column labels, as negative numbers do not appear in the data matrix**). The sum of squares, mean squares and F ratio calculated from this new data matrix are the A_{comp} and $A_{comp \times S}$ effects: $F_{Acomp} = MS_{Acomp} \div MS_{Acomp \times S}$. A warning label was tacked on to these calculations to alert you to the fact that both MS estimates may be inflated by the same factor (e.g., $\Sigma c_i^2 \div 2$), which resulted from doing the analysis of variance after transforming the original scores. If in reporting the results of the statistical test of a given analytical comparison you need to provide the estimate of the actual error variance ($MS_{Acomp \times S}$), then you will need to **undo** the transformation operation by dividing the error variance you calculated by a correction factor—i.e., $MS_{Acomp \times S}$ (corrected) = $MS_{Acomp \times S} \div (\Sigma c_i^2 \div 2)$.

11.1: Model Problem 1. Assume the data below are from a class experiment in which the instructor gave her students (n = 10) a 10-word vocabulary quiz on two occasions (a = 2). For one experimental condition, the quiz was announced 24 hours in advance (a_1), and for the other condition, it was a surprise quiz (unannounced) (a_2). The scores below are number of correct responses out of a maximum score of 10. We will use these data to test H_0 that $a_1 = a_2$ three times: once for $F_{.05}$, once for $t_{.05}$, and once using $CI_{.95}$.

Fictitious Data for Announced (An) vs. Surprise (Su) Vocabulary Tests

	a_1 (An)	a_2 (Su)	ΣS_i
s_1	8	7	**15**
s_2	5	5	**10**
s_3	10	9	**19**
s_4	6	5	**11**
s_5	7	3	**10**
s_6	5	2	**7**
s_7	7	4	**11**
s_8	10	6	**16**
s_9	6	3	**9**
s_{10}	6	6	**12**
ΣA_j	**70**	**50**	

A. Single-Factor ANOVA (F):

Step 1. The Basic Ratios:

$[Y] = \Sigma Y_i^2 = 8^2 + 5^2 + \ldots + 6^2 = \mathbf{810.00}$

$[T] = (\Sigma Y_i)^2 \div (a)(n) = (120)^2 \div 20 = \mathbf{720.00}$

$[A] = \Sigma A_j^2 \div n = (70^2 + 50^2) \div 10 = \mathbf{740.00}$

$[S] = \Sigma S_i^2 \div a = (15^2 + 10^2 + \ldots + 12^2) \div 2 = 1{,}558 \div 2 = \mathbf{779.00}$

Step 2. The Sums of Squares (SS):

$SS_A = [A] - [T] = 740.00 - 720.00 = \mathbf{20.00}$

$SS_S = [S] - [T] = 779.00 - 720.00 = \mathbf{59.00}$

$SS_{A \times S} = [Y] - [A] - [S] + [T] = 810.00 - 740.00 - 779.00 + 720.00 = \mathbf{11.00}$

Step 3. Degrees of Freedom (df):

$$df_A = a - 1 = 2 - 1 = \mathbf{1}$$

$$df_S = (n - 1) = (10 - 1) = \mathbf{9}$$

$$df_{A\times S} = (a - 1)(n - 1) = (1)(9) = \mathbf{9}$$

Step 4. The Variances (MS):

$$MS_A = SS_A \div df_A = 20.00 \div 1 = \mathbf{20.00}$$

$$MS_S = SS_S \div df_S = 59.00 \div 9 = \mathbf{6.56}$$

$$MS_{A\times S} = SS_{A\times S} \div df_{A\times S} = 11.00 \div 9 = \mathbf{1.22}$$

Step 5. The F ratio:

$$F = MS_A \div MS_{A\times S} = 20.00 \div 1.22 = \mathbf{16.39}$$

Step 6. The ANOVA summary table:

Analysis of Variance Summary Table

Source	SS	df	MS	F	
A	20.00	1	20.00	16.39	$p < .05$
S	59.00	9	6.56		
A × S	11.00	9	1.22		

B. Paired-Sample t: Since $F = t^2$, we already know that within rounding error, t will be the square root of F: $t = \sqrt{16.39} = 4.05$; and further, that this value of t will be significant at $p \leq .05$. The formula for the paired-sample t is

$$t = (\bar{Y}_{A1} - \bar{Y}_{A2}) \div \sqrt{(MS_D \div n)}.$$

The $a_1 - a_2$ scores (D_i) for the 10 subjects are $D_1 = 1$, $D_2 = 0$, $D_3 = 1$, $D_4 = 1$, $D_5 = 4$, $D_6 = 3$, $D_7 = 3$, $D_8 = 4$, $D_9 = 3$, and $D_{10} = 0$.

$$\bar{Y}_{A1} = 7.00,\ \bar{Y}_{A2} = 5.00, \text{ and } \bar{Y}_{A1} - \bar{Y}_{A2} = 7.00 - 5.00 = \mathbf{2.00}$$

$$SS_D = [D] - [T] = (1^2 + 0^2 + 1^2 + 1^2 + 4^2 + 3^2 + 3^2 + 4^2 + 3^2 + 0^2) - (20^2 \div 10) = 62.00 - 40.00 = \mathbf{22.00}$$

$MS_D = SS_D \div (n - 1) = 22.00 \div 9 = \mathbf{2.44}$;

$t = 2.00 \div \sqrt{(2.44 \div 10)} = 2.00 \div \sqrt{.24} = 2.00 \div .4940 = \mathbf{4.05}, p < .05$.

C. The .95 Confidence Interval: Given that $\bar{Y}_{A1}$ is "significantly" larger than $\bar{Y}_{A2}$, we know that the corresponding $CI_{.95}$ must specify an interval that does not include 0.

$CI_{.95}$: $(\bar{Y}_{A1} - \bar{Y}_{A2}) - (t_{.05})\sqrt{(MS_D \div n)} \leq \mu_2 - \mu_1 \leq (\bar{Y}_{A1} - \bar{Y}_{A2}) + (t_{.05})\sqrt{(MS_D \div n)}$

The critical $t_{.05}(9)$ from Table A–3 is **2.26**; $(\bar{Y}_{A1} - \bar{Y}_{A2}) = \mathbf{2.00}$, and $MS_D = \mathbf{2.44}$. Therefore,

$CI_{.95}$: $2.00 - (2.26)\sqrt{(2.44 \div 10)} \leq \mu_1 - \mu_2 \leq (2.00) + (2.26)(\sqrt{(2.44 \div 10)}$

$CI_{.95}$: $2.00 - (2.26)\sqrt{.244} \leq \mu_1 - \mu_2 \leq 2.00 + (2.26)\sqrt{.244}$

$CI_{.95}$: $2.00 - (2.26)(.4940) \leq \mu_1 - \mu_2 \leq 2.00 + (2.26)(.4940)$

$CI_{.95}$: $2.00 - 1.12 \leq \mu_1 - \mu_2 \leq 2.00 + 1.12$

$\mathbf{CI_{.95}: .88 \leq \mu_1 - \mu_2 \leq 3.12}$, which does not include 0, a result consistent with the significant difference between the a_1 and a_2 conditions reported with the ANOVA and paired sample t test.

11.2: Model Problem 2. For this problem we will assume a researcher is interested in tracking performance on a standard IQ test for seven underprivileged children who participated in a pre-school learning enrichment program. The students were tested on the first day they started the program (a_1), after they had been in the program for six months (a_2), and after they had been in the program for 12 months. Assume the scores in the following Table are from three comparable IQ tests for each of the seven children.

A. Is there a significant difference among the three conditions of this experiment?

	a_1(Test 1)	a_2(Test 2)	a_3(Test 3)	ΣS_j
S_1	85	90	90	**265**
S_2	90	90	80	**260**
S_3	70	90	100	**260**
S_4	110	100	110	**320**
S_5	75	80	85	**240**
S_6	95	100	110	**305**
S_7	100	85	100	**285**
ΣA_i	**625**	**635**	**675**	

The step-wise computations necessary for calculating F are as follows:

Step 1. The Basic Ratios:

$[Y] = \Sigma Y_i^2 = 85^2 + 90^2 + \ldots + 100^2 = \mathbf{180{,}825.00}$

$[T] = (\Sigma Y_i)^2 \div (a)(n) = (1{,}935)^2 \div 21 = \mathbf{178{,}296.43}$

$[A] = \Sigma A_j^2 \div n = (625^2 + 635^2 + 675^2) \div 7 = \mathbf{178{,}496.43}$

$[S] = \Sigma S_i^2 \div a = (265^2 + 260^2 + 260^2 + 320^2 + 240^2 + 305^2 + 285^2) \div 3$

$[S] = \mathbf{179{,}891.67}$

Step 2. The Sums of Squares (SS):

$SS_A = [A] - [T] = 178{,}496.43 - 178{,}296.43 = \mathbf{200.00}$

$SS_S = [S] - [T] = 179{,}891.67 - 178{,}296.43 = \mathbf{1{,}595.24}$

$SS_{A\times S} = [Y] - [A] - [S] + [T] = 180{,}825.00 - 178{,}496.43 - 179{,}891.67 + 178{,}296.43 = \mathbf{733.33}$

Step 3. Degrees of Freedom (df):

$df_A = 3 - 1 = \mathbf{2}$; $df_S = n - 1 = 7 - 1 = \mathbf{6}$; $df_{A\times S} = (a - 1)(n - 1) = (2)(6) = \mathbf{12}$

Step 4. The Variances (MS):

$MS_A = SS_A \div df_A = 200.00 \div 2 = \mathbf{100.00}$

$MS_S = SS_S \div df_S = 1{,}595.24 \div 6 = \mathbf{265.87}$

$MS_{A\times S} = SS_{A\times S} \div df_{A\times S} = 733.33 \div 12 = \mathbf{61.11}$

Step 5. The ANOVA Summary Table and F ratio:

Source	SS	df	MS	F	
A	200.00	2	100.00	1.64	not significant
S	1,595.24	6	265.87		
A × S	733.33	12	61.11		

B. Is there a significant difference in IQ scores before the program started (a_1) and one year later (a_3)? Even though the difference among these three conditions was not significant, for practice we will do an analytical comparison for this maximum "before" and "after" difference. We will follow the two-step procedure of transforming individual scores to get the data ready for an ANOVA with a = 2. (This is the "long" solution method for this particular comparison, as it was shown in the text that a comparison between two intact groups from an original data set can be done by extracting the data from the two groups under consideration and doing an ANOVA on the extracted data set.) Individual scores will be transformed by multiplying them by the appropriate weighting coefficients: for an a_1 vs. a_3 contrast, set $c_1 = +1$, $c_2 = 0$ and $c_3 = -1$.

Step 1: The Transformed Data Matrix

	a_1(Test 1)	a_2(Test 2)	a_3(Test 3)
s_1	85	0	-90
s_2	90	0	-80
s_3	70	0	-100
s_4	110	0	-110
s_5	75	0	-85
s_6	95	0	-110
s_7	100	0	-100

Step 2: Combine Same-Signed Transformed Scores to Create Two Conditions (at this point it should be obvious that this procedure is comparable to extracting the data from the two conditions under consideration, as the +Sign and -Sign columns are identical to the a_1 and a_3 scores in the original data set.)

	+Sign	-Sign	ΣS_j
s_1	85	90	**175**
s_2	90	80	**170**
s_3	70	100	**170**
s_4	110	110	**220**
s_5	75	85	**160**
s_6	95	110	**205**
s_7	100	100	**200**
ΣA_i	**625**	**675**	

The ANOVA:

Step 1. The Basic Ratios:

$$[Y] = \Sigma Y_i^2 = 85^2 + 90^2 + \ldots + 100^2 = \mathbf{122{,}900.00}$$

$[T] = (\Sigma Y_i)^2 \div (a)(n) = (1{,}300)^2 \div 14 = \mathbf{120{,}714.29}$

$[A] = \Sigma A_j^2 \div n = (625^2 + 675^2) \div 7 = \mathbf{120{,}892.86}$

$[S] = (175^2 + 170^2 + 170^2 + 220^2 + 160^2 + 205^2 + 200^2) \div 2 = \mathbf{122{,}225.00}$

Step 2. The Sums of Squares (SS):

$SS_A = [A] - [T] = 120{,}892.86 - 120{,}714.29 = \mathbf{178.57}$

$SS_S = [S] - [T] = 122{,}225.00 - 120{,}714.29 = \mathbf{1{,}510.71}$

$SS_{A\times S} = [Y] - [A] - [S] + [T] = 122{,}900.00 - 120{,}892.86 - 122{,}225.00 + 120{,}714.29 = \mathbf{496.43}$

Step 3. Degrees of Freedom (df):

$df_A = 2 - 1 = \mathbf{1}$; $df_S = n - 1 = 7 - 1 = \mathbf{6}$; $df_{A\times S} = (a - 1)(n - 1) = (1)(6) = \mathbf{6}$

Step 4. The Variances (MS):

$MS_A = SS_A \div df_A = 178.57 \div 1 = \mathbf{178.57}$

$MS_S = SS_S \div df_S = 1{,}510.71 \div 6 = \mathbf{251.78}$

$MS_{A\times S} = SS_{A\times S} \div df_{A\times S} = 496.43 \div 6 = \mathbf{82.74}$

Step 5. The ANOVA Summary Table and F_{comp}:

Source	SS	df	MS	F	
A_{comp}	178.57	1	178.57	2.16	not significant
S	1,510.71	6	251.78		
$A_{comp} \times S$	496.43	6	82.74		

11.3: Practice Problem 1. Tables 6.7 and 6.8 from Chapter 6 (pages 106 and 107) present data and an ANOVA summary table, respectively, for a single-factor, between-subject design. These two tables are copied below, and your task is to re-analyze the data as if they came from a single-factor, within-subject design, presenting the results of your analysis in an ANOVA summary table.

From Table 6.7: Data Matrix

a_1	a_2	a_3	a_4
3	7	6	7
5	5	7	4
2	9	11	8
4	6	10	8
7	6	8	5
5	9	6	3
1	8	5	6
4	4	9	7
6	5	9	6
2	5	10	8

From Table 6.8: Analysis of Variance Summary Table

Source	SS	df	MS	F	
A	89.30	3	29.77	8.53	$p < .05$
S/A	125.80	36	3.49		

11.4: Practice Problem 2. Assume the data in Table 11.2 (page 214) came from a single-factor, within-subject experiment involving a **quantitative** independent variable with equally spaced intervals. The ANOVA was done for these data, so for this problem you are asked to do a trend analysis: (1) Is there a significant linear trend? (2) Is there a significant quadratic trend?

11.5: Practice Problem 3. A student did an experiment with 5 subjects ($n = 5$), with each subject serving in all 3 conditions of her experiment ($a = 3$). The data are presented below.

	a_1	a_2	a_3	ΣS_j
s_1	4	8	5	**17**
s_2	5	4	3	**12**
s_3	10	6	5	**21**
s_4	6	5	3	**14**
s_5	4	2	4	**10**
ΣA_i	**29**	**25**	**20**	

1. Is there a significant difference at $p \leq .05$ among these three conditions?
2. Determine whether F_{Acomp1} involving the complex comparison contrasting a_1 and a_2 as a "single condition" with the a_3 condition ($a_1 + a_2$ vs. a_3) is significant at $p \leq .05$.
3. What is the corrected value of the $MS_{Acomp \times S}$ for the "$a_1 + a_2$ vs. a_3" analytical comparison in part 2?

11.6: Practice Problem 4. A social psychologist was interested in the accumulative impact of an "egalitarian" mode of interaction between teachers and students on academic achievement. Six randomly selected students (s_1 through s_6) were observed over five one-week sessions. Their respective achievement scores appear in the data matrix below:

Weeks (Factor A)

	a_1	a_2	a_3	a_4	a_5	ΣS_j
s_1	8	9	7	7	10	**41**
s_2	4	3	6	5	9	**27**
s_3	7	4	4	4	7	**26**
s_4	3	3	2	1	4	**13**
s_5	9	8	7	5	10	**39**
s_6	6	6	4	3	9	**28**
ΣA_i	**37**	**33**	**30**	**25**	**49**	

A. Are the observed differences in achievement scores over the five-week period significant at $p \leq .05$?

B. Do the appropriate follow-up contrasts to determine if the linear trend ($F_{ALinear}$) and the quadratic trend ($F_{Aquadratic}$) are significant at $\alpha \leq .05$.

Chapter 12

Two-Factor Designs with One or Two Within-Subject Factors

Do not put faith in what statistics say until you have carefully considered what they do not say.

—Watt

I. Three Variations for Two-Factor Designs

With two independent variables manipulated in an experiment and the possibility of either between-subject or within-subject procedures for representing levels of independent variables, three design variations may be considered. The two-factor design was introduced in Chapter 9 in a between-subject format (subjects randomly assigned to only one a_ib_j group). A single-factor within-subject design, for which each subject participates once in each condition of the experiment, was introduced in Chapter 11. These two variations in how subjects participate in an experiment actually give rise to three different design combinations with two independent variables. In addition to the two-factor random groups design introduced in Chapter 9, we may now add two new variations: (1) a within-subject two-factor design in which each subject serves in every combination of a_ib_j conditions, and (2) a "mixed" two-factor design in which different subjects are randomly assigned to one and only one condition defined by one of the two independent variables (e.g., Factor A is a "between-subject" factor)—however, each subject participates in every condition defined by the second independent variable (e.g., Factor B is a "within-subject" factor). Assigning subjects to conditions with this procedure defines a two-factor *mixed design*, as the design is a mixture of one between-subject independent variable and one within-subject independent variable. Not surprisingly, these variations in design are associated with different analysis of variance procedures. The purpose of this chapter is to introduce procedures for calculating F ratios for two independent variables (Factor A and Factor B) and the interaction (A × B) for (1) the two-factor mixed design and (2) the two-factor within-subject design.

Whenever subjects serve in all experimental conditions defined by an independent variable, variability due to subjects can be removed from the estimate of random error in the denominator of the appropriate F ratio. The analysis of variance procedures for the two-factor between-subject design were described in Chapter 9. With two more variations on this design, it is important to recognize the distinguishing features of each. Three brief descriptions of experiments follow. The one that indicates that the different levels of both Factor A and Factor B are represented by different groups of individuals is a between-subject design, the one that indicates that the same individuals participate in all a_ib_j conditions is a complete two-factor within-subject design, and the one with different groups of subjects for the different levels of one independent variable but with each subject participating in all conditions defined by the other independent variable is a mixed two-factor design.

Experimental Brief #1: Factor A consists of four levels of shock intensity ($a = 4$), and Factor B consists of three discrimination tasks of differing difficulty ($b = 3$). Sixty rats are randomly assigned to 12 experimental groups ($4 \times 3 = 12$), such that five different rats ($n = 5$) are in each a_ib_j group. Is this a two-factor between-subject design, a two-factor within-subject design, or a two-factor mixed design? This brief description informs us that 60 different rats were separated into 12 groups, with five different rats in each group. With $n = 5$, this has to be a two-factor, between-subject design. How would you change the description of shock intensity (Factor A) and task difficulty (Factor B) to make this a two-factor within-subject example or a two-factor mixed design example?

Experimental Brief #2: The following is a brief description adapted from a practice problem taken from Keppel and Wickens (2004, page 429). An instructor did an experiment on stereotyping. She prepared six biographical sketches describing a male (a_1) or female (a_2) working as a university professor (b_1), a CEO of a small company (b_2), or a State politician (b_3). A booklet with these six sketches was distributed to all of the students in her beginning psychology class ($n = 50$), and each student in the class was asked to rate how much he or she admired the individual described in the biographical sketch. Is this a two-factor between-subject design, a two-factor within-subject design, or a two-factor mixed design? This brief description informs us that 50 students in a class rated six biographical sketches (e.g., one for a male professor, one for a female professor, one for a male CEO of a small company, etc.). That is, each of the 50 students provided a rating for each of the six experimental conditions created by the 2 (sex) by 3 (profession) combination of Factor A and Factor B. Thus, this brief describes a two-factor within-subject design. How would you change the description to make gender and profession a two-factor between-subject example or a two-factor mixed design example?

Experimental Brief #3: Subjects performed a vigilance task involving detection of targets on a radar screen during a 40-minute test session. Factor A consisted of two types of displays: white targets on a black background (a_1) or black targets on a white background (a_2). A total of 60 students participated in this experiment, with 30 students randomly assigned to the "white-on-black" display group and the other 30 students randomly assigned to the "black-on-white" display group. The experimenter was also interested in the effects of practice on detection accuracy. Each subject in the white-on-black group and each subject in the black-on-white group was tested on four successive days (b = 4), with the assumption that practice effects, if any, would improve performance across the four days. Is this a two-factor between-subject design, a two-factor within-subject design, or a two-factor mixed design? This brief description informs us that 60 students were separated into two groups, with half viewing white-on-black displays (Group a_1) and half viewing black-on-white displays (Group a_2)—clearly a between-subject manipulation. However, all subjects had four practice trials, indicating that Factor B was a within-subject variable. How would you change the description of target displays and practice to make this a two-factor between-subject example or a two-factor within-subject example? An experiment with practice as a between-subject variable would be quite unusual; however, it can be done.

An important difference for independent variables manipulated between- vs. within-subjects is that there are multiple scores for each subject across the different levels of a within-subject variable, and it is meaningful to combine scores for the same subject when assessing subject variability. For the analysis of variance this means that there will be one or more meaningful ways to sum scores for subjects to calculate basic ratios such as [AS] for a mixed design; and [S], [AS], and [BS] for a complete within-subject design. For all three variations, an analysis of variance provides statistical tests of the null hypothesis for Factor A, Factor B, and the A × B interaction; however, **different error terms** must be calculated for the different designs, reflecting whether variation due to individual differences among subjects can be removed from the appropriate estimate of random error.

II. The ANOVA for the Mixed Two-Factor Design

To illustrate computational procedures for the mixed two-factor design we will consider an experiment for which subjects serve in only one Factor A condition (a between-subject variable), but in all Factor B conditions (a within-subject variable). This combination means that a subject who is randomly assigned to the a_1 group will not experience other levels of Factor A; however, each subject in an a_i group will provide multiple scores (one score for each of the b_j conditions): i.e., an a_1 subject will provide an a_1b_1 score, an a_1b_2 score, an a_1b_3 score, etc. Thus, one new calculation (compared to calculations required for the two-factor, between-subject ANOVA from Chapter 9) is required for the two-factor mixed design. A basic ratio based on a total score for each individual subject can now be computed, and with repeated measures on Factor B, this basic ratio is [AS] (e.g., a_1s_1 is the combination of $a_1b_1s_1$ + $a_1b_2s_1$ + … + $a_1b_js_1$). This calculation is used to estimate two error terms. The first error term is for the between-subject factor (Factor A); this estimate of random error includes variation due to individual differences (subjects "nested" within the different A groups, hence the subject factor in a mixed two-factor design is coded as **S/A**). The second error term is for evaluating Factor B and the A × B interaction. This estimate has subject variation removed from random error (this random error estimate is the interaction of the within-subject Factor B and the S/A subject factor, with the variance estimate coded as $\mathbf{MS}_{\mathbf{B \times S/A}}$).

You should notice as we go through the computational steps for the two-factor mixed ANOVA that in addition to calculating basic ratios for [A], [B], [T], [AB], and [Y] as was done with the two-factor between-subject design, a new [AS] basic ratio appears. Three sums of squares, SS_A, SS_B, and $SS_{A \times B}$, are computed as before; however, $SS_{S/AB}$ (sum of squares for the error term in the two-factor between-subject design) is replaced by two new sums of squares: $SS_{S/A}$, the error term (denominator for F) for the between-subject factor and $SS_{B \times S/A}$, the error term for the within-subject factor and the A × B interaction. Using data from the two-factor between-subject design in Chapter 10 for this illustration has a couple of advantages: (1) most of the computational work has already been done (e.g., squaring and summing each individual score to calculate [Y]) and (2) by using the same data we can see that the sum of squares for random error in the between-subject design ($SS_{S/AB}$) is actually divided into two component parts ($SS_{S/AB} = SS_{S/A} + SS_{B \times S/A}$) in the two-factor mixed design. The data from Table 10.4 (page 193) are reorganized to accommodate the change from a two-factor between-subject design to a two-factor mixed design. These data are presented in Table 12.1, showing that every subject (s_1 through s_{12}) in the experiment has a score for all Factor B levels (different subscripts for "s" are used to denote different subjects).

Table 12.1: Fictitious Data for a 3 × 3 Mixed Design

	a_1b_1	a_1b_2	a_1b_3	$\Sigma\Sigma A_1S_k$		a_2b_1	a_2b_2	a_2b_3	$\Sigma\Sigma A_2S_k$		a_3b_1	a_3b_2	a_3b_3	$\Sigma\Sigma A_3S_k$
s_1	3	4	2	**9**	s_5	9	6	5	**20**	s_9	7	6	10	**23**
s_2	6	4	6	**16**	s_6	7	7	7	**21**	s_{10}	2	7	9	**18**
s_3	2	8	5	**15**	s_7	5	8	1	**14**	s_{11}	3	6	13	**22**
s_4	4	4	7	**15**	s_8	9	4	5	**18**	s_{12}	8	9	8	**25**
Σ	**15**	**20**	**20**	**55**		**30**	**25**	**18**	**73**		**20**	**28**	**40**	**88**

AxB	**a_1**	**a_2**	**a_3**	
b_1	15	30	20	**65**
b_2	20	25	28	**73**
b_3	20	18	40	**78**
	55	73	88	**216**

Step 1: Calculate the Six Basic Ratios (make sure you understand where each squared sum and the denominators in the following formulas came from—e.g., the "55" in [A] is the sum of the 12 a_1 scores (3 + 6 + … + 5 + 7 = 55), and the denominator, b × n (3 × 4 = 12) is the number of individual a_1 scores that were summed.)

$[A] = \Sigma A_i^2 \div (b)(n) = (55^2 + 73^2 + 88^2) \div (3)(4) = 16{,}098 \div 12 = \mathbf{1{,}341.50}$

$[B] = \Sigma B_j^2 \div (a)(n) = (65^2 + 73^2 + 78^2) \div (3)(4) = 15{,}638 \div 12 = \mathbf{1{,}303.17}$

$[T] = (\Sigma\Sigma A_i B_j)^2 \div (a)(b)(n) = 216^2 \div (3)(3)(4) = 46{,}656 \div 36 = \mathbf{1{,}296.00}$

$[AB] = \Sigma\Sigma\, (A_i B_j)^2 \div n = (15^2 + 20^2 + \ldots + 40^2) \div 4 = 5{,}658 \div 4 = \mathbf{1{,}414.50}$

$[Y] = \Sigma\Sigma\Sigma Y_{i,j,k}^{\ 2} = 3^2 + 6^2 + \ldots + 8^2 = \mathbf{1{,}534.00}$

For the "new" basic ratio (1) sum all a_i scores for each s_k subject; (2) square each of the summed $A_i S_k$ scores; (3) sum the squared $A_i S_k$ scores; and (4) divide the resulting sum of squared scores by the number of b_j scores that had to be combined to get each a_i sum. $[AS] = \Sigma\Sigma\, (A_i S_k)^2 \div b$. To calculate [AS] it is helpful to reorganize the data in an AS matrix as shown below, where $a_1 s_1 = 3 + 4 + 2 = 9$, $a_1 s_2 = 6 + 4 + 6 = 16 \ldots$ and $a_3 s_{12} = 8 + 9 + 8 = 25$.

	a_1		a_2		a_3
s_1	9	s_5	20	s_9	23
s_2	16	s_6	21	s_{10}	18
s_3	15	s_7	14	s_{11}	22
s_4	15	s_8	18	s_{12}	25

$[AS] = \Sigma(A_i S_k)^2 \div b = (9^2 + 16^2 + \ldots + 25^2) \div 3 = 4{,}110 \div 3 = \mathbf{1{,}370.00}$

Step 2: Calculate the five Sums of Squares (Note the new formulas for $SS_{S/A}$ and for $SS_{B\times S/A}$)

$SS_A = [A] - [T] = 1{,}341.50 - 1{,}296.00 = \mathbf{45.50}$

$SS_B = [B] - [T] = 1{,}303.17 - 1{,}296.00 = \mathbf{7.17}$

$SS_{AxB} = [AB] - [A] - [B] + [T] = 1{,}414.50 - 1{,}341.50 - 1{,}303.17 + 1{,}296.00 = \mathbf{65.83}$

$SS_{S/A} = [AS] - [A] = 1{,}370.00 - 1{,}341.50 = \mathbf{28.50}$

$SS_{B\times S/A} = [Y] - [AB] - [AS] + [A] = 1{,}534.00 - 1{,}414.50 - 1{,}370.00 + 1{,}341.50 = \mathbf{91.00}$

Step 3: Degrees of Freedom: $df_A = a - 1 = 3 - 1 = \mathbf{2}$; $df_{S/A} = (a)(n - 1) = (3)(4 - 1) = \mathbf{9}$; $df_B = b - 1 = 3 - 1 = \mathbf{2}$; $df_{A\times B} = (a - 1)(b - 1) = (2)(2) = \mathbf{4}$; $df_{B\times S/A} = (b - 1)(a)(n - 1) = (2)(3)(3) = \mathbf{18}$

Step 4: Calculate the five Mean Squares

$MS_A = SS_A \div df_A = 45.50 \div 2 = \mathbf{22.75}$

$MS_B = SS_B \div df_B = 7.17 \div 2 = \mathbf{3.58}$

$MS_{A\times B} = SS_{AxB} \div df_{AxB} = 65.83 \div 4 = \mathbf{16.46}$

$MS_{S/A} = SS_{S/A} \div df_{S/A} = 28.50 \div 9 = \mathbf{3.17}$

$MS_{B\times S/A} = SS_{B\times S/A} \div df_{B\times S/A} = 91.00 \div 18 = \mathbf{5.06}$

Step 5: Calculate the three F ratios by dividing each mean square by the appropriate error term ($F_A = MS_A \div MS_{S/A} = 22.75 \div 3.17 = 7.18$; $F_B = MS_B \div MS_{B\times S/A} = 3.58 \div 5.06 = .71$; $F_{A\times B} = MS_{A\times B} \div MS_{B\times S/A} = 16.46 \div 5.06 = 3.25$)

Step 6: Report the results of your analysis in an analysis of variance summary table.

ANOVA Summary Table

Source	SS	df	MS	F	F_{crit}
A	45.50	2	22.75	7.18 $p < .05$	4.26
S/A (error)	28.50	9	3.17		
B	7.17	2	3.58	.71	3.55
A × B	65.83	4	16.46	3.25 $p < .05$	2.93
B × S/A (error)	91.00	18	5.06		

Inspection of the between-subject ANOVA for these data (Table 10.5, page 193) reveals that MS_A, MS_B, and $MS_{A\times B}$ are unchanged, and $SS_{S/AB} = SS_{S/A} + SS_{B\times S/A}$ (119.50 = 28.50 + 91.00; as does $df_{S/AB} = df_{S/A} + df_{B\times S/A}$). This example problem is based on made-up numbers, with only a change in the storyline to illustrate computations for a mixed two-factor analysis of variance problem. Real data would most likely show a stronger relationship among scores from the same subject, and hence a smaller error estimate for Factor B and the A × B interaction compared to the error estimate for Factor A (i.e., $MS_{B\times S/A}$ would likely be smaller than $MS_{S/A}$).

III. The ANOVA for the Within-Subject Two-Factor Design

With the two-factor within-subject design the same subjects participate in all conditions of the experiment (each subject is in every a_ib_j condition). The analysis of variance for the two-factor within-subject design may be set in the context of procedures introduced for conducting an analysis of variance for the two-factor between-subject design and two-factor mixed design. From the computational perspective, what is old and what is new for the two-factor within-subject design? For both of the previous designs considered, [A], [B], [T], [AB], and [Y] had to be calculated, and the same is true for the two-factor within-subject design. Also, for the two-factor mixed design, [AS] had to be calculated, and this basic ratio will again have to be computed in the same way for the two-factor, within-subject design. These "old" basic ratio computations are necessary for determining SS_A, SS_B, and $SS_{A\times B}$, and they are calculated as before. Two new basic ratios ([S] and [BS]) and four new sums of squares (SS_S, $SS_{A\times S}$, $SS_{B\times S}$, and $SS_{A\times B\times S}$) are calculated in the process of completing a two-factor within-subject analysis of variance. In this design, a total score may be obtained for each subject (for this example data set when viewed as a within subject design, each subject contributes nine scores—e.g., the sum of the nine scores for the first subject is $\Sigma S_1 = 3 + 4 + 2 + 9 + 6 + 5 + 7 + 6 + 10 = 52$), which provides the basis for calculating the basic ratio for S [S]. You will recall from the two-factor mixed design that a new basic ratio [AS] was computed, as each subject contributed three scores to each a_i condition (e.g., $\mathbf{a_1}b_1s_1 + \mathbf{a_1}b_2s_1 + \mathbf{a_1}b_3s_1 = a_1s_1 = 3 + 4 + 2 = 9$). This basic ratio [AS] can also be computed for the two-factor within-subject design. In addition, a new, similar basic ratio [BS] can be calculated that involves Factor B, as each subject now provides a total score restricted for specific b_j conditions (e.g., $a_1\mathbf{b_1}s_1 + a_2\mathbf{b_1}s_1 + a_3\mathbf{b_1}s_1 = b_1s_1 = 3 + 9 + 7 = 19$). The formulas for the new sums of squares appear below:

$$SS_S = [S] - [T], \text{ where } [S] = \Sigma S_k^2 \div (a)(b)$$

$$SS_{A\times S} = [AS] - [A] - [S] + [T], \text{ where } [AS] = \Sigma\Sigma(A_iS_k)^2 \div b$$

$$SS_{B\times S} = [BS] - [B] - [S] + [T], \text{ where } [BS] = \Sigma\Sigma(B_jS_k)^2 \div a$$

$$SS_{A\times B\times S} = [Y] - [AB] - [AS] - [BS] + [A] + [B] + [S] - [T]$$

You may recall from the single-factor within-subject analysis of variance introduced in Chapter 11 that the error term for evaluating the effect of Factor A was the interaction between Factor A and the subject factor ($F_A = MS_A \div MS_{A\times S}$). For the two-factor, within-subject analysis of variance, each effect has a different error term based on the interaction of the effect in question with the subject factor. The error term (denominator in the F ratio) for Factor A is $MS_{A\times S}$, for Factor B the corresponding error term is $MS_{B\times S}$, and for the A × B interaction, the appropriate error term is $MS_{A\times B\times S}$ (in summary, $F_A = MS_A \div MS_{A\times S}$, $F_B = MS_B \div MS_{B\times S}$, and $F_{A\times B} = MS_{A\times B} \div MS_{A\times B\times S}$). A key to completing the ANOVA successfully lies with reorganizing data appropriately for each basic ratio that must be calculated. As we consider an example problem, an appropriate reorganized data matrix for calculating [BS] will accompany the computational steps for calculating the basic ratio. As was the case for the single-factor, within-subject design, there is only one score in each $a_ib_js_k$ cell, hence $SS_{S/ABS} = [Y] - [ABS] = 0$ (the same set of numbers are squared and summed in calculating both [Y] and [ABS]). The number set in Table 12.2 will be used to illustrate computational procedures for the two-factor within-subject analysis of variance. This number set should be familiar by now; it is an appropriately reorganized version of the data from Table 12.1. However, in this table each row in the data matrix represents scores from just one of only four different participants in the experiment (s_1 through s_4).

Table 12.2: Fictitious Data for a 3 × 3 Within-Subject Design

	a_1b_1	a_1b_2	a_1b_3	$\Sigma\Sigma A_1S_k$	a_2b_1	a_2b_2	a_2b_3	$\Sigma\Sigma A_2S_k$	a_3b_1	a_3b_2	a_3b_3	$\Sigma\Sigma A_3S_k$	ΣS_k
s_1	3	4	2	**9**	9	6	5	**20**	7	6	10	**23**	**52**
s_2	6	4	6	**16**	7	7	7	**21**	2	7	9	**18**	**55**
s_3	2	8	5	**15**	5	8	1	**14**	3	6	13	**22**	**51**
s_4	4	4	7	**15**	9	4	5	**18**	8	9	8	**25**	**58**
Σ	**15**	**20**	**20**	**55**	**30**	**25**	**18**	**73**	**20**	**28**	**40**	**88**	**216**

With each of the four subjects serving in all a_i conditions, it is now meaningful to sum a given subject's scores at each level of Factor B (e.g., $b_1s_1 = a_1b_1s_1 + a_2b_1s_1 + a_3b_1s_1 = 3 + 9 + 7 = 19$). The sums for B_j are $\mathbf{\Sigma Y_{B1} = 65, \Sigma Y_{B2} = 73, \Sigma Y_{B3} = 78}$.

Step 1: Calculate the Eight Basic Ratios

$[A] = \Sigma A_i^2 \div (b)(n) = (55^2 + 73^2 + 88^2) \div (3)(4) = 16{,}098 \div 12 = \mathbf{1{,}341.50}$

$[B] = \Sigma B_j^2 \div (a)(n) = (65^2 + 73^2 + 78^2) \div (3)(4) = 15{,}638 \div 12 = \mathbf{1{,}303.17}$

$[T] = (\Sigma\Sigma A_iB_j)^2 \div (a)(b)(n) = 216^2 \div (3)(3)(4) = 46{,}656 \div 36 = \mathbf{1{,}296.00}$

$[S] = \Sigma S_k^2 \div (a)(b) = (52^2 + 55^2 + 51^2 + 58^2) \div 9 = 11{,}694 \div 9 = \mathbf{1{,}299.33}$

$[AB] = \Sigma\Sigma(A_iB_j)^2 \div n = (15^2 + 20^2 + \ldots + 40^2) \div 4 = 5{,}658 \div 4 = \mathbf{1{,}414.50}$

$[AS] = \Sigma\Sigma(A_iS_k)^2 \div b = (9^2 + 16^2 + \ldots + 25^2) \div 3 = 4{,}110 \div 3 = \mathbf{1{,}370.00}$

$[Y] = \Sigma\Sigma\Sigma Y_{i,j,k}^{\ 2} = 3^2 + 6^2 + \ldots + 8^2 = \mathbf{1{,}534.00}$

$[BS] = \Sigma\Sigma(B_jS_k)^2 \div a$; to calculate [BS] it may be helpful to reorganize the data into a BS matrix so that the various b_js_k sums are correctly identified, e.g., $a_1b_1s_1 + a_2b_1s_1 + a_3b_1s_1 = 3 + 9 + 7 = 19$, which is the entry for b_1s_1 in the reorganized BS data matrix:

	b_1	b_2	b_3
s_1	19	16	17
s_2	15	18	22
s_3	10	22	19
s_4	21	17	20

$[BS] = \Sigma\Sigma(B_jS_k)^2 \div a = (19^2 + 15^2 + \ldots + 20^2) \div 3 = 4{,}014 \div 3 = \mathbf{1{,}338.00}$

Step 2: Calculate the Seven Sums of Squares

$SS_A = [A] - [T] = 1{,}341.50 - 1{,}296.00 = \mathbf{45.50}$

$SS_B = [B] - [T] = 1{,}303.17 - 1{,}296.00 = \mathbf{7.17}$

$SS_S = [S] - [T] = 1{,}299.33 - 1{,}296.00 = \mathbf{3.33}$

$SS_{A\times B} = [AB] - [A] - [B] + [T] = 1{,}414.50 - 1{,}341.50 - 1{,}303.17 + 1{,}296.00 = \mathbf{65.83}$

$SS_{A\times S} = [AS] - [A] - [S] + [T] = 1{,}370.00 - 1{,}341.50 - 1{,}299.33 + 1{,}296.00 = \mathbf{25.17}$

$SS_{B\times S} = [BS] - [B] - [S] + [T] = 1{,}338.00 - 1{,}303.17 - 1{,}299.33 + 1{,}296.00 = \mathbf{31.5}$

$SS_{A\times B\times S} = [Y] - [AB] - [AS] - [BS] + [A] + [B] + [S] - [T] = 1{,}534.00 - 1{,}414.50 - 1{,}370.00 - 1{,}338.00 + 1{,}341.50 + 1{,}303.17 + 1{,}299.33 - 1{,}296.00 = \mathbf{59.50}$

Step 3: Degrees of Freedom: $df_S = n - 1 = 4 - 1 = \mathbf{3}$; $df_A = a - 1 = 3 - 1 = \mathbf{2}$; $df_{A\times S} = (a - 1)(n - 1) = (3 - 1)(4 - 1) = \mathbf{6}$; $df_B = b - 1 = 3 - 1 = \mathbf{2}$; $df_{B\times S} = (b - 1)(n - 1) = (3 - 1)(4 - 1) = \mathbf{6}$; $df_{A\times B} = (a - 1)(b - 1) = (3 - 1)(3 - 1) = \mathbf{4}$; $df_{A\times B\times S} = (a - 1)(b - 1)(n - 1) = (3 - 1)(3 - 1)(4 - 1) = \mathbf{12}$

Step 4: Calculate the Seven Mean Squares

$MS_A = SS_A \div df_A = 45.50 \div 2 = \mathbf{22.75}$

$MS_B = SS_B \div df_B = 7.17 \div 2 = \mathbf{3.58}$

$MS_{A\times B} = SS_{A\times B} \div df_{A\times B} = 65.83 \div 4 = \mathbf{16.46}$

$MS_S = SS_S \div df_S = 3.33 \div 3 = \mathbf{1.11}$

$MS_{A\times S} = SS_{A\times S} \div df_{A\times S} = 25.17 \div 6 = \mathbf{4.20}$

$MS_{B\times S} = SS_{B\times S} \div df_{B\times S} = 31.50 \div 6 = \mathbf{5.25}$

$MS_{A\times B\times S} = SS_{A\times B\times S} \div df_{A\times B\times S} = 59.50 \div 12 = \mathbf{4.96}$

Step 5: Present the results in an analysis of variance summary table:

ANOVA Summary Table

Source	SS	df	MS	F	F_{crit}
A	45.50	2	22.75	5.42 $p < .05$	5.14
A × S (error)	25.17	6	4.20		
B	7.17	2	3.58	.68	5.14
B × S (error)	31.50	6	5.25		
A × B	65.83	4	16.46	3.32 $p < .05$	3.26
A × B × S (error)	59.50	12	4.96		
S	3.33	3	1.11	**cannot be evaluated**	

The total sum of squares for random error and subject effects (for two-factor designs: $SS_{S/AB}$ for the between-subject design, $SS_{S/A} + SS_{B×S/A}$ for the mixed design, and $SS_S + SS_{A×S} + SS_{B×S} + SS_{A×B×S}$ for the within-subject design) is the same value for all three design variations since all three illustrations used this common set of 36 numbers. The $SS_{S/AB}$ error term from the two-factor between-subject design is simply partitioned into meaningful components in the mixed and within-subject designs by removing portions of individual-difference variability appropriately from random error.

IV. Using SPSS for Mixed and Within-Subject Two-Factor Designs

Doing ANOVA problems with a hand calculator for mixed and within-subject two-factor designs should leave even technologically challenged individuals looking to the computer for relief. You will recall that the task of entering data in an SPSS spreadsheet for single-factor and two-factor between-subject designs required creating "dummy" codes that were unique for each level of an independent variable (e.g., "1" for each a_1 level, "2" for each a_2 level, "3" for each a_3 level, etc.). Entering data in an SPSS spreadsheet for a within-subject factor simply required setting up a column of scores for each level of that factor (e.g., all a_1 scores are entered in the first column of the SPSS spreadsheet, all a_2 scores are entered in the second column, etc.) and organizing it so that each row in the spreadsheet corresponded to scores for a given subject. The same rules apply for entering data for two-factor mixed designs (a dummy code for the between-subject factor and separate score columns for each level of the within-subject factor) and for two-factor within-subject designs (each column corresponds to scores for one a_ib_j condition, with rows representing individual subjects). The original data from Table 12.1 are repeated in Table 12.3 as they would appear in an SPSS spreadsheet for a two-factor mixed analysis of variance, and they are repeated in Table 12.4 as they would appear in an SPSS spreadsheet for a two-factor within-subject analysis of variance.

Table 12.3: Data View Screen for SPSS for a Two-Factor Mixed ANOVA

	Factor A	b_1	b_2	b_3
1	1.00	3.00	4.00	2.00
2	1.00	6.00	4.00	6.00
3	1.00	2.00	8.00	5.00
4	1.00	4.00	4.00	7.00
5	2.00	9.00	6.00	5.00
6	2.00	7.00	7.00	7.00
7	2.00	5.00	8.00	1.00
8	2.00	9.00	4.00	5.00
9	3.00	7.00	6.00	10.00
10	3.00	2.00	7.00	9.00
11	3.00	3.00	6.00	13.00
12	3.00	8.00	9.00	8.00

Table 12.4: Data View Screen for SPSS for a Two-Factor, Within-Subject ANOVA

	a_1b_1	a_1b_2	a_1b_3	a_2b_1	a_2b_2	a_2b_3	a_3b_1	a_3b_2	a_3b_3
1	3.00	9.00	7.00	4.00	6.00	6.00	2.00	5.00	10.00
2	6.00	7.00	2.00	4.00	7.00	7.00	6.00	7.00	9.00
3	2.00	5.00	3.00	8.00	8.00	6.00	5.00	1.00	13.00
4	4.00	9.00	8.00	4.00	4.00	9.00	7.00	5.00	8.00

To perform an ANOVA with one or more within-subject variables (repeated measures), you begin by going to "Analyze" on the tool bar at the top of the screen. Scroll down to "General Linear Model," then click on "**Repeated Measures.**" This action will open a dialog box similar to the one in Table 12.5 (same as Table 11.4, page 216).

Table 12.5: SPSS Dialog Box for Repeated Measures ANOVA

Number of levels

3

Add

Define

Within-subject variables

factor 1

factor1(3)

For the two-factor mixed design example problem, Factor B is the within-subject variable, and it has three levels, so you will need to type in the number "3" in the box below "Number of levels" (the "3" is shown as already entered in this box). Then click on the "Add" box and "Factor 1(3)" will appear as shown in the box below "within-subject variables." In order to differentiate the between-subject and within-subject variables, move your cursor to the "Define" box, and click on it. A new dialog box will appear, similar to the one in Table 12.6 (same as Table 11.5, page 217).

Table 12.6: Completed SPSS Dialog Box for Two-Factor Mixed ANOVA

Factor A
b1
b2
b3

Within-subject variables

→

Between-subject variables

→

Options

OK

Covariates

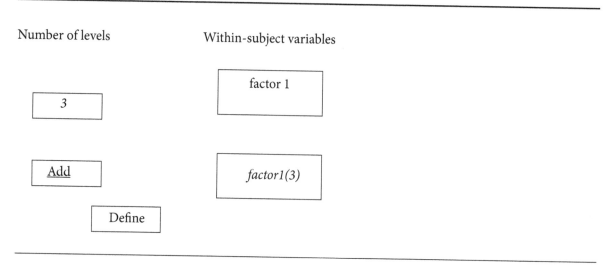

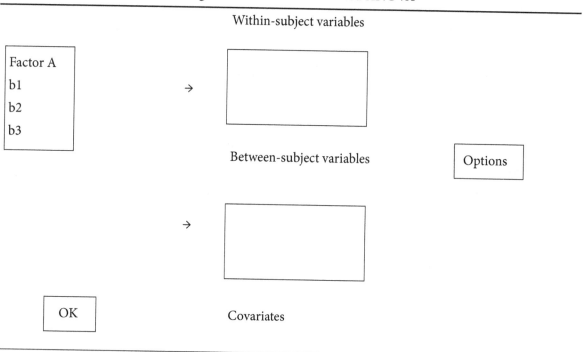

Since Factor B is the within-subject variable, you need to highlight "b1, b2, and b3" and click on the arrow to the left of the "within-subject variables" box. This will move "b1, b2, and b3" to that box. Similarly, highlight "factor A" and click on the arrow to the left of the "between-subject variables" box. This will move "factor A" to that box. Move your cursor to click on "OK" to produce a completed ANOVA summary table. The essential part of the SPSS ANOVA table is similar to what has been presented before. The ANOVA results for the within-subject factor and the A × B interaction appear in the table labeled, "Tests of Within-Subjects Effects." Results for the between-subject factor appear in the table labeled, "Tests of Between-Subjects Effects." These results are copied in Table 12.7, and they confirm the results of the analysis of variance using a hand calculator reported earlier.

Table 12.7: SPSS ANOVA Summary Table for Within- and Between-Subject Factors

Tests of Within-Subjects Effects

Source	Type III Sum of Squares	df	Mean Square	F	Sig
Factor 1 (B) Sphericity Assumed	7.167	2	3.583	**.709**	.505
Factor 1 × A	65.833	4	16.458	**3.255**	.036
Error	91.000	18	5.056		

Tests of Between-Subjects Effects

Source	Type III Sum of Squares	df	Mean Square	F	Sig
Factor A	45.500	2	22.750	**7.184**	.014
Error	28.500	9	3.167		

The second SPSS illustration is appropriate for a two-factor within-subject design after the data have been entered in the spreadsheet that appeared in Table 12.4 on page 243 Since both Factors A and B are manipulated within subjects for the two-factor, within-subject design, only the "within-subjects variables" box is used. So again, go to the tool bar at the top of the screen and follow the same sequence of "Analyze," "General Linear Model," and "Repeated Measures" to open the first dialog box. Type "3" in the "number of levels" box, then click on the "add" box to move the first factor to the "within-subjects variables" window. Then in the "within-subject factor" box, replace "factor1" with "factor2," then type "3" again (number of levels for factor 2) since Factor B also has three levels, and click on the "add" box. Both factor 1 (our Factor A) and factor 2 (our Factor B) will appear in the "within-subjects variables" window. Now move your cursor to the "define" box, and click on it to bring up the second dialog box. All you need to do is highlight each of the nine "a_ib_j" lines and click on the arrow to the left of the "within-subjects variables" box to move them. Next click on "OK," and completed ANOVA tables will appear on the screen. The essential ANOVA components appear in the table labeled "Tests of Within-Subjects Effects," a facsimile of which appears in Table 12.8. The "A," "B," and "AxB" in parentheses and italics have been added to indicate that factor 1 is A, factor 2 is B and factor 1 * factor 2 is the A×B interaction. The sum of squares for subjects (SS_S) does not appear in this SPSS summary table.

Table 12.8: SPSS ANOVA Summary Table for Two-Factor, Within-Subject Example

Tests of Within-Subjects Effects

Source	Type III Sum of Squares	df	Mean Square	F	Sig
factor1(*A*) Sphericity Assumed	7.167	2	3.583	.683	.541
error(factor1)	31.500	6	5.250		
factor2(*B*)	45.500	2	22.750	5.424	.045
error(factor2)	25.167	6	4.194		
factor1*factor2(*A*×*B*)	65.833	4	16.458	3.319	.048
error(factor1*factor2)	59.500	12	4.958		

Both SPSS analyses confirm, within rounding error, the results calculated by hand and reported earlier in this chapter. Since the same data were used for the three two-factor ANOVA example problems, comparing the ANOVA Summary Tables for these three variations reveals that repeated measures (within-subject variables) impact both SS and df for error estimates; however, the sums of squares and degrees of freedom for the independent variables (Factor A and Factor B) and the A × B interaction remain the same across all three design variations. Comparing the three ANOVA summary tables, you can see that SS_A, SS_B, and $SS_{A\times B}$ are identical regardless of design variation. Further, the estimate of random error in the two-factor between-subject ANOVA ($SS_{S/AB}$) equals the sum of the two error terms ($SS_{S/A} + SS_{B\times S/A}$) in the two-factor mixed ANOVA, and $SS_{S/AB}$ equals the sum of squares for subjects (SS_S) plus the sum of squares for the three error terms, $SS_S + SS_{A\times S} + SS_{B\times S} + SS_{A\times B\times S}$, in the complete two-factor within-subject ANOVA (119.50 = 3.33 + 25.17 + 31.50 + 59.50).

V. Analytical Comparisons for Mixed and Within-Subject Two-Factor Designs

Analytical comparisons following an analysis of variance enable the researcher to probe more deeply into significant effects. There are three possible effects in two-factor designs (main effects for Factor A and Factor B, and an A × B interaction). If the interaction is significant, that effect becomes the focus of attention for any follow-up analysis (simple effects or interaction contrasts). If the interaction is not significant, then significant main effects for Factor A and/or Factor B may be considered for more detailed analyses (main comparisons). Recall from Chapter 10 that for between-subject designs, specific contrasts (e.g., the simple effect of Factor A at b_2) were tested with the error term from the overall analysis of variance as the denominator in the corresponding F ratio. For the single-factor within-subject design described in Chapter 11, analytical comparisons to test specific contrasts (e.g., an A_{comp} comparing a_2 vs. a_3) required calculating a new error term based on the interaction of subjects with the conditions involved in the specific A_{comp} tested (e.g., $MS_{A comp \times S}$).

Analytical comparisons for the two-factor mixed and two-factor within-subject designs obey these same principles. For the mixed design, the error term from the ANOVA (e.g., $MS_{S/A}$) may be used for main comparisons done on the between-subject variable (e.g., Factor A). **For the within-subject**

variable (e.g., Factor B) and the A × B interaction, and for all three effects in the two-factor within-subject design, new error terms need to be calculated for each contrast tested. For example, $MS_{B \times S/A}$ provides the error estimate in the two-factor mixed design for Factor B and for the A × B interaction. A simple-effect analysis of this interaction for a between-subject variable contrast (e.g., Factor A at b_2) requires a new error term ($MS_{S/A \text{ at } b2}$), as does a simple-effect analysis for a within-subject contrast (e.g., $MS_{B \times S \text{ at } a3}$ is the error term for the simple effect of B at a_3). Similarly, $MS_{A \times B \times S}$ provides the error estimate in the two-factor within-subject design for the A × B interaction. Thus, new error terms must be calculated to test simple effects (e.g., $MS_{A \times S \text{ at } b1}$ for the simple effect of A at b_1) and to test interaction contrasts (e.g., $MS_{Acomp \times Bcomp \times S}$ for the $A_{comp} \times B_{comp}$ interaction).

While keeping track of all of these different error terms and subscripts may seem overwhelming, things do fall into place once conditions relevant to contrasts of interest are isolated. Consider a simple-effect analysis to test the significance of Factor B (the within-subject factor) at level 3 of Factor A for the mixed two-factor design (procedures for doing interaction contrasts are illustrated in Model Problem 1, Part 3 at the end of this chapter). The subset of B at a_3 data from Table 12.1 are presented in Table 12.9. The data in this table represent a single-factor, within-subject design; a simple-effects analysis of Factor B at a_3 only requires doing a single-factor within-subject ANOVA using this extracted data set. The two relevant sums of squares, $SS_{B \text{ at } a3}$ and $SS_{B\times S \text{ at } a3}$, are simply "$SS_B$" and "$SS_{B\times S}$" for this reconfigured subset of data that identify the contrast of interest. Therefore, merely doing a single-factor within-subject ANOVA for the a_3b_j data accomplishes the simple-effect analysis for B at a_3. If this simple effect were calculated for the complete two-factor within-subject design, all that would change in Table 12.9 are the four subscripts for 's' (9, 10, 11, and 12 would be replaced by 1, 2, 3, and 4, respectively). Thus, all SS and MS computations would be the same. The sum of squares for Factor B for the data in Table 12.9 is actually $SS_{B \text{ at } a3}$, and $SS_{B\times S}$ for the data in this table is actually $SS_{B\times S \text{ at } a3}$.

Table 12.9: Factor B at a_3 Data Extracted from Table 12.1

	a_3b_1	a_3b_2	a_3b_3	ΣS_k
s_9	7	6	10	**23**
s_{10}	2	7	9	**18**
s_{11}	3	6	13	**22**
s_{12}	8	9	8	**25**
$\Sigma\Sigma A_3B_j$	**20**	**28**	**40**	

Step 1: Calculate Basic Ratios

$[B] = (20^2 + 28^2 + 40^2) \div n = 2{,}784 \div 4 = \mathbf{696.00}$

$[S] = (23^2 + 18^2 + 22^2 + 25^2) \div b = 1{,}962 \div 3 = \mathbf{654.00}$

$[T] = 88^2 \div (b)(s) = 7{,}744 \div 12 = \mathbf{645.33}$

$[Y] = [BS] = 7^2 + 2^2 + \ldots + 8^2 = \mathbf{742.00}$

Step 2: Calculate Sums of Squares

$SS_B = [B] - [T] = 696.00 - 645.33 = \mathbf{50.67}$

$SS_S = [S] - [T] = 654.00 - 645.33 = \mathbf{8.67}$

$SS_{B\times S} = [Y] - [B] - [S] + [T] = 742.00 - 696.00 - 654.00 + 645.33 = \mathbf{37.33}$

Step 3: Determine Degrees of Freedom

$df_B = b - 1 = 3 - 1 = \mathbf{2}$; $df_S = n - 1 = 4 - 1 = \mathbf{3}$; $df_{B\times S} = (b - 1)(n - 1) = (2)(3) = \mathbf{6}$

Step 4: Calculate Mean Squares

$MS_B = SS_B \div df_B = 50.67 \div 2 = \mathbf{25.34}$

$MS_S = SS_S \div df_S = 8.67 \div 3 = \mathbf{2.89}$

$MS_{B\times S} = SS_{B\times S} \div df_{B\times S} = 37.33 \div 6 = \mathbf{6.22}$

Step 5: Report the results in an analysis of variance summary table

ANOVA Summary Table

Source	SS	df	MS	F	F_{crit}
B at a_3	50.67	2	25.34	4.07 (ns)	5.14
S	8.67	3	2.89		
B × S at a_3	37.33	6	6.22		

As you can see, analytical comparisons for two-factor designs involving a within-subject variable require calculating a sum of squares for the contrast of interest and calculating a sum of squares for an error term appropriate for evaluating the contrast of interest. An important first step in performing this task is to isolate the relevant data by either combining scores for main comparisons (e.g., a_1 vs. a_2 involves a combination a_1s_1 score derived by summing $a_1b_1s_1 + a_1b_2s_1 + a_1b_3s_1$, etc.) or extracting subsets of data for simple effects as was done for the preceding example problem. In both cases, the task is completed by doing a single-factor within-subject analysis of variance on the new combined or restricted data set.

Now consider a simple effects illustration for Factor A: a within-subject variable in the complete two-factor within-subject design, but a between-subject factor in the two-factor mixed design. Recall that the error term for the repeated factor is $MS_{A\times S\text{ at }b3}$ and for the non-repeated factor the error term is

$MS_{S/A\ at\ b3}$. Two data sets are shown in Table 12.10, with the left-hand side showing the A at b_3 data for an A × (B × S) design and the right-hand side showing the same data for an (A × B × S) design.

Table 12.10: Data for the b_3 Groups for Determining Simple Effects of A at b_3

The A × (B × S) Design

	a_1b_3		a_2b_3		a_3b_3
s_1	2	s_5	5	s_9	10
s_2	6	s_6	7	s_{10}	9
s_3	5	s_7	1	s_{11}	13
s_4	7	s_8	5	s_{12}	8
ΣS	**20**		**18**		**40**

The (A × B × S) Design

	a_1b_3	a_2b_3	a_3b_3	$\Sigma\Sigma A_iS_k$
s_1	2	5	10	**17**
s_2	6	7	9	**22**
s_3	5	1	13	**19**
s_4	7	5	8	**20**
	20	**18**	**40**	**78**

The difference between these two extracted data sets involves 12 subscripts for subjects on the left and four subscripts for four subjects on the right, as there are only four subjects with this design. In addition, there is a final summation column for (A × B × S), as it is possible to sum the three scores for each of the four subjects with this design. Appropriate ANOVAs for each restricted data set complete the simple effect analyses for A at b_3.

For A × (B × S):

Step 1: Basic Ratios

$[A] = (20^2 + 18^2 + 40^2) \div 4 = 2{,}324 \div 4 = \mathbf{581.00}$

$[T] = 78^2 \div 12 = \mathbf{507.00}$

$[Y] = 2^2 + 6^2 + \ldots + 8^2 = \mathbf{628.00}$

Step 2: Sums of Squares

$SS_A = [A] - [T] = 581.00 - 507.00 = \mathbf{74.00}$

$SS_{S/A} = [Y] - [A] = 628.00 - 581.00 = \mathbf{47.00}$

Step 3: Degrees of Freedom

$df_A = a - 1 = 3 - 1 = \mathbf{2}$, and $df_{S/A} = (a)(n - 1) = 3(3) = \mathbf{9}$

Step 4: Mean Squares

$MS_A = SS_A \div df_A = 74.00 \div 2 = \mathbf{37.00}$

$MS_{S/A} = SS_{S/A} \div df_{S/A} = 47.00 \div 9 = \mathbf{5.22}$

Step 5: ANOVA Summary Table

Source	SS	df	MS	F	F_{crit}
A at b_3	74.00	2	37.00	7.09 $p < .05$	4.26
S/A at b_3	47.00	9	5.22		

For (A × B × S):

Step 1: Basic Ratios

$[A] = (20^2 + 18^2 + 40^2) \div 4 = 2{,}324 \div 4 = \mathbf{581.00}$

$[S] = (17^2 + 22^2 + 19^2 + 20^2) \div 3 = \mathbf{511.33}$

$[T] = 78^2 \div 12 = \mathbf{507.00}$

$[Y] = 2^2 + 6^2 + \ldots + 8^2 = \mathbf{628.00}$

Step 2: Sums of Squares

$SS_A = [A] - [T] = 581.00 - 507.00 = \mathbf{74.00}$

$SS_{A\times S} = [Y] - [A] - [S] + [T] = 628.00 - 581.00 - 511.33 + 507.00 = \mathbf{42.67}$

Step 3: Degrees of Freedom

$df_A = a - 1 = 3 - 1 = \mathbf{2}$, and $df_{A\times S} = (a - 1)(n - 1) = 2(3) = \mathbf{6}$

Step 4: Mean Squares

$MS_A = SS_A \div df_A = 74.00 \div 2 = \mathbf{37.00}$

$MS_{A\times S} = SS_{A\times S} \div df_{A\times S} = 42.67 \div 6 = \mathbf{7.11}$

Step 5: ANOVA Summary Table

Source	SS	df	MS	F	F_{crit}
A at b_3	74.00	2	37.00	5.20 $p < .05$	5.14
A × S at b_3	42.67	6	7.11		

Given that different error terms (the appropriate denominator for various F values) are required depending on design and whether Factor A and/or Factor B are non-repeated or repeated factors, the summary provided in Table 12.11 may provide a helpful reference.

Table 12.11: Appropriate MS_{error} for Simple Effects with A × (B × S), and (A × B × S) Designs

A × (B × S) Design		(A × B × S) Design	
Simple Effect	Error Term	Simple Effect	Error Term
For A at bj	$MS_{S/A \text{ at } bj}$	For A at bj	$MS_{A\times S \text{ at } bj}$
For B at ai	$MS_{B\times S \text{ at } ai}$	For B at ai	$MS_{B\times S \text{ at } ai}$

VI. Summary

Independent variables in an experiment may be manipulated between-subjects (individuals randomly assigned to participate in only one group) or within-subjects (individuals participate in all experimental conditions). For experiments with two independent variables, one or both may be within-subject variables: a two-factor mixed design or a two-factor within-subject design, respectively. Analysis of variance procedures for two-factor mixed and within-subject designs were described in this chapter.

Variance due to individual differences can be estimated whenever two or more scores are obtained for each subject, which in effect removes subject differences from estimates of random error. Thus, within-subject variables normally represent more powerful or sensitive manipulations of independent variables than between-subject variables. For computational purposes, the subject variable is treated like an additional factor (coded as S/A in the two-factor mixed design, and as S in the two-factor within-subject design) that interacts with the independent variables (B × S/A; or A × S, B × S, and A × B × S for two-factor mixed and two-factor within-subject designs, respectively). When participants provide more than one score in a data set, additional combinations of data (scores common to a given subject) are possible, resulting in additional sources of variation (variability due to subjects and due to interactions between subjects and one or both independent variables). The different combination possibilities are summarized in the side-by-side-by-side comparisons shown in Table 12.10.

Table 12.12: Comparisons among A × B, A × (B × S), and (A × B × S) Designs

A × B Design		A × (B × S) Design		(A × B × S) Design	
Source	F	Source	F	Source	F
A	$MS_A \div MS_{S/AB}$	A	$MS_A \div MS_{S/A}$	A	$MS_A \div MS_{A\times S}$
B	$MS_B \div MS_{S/AB}$	S/A		A × S	
A × B	$MS_{A\times B} \div MS_{S/AB}$	B	$MS_B \div MS_{B\times S/A}$	B	$MS_B \div MS_{B\times S}$
S/AB		A × B	$MS_{A\times B} \div MS_{B\times S/A}$	B × S	
		B × S/A		A × B	$MS_{A\times B} \div MS_{A\times B\times S}$
				A × B × S	
				S	

As with all designs with significant treatment effects and $df \geq 2$, more detailed analytical comparisons may be carried out. For within-subject variables, analytical comparisons require calculating sums of squares for the contrast of interest **and** for error terms restricted to the data relevant for the contrast of interest. Thus, the data must first be reorganized to fit the question being asked (e.g., a simple-effects question about the possible significance of Factor A at b_1 requires a first step of extracting only the b_1 data from the original data matrix). Calculating SS_A from the extracted data set is actually $SS_{A \text{ at } b1}$, and calculating $SS_{A\times S}$ from this extracted data set is actually $SS_{A\times S \text{ at } b1}$.

12.1: Model Problem 1. Assume the data in the table below are from an A × (B × S) **two-factor mixed design** with a = 2, b = 3, and n = 6. First, test the significance of the three effects that can be evaluated with this design: Factor A, Factor B, and the A × B interaction. Second, given the significant interaction, test the simple effects of Factor B at a_1 and Factor B at a_2.

Data Set for 2 × 3 Mixed Factorial Design

Subjects	a_1b_1	a_1b_2	a_1b_3	ΣS	Subjects	a_2b_1	a_2b_2	a_2b_3	ΣS
s_1	9	6	5	**20**	s_7	15	10	8	**33**
s_2	11	9	6	**26**	s_8	15	9	7	**31**
s_3	11	8	5	**24**	s_9	14	6	4	**24**
s_4	12	10	10	**32**	s_{10}	17	13	11	**41**
s_5	12	12	9	**33**	s_{11}	11	7	5	**23**
s_6	11	11	7	**29**	s_{12}	16	10	8	**34**
Σ	**66**	**56**	**42**	**164**		**88**	**55**	**43**	**186**

1. Two-Factor Mixed ANOVA:

Step 1: Basic Ratios

$[A] = (164^2 + 186^2) \div (b)(n) = 61{,}492 \div (3)(6) = 61{,}492 \div 18 = \mathbf{3{,}416.22}$

$[B] = ([66{+}88]^2 + [56{+}55]^2 + [42{+}43]^2) \div (a)(n) = (154^2 + 111^2 + 85^2) \div (2)(6) = 43{,}262 \div 12 =$ $\mathbf{3{,}605.17}$

$[T] = 350^2 \div (a)(b)(n) = 122{,}500 \div (2)(3)(6) = 122{,}500 \div 36 = \mathbf{3{,}402.78}$

$[AB] = (66^2 + 56^2 + \ldots + 43^2) \div n = 21{,}874 \div 6 = \mathbf{3{,}645.67}$

$[AS] = (20^2 + 26^2 + \ldots + 34^2) \div (b) = 10{,}598 \div 3 = \mathbf{3{,}532.67}$

$[Y] = 9^2 + 11^2 + \ldots + 8^2 = \mathbf{3{,}780.00}$

Step 2: Sums of Squares

$SS_A = [A] - [T] = 3{,}416.22 - 3{,}402.78 = \mathbf{13.44}$

$SS_{S/A} = [AS] - [A] = 3{,}532.67 - 3{,}416.22 = \mathbf{116.45}$

$SS_B = [B] - [T] = 3{,}605.17 - 3{,}402.78 = \mathbf{202.39}$

$SS_{A \times B} = [AB] - [A] - [B] + [T] = 3{,}645.67 - 3{,}416.22 - 3{,}605.17 + 3{,}402.78 = \mathbf{27.06}$

$SS_{B \times S/A} = [Y] - [AB] - [AS] + [A] = 3{,}780.00 - 3{,}645.67 - 3{,}532.67 + 3{,}416.22 = \mathbf{17.88}$

Step 3: Degrees of Freedom

$df_A = (a - 1) = (2 - 1) = \mathbf{1}$; $df_{S/A} = (a)(n - 1) = (2)(6 - 1) = \mathbf{10}$; $df_B = (b - 1) = (3 - 1) = \mathbf{2}$; $df_{A \times B} = (a - 1)(b - 1) = (1)(2) = \mathbf{2}$; and $df_{B \times S/A} = (b - 1)(a)(n - 1) = (2)(2)(5) = \mathbf{20}$

Step 4: Mean Squares

$MS_A = SS_A \div df_A = 13.44 \div 1 = \mathbf{13.44}$

$MS_{S/A} = SS_{S/A} \div df_{S/A} = 116.45 \div 10 = \mathbf{11.64}$

$MS_B = SS_B \div df_B = 202.39 \div 2 = \mathbf{101.20}$

$MS_{A \times B} = SS_{A \times B} \div df_{A \times B} = 27.06 \div 2 = \mathbf{13.53}$

$MS_{B \times S/A} = SS_{B \times S/A} \div df_{B \times S/A} = 17.88 \div 20 = \mathbf{.89}$

Step 5: ANOVA Summary Table

Source	SS	df	MS	F	F_{crit}
A	13.44	1	13.44	1.15	4.96
S/A	116.45	10	11.64		
B	202.39	2	101.20	113.71 $p < .05$	3.49
A × B	27.06	2	27.06	30.40 $p < .05$	3.49
B × S/A	17.88	20	.89		

2. Simple-Effects Analyses: Is Factor B significant at level 1 of Factor A? Factor B is the within-subject factor, and the restricted a_1 data set represents a single-factor, within-subject design (the same is true for the simple effect of B at a_2).

   a. Simple Effect of B at a_1 (the left half of the data matrix)

$[B \text{ at } a_1] = (66^2 + 56^2 + 42^2) \div 6 = 9{,}256 \div 6 = \mathbf{1{,}542.67}$

$[S \text{ at } a_1] = (20^2 + 26^2 + \ldots + 29^2) \div 3 = 4{,}606 \div 3 = \mathbf{1{,}535.33}$

$[T \text{ at } a_1] = 164^2 \div 18 = \mathbf{1{,}494.22}$

$[Y \text{ at } a_1] = 9^2 + 11^2 + \ldots + 7^2 = \mathbf{1{,}594.00}$

$SS_{B \text{ at } a1} = 1{,}542.67 - 1{,}494.22 = \mathbf{48.45}$

$SS_{S \text{ at } a1} = 1{,}535.33 - 1{,}494.22 = \mathbf{41.11}$

$SS_{B\times S \text{ at } a1} = 1{,}594.00 - 1{,}542.67 - 1{,}535.33 + 1{,}494.22 = \mathbf{10.22}$

$df_{B \text{ at } a1} = (b - 1) = \mathbf{2}$, and $SS_{B\times S \text{ at } a1} = (b - 1)(n - 1) = (2)(5) = \mathbf{10}$

$MS_{B \text{ at } a1} = 48.45 \div 2 = \mathbf{24.22}$; $MS_{B\times S \text{ at } a1} = 10.22 \div 10 = \mathbf{1.02}$; and $F_{B \text{ at } a1}(2, 10) = 24.22 \div 1.02 = \mathbf{23.75}$, $p < .05$

b. Simple Effect of B at a_2 (the right half of the data matrix)

$[B \text{ at } a_2] = (88^2 + 55^2 + 43^2) \div 6 = 12, 186 \div 6 = \mathbf{2{,}103.00}$

$[S \text{ at } a_2] = (33^2 + 31^2 + \ldots + 34^2) \div 3 = 5{,}992 \div 3 = \mathbf{1{,}997.33}$

$[T \text{ at } a_2] = 186^2 \div 18 = \mathbf{1{,}922.00}$

$[Y \text{ at } a_2] = 15^2 + 15^2 + \ldots + 8^2 = \mathbf{2{,}186.00}$

$SS_{B \text{ at } a2} = 2{,}103.00 - 1{,}922.00 = \mathbf{181.00}$

$SS_{S \text{ at } a2} = 1{,}997.33 - 1{,}922.00 = \mathbf{75.33}$

$SS_{B\times S \text{ at } a2} = 2{,}186.00 - 2{,}103.00 - 1{,}997.33 + 1{,}922.00 = \mathbf{7.67}$

$df_{B \text{ at } a2} = (b - 1) = \mathbf{2}$, and $SS_{B\times S \text{ at } a2} = (b - 1)(n - 1) = (2)(5) = \mathbf{10}$

$MS_{B \text{ at } a2} = 181.00 \div 2 = \mathbf{90.50}$; $MS_{B\times S \text{ at } a2} = 7.67 \div 10 = \mathbf{.77}$; and $F_{B \text{ at } a2}(2, 10) = 90.50 \div .77 = \mathbf{117.53}$, $p < .05$

3. $A_{comp} \times B_{comp}$ Interaction Contrast:

For an interaction contrast involving levels 1 and 2 for Factor A and levels 1 and 2 of factor B, the relevant data are found in columns 1, 2, 4, and 5 of the table for Model Problem 1 (a_1b_1, a_2b_1, a_1b_2, and a_2b_2). The procedures for an interaction contrast are identical to the overall ANOVA, except now we have a simpler data set that involves only four of the original six groups.

Step 1: Basic Ratios

$[A] = (122^2 + 143^2) \div (b)(n) = 35{,}333 \div 12 = \mathbf{2{,}944.42}$

$[B] = (154^2 + 111^2) \div (a)(n) = 36{,}037 \div 12 = \mathbf{3{,}003.08}$

$[T] = 265^2 \div (a)(b)(n) = 70{,}225 \div 24 = \mathbf{2{,}926.04}$

$[AB] = (66^2 + 56^2 + 88^2 + 55^2) \div (n) = 18{,}261 \div 6 = \mathbf{3{,}043.50}$

$[AS] = (15^2 + 20^2 + \ldots + 26^2) \div (b) = 6{,}031 \div 2 = \mathbf{3{,}015.50}$

$[Y] = 9^2 + 11^2 + \ldots + 10^2 = \mathbf{3{,}125.00}$

Step 2: Sums of Squares (All that is needed to calculate the $A_{comp} \times B_{comp}$ interaction is $SS_{A\times B}$ and $SS_{B\times S/A}$ for this restricted data set)

$SS_{A\times B} = [AB] - [A] - [B] + [T] = 3{,}043.50 - 2{,}944.42 - 3{,}003.08 + 2{,}926.04 = \mathbf{22.04}$

$SS_{B\times S/A} = [Y] - [AB] - [AS] + [A] = 3{,}125.00 - 3{,}043.50 - 3{,}015.50 + 2{,}944.42 = \mathbf{10.42}$

Step 3: Degrees of Freedom

$df_{A\times B} = (a - 1)(b - 1) = \mathbf{1}$ and $df_{B\times S/A} = (b - 1)(a)(n - 1) = (1)(2)(5) = \mathbf{10}$

Step 4: Mean Squares

$MS_{A\times B} = SS_{A\times B} \div df_{A\times B} = 22.04 \div 1 = \mathbf{22.04}$

$MS_{B\times S/A} = SS_{B\times S/A} \div df_{B\times S/A} = 10.42 \div 10 = \mathbf{1.04}$

Step 5: ANOVA Summary Table for this $A_{comp} \times B_{comp}$ Interaction

Source	SS	df	MS	F	F_{crit}
A × B	22.04	1	22.04	21.19 $p < .05$	4.96
B × S/A	10.42	10	1.04		

12.2: Model Problem 2. Repeat Model Problem 1; however, for this problem assume the data came from a two-factor within-subject design (treat each row of scores as if they were scores from the same subject; i.e., s_1 and s_7 are now the same subject, and the same for s_2 and s_8, etc.).

1. Two-Factor Within-Subject ANOVA

Step 1: Basic Ratios

[A], [B], [T], [AB], [AS], and [Y] are the same as they were for the two-factor mixed design. However, two new basic ratios must be calculated: a basic ratio [S] based on the sum of all $a_i b_j$ scores for each subject and a basic ratio [BS] based on the sum of the b_j scores for each subject.

$[A] = (164^2 + 186^2) \div (b)(n) = \mathbf{3{,}416.22}$

$[B] = ([66+88]^2 + [56+55]^2 + [42+43]^2) \div (a)(n) = \mathbf{3{,}605.17}$

$[T] = 350^2 \div (a)(b)(n) = \mathbf{3{,}402.78}$

$[AB] = (66^2 + 56^2 + \ldots + 43^2) \div n = \mathbf{3{,}645.67}$

$[AS] = (20^2 + 26^2 + \ldots + 34^2) \div (b) = \mathbf{3{,}532.67}$

$[Y] = 9^2 + 11^2 + \ldots + 8^2 = \mathbf{3{,}780.00}$

$[S] = (53^2 + 57^2 + \ldots + 63^2) \div (a)(b) = 20{,}796 \div 6 = \mathbf{3{,}466.00}$

$[BS] = (24^2 + 26^2 + \ldots + 15^2) \div (a) = 7{,}364 \div 2 = \mathbf{3{,}682.00}$

Step 2: Sums of Squares

$SS_A = [A] - [T] = 3{,}416.22 - 3{,}402.78 = \mathbf{13.44}$

$SS_{A\times S} = [AS] - [A] - [S] + [T] = 3{,}532.67 - 3{,}416.22 - 3{,}466.00 + 3{,}402.78 = \mathbf{53.23}$

$SS_B = [B] - [T] = 3{,}605.17 - 3{,}402.78 = \mathbf{202.39}$

$SS_{B\times S} = [BS] - [B] - [S] + [T] = 3{,}682.00 - 3{,}605.17 - 3{,}466.00 + 3{,}402.78 = \mathbf{13.61}$

$SS_{A\times B} = [AB] - [A] - [B] + [T] = 3{,}645.67 - 3{,}416.22 - 3{,}605.17 + 3{,}402.78 = \mathbf{27.06}$

$SS_{A\times B\times S} = [Y] - [AB] - [AS] - [BS] + [A] + [B] + [S] - [T]$

$SS_{A\times B\times S} = 3{,}780.00 - 3{,}645.67 - 3{,}532.67 - 3{,}682.00 + 3{,}416.22 + 3{,}605.17 + 3{,}466.00 - 3{,}402.78$
$= \mathbf{4.27}$

$SS_S = [S] - [T] = 3{,}466.00 - 3{,}402.78 = \mathbf{63.22}$

Step 3: Degrees of Freedom

df_A = **1**; $df_{A\times S} = (a-1)(n-1) = (1)(5) =$ **5**; df_B = **2**; $df_{B\times S} = (b-1)(n-1) = (2)(5) = 10$; $df_{A\times B}$ = **2**; $df_{A\times B\times S}$ $= (a-1)(b-1)(n-1) = (1)(2)(5) =$ **10**; and $df_S = (n-1) =$ **5**

Step 4: Mean Squares

$MS_A = SS_A \div df_A = 13.44 \div 1 = \mathbf{13.44}$

$MS_{A\times S} = SS_{A\times S} \div df_{A\times S} = 53.23 \div 5 = \mathbf{10.65}$

$MS_B = SS_B \div df_B = 202.39 \div 2 = \mathbf{101.20}$

$MS_{B\times S} = SS_{B\times S} \div df_{B\times S} = 13.61 \div 10 = \mathbf{1.36}$

$MS_{A\times B} = SS_{A\times B} \div df_{A\times B} = 27.06 \div 2 = \mathbf{13.53}$

$MS_{A\times B\times S} = SS_{A\times B\times S} \div df_{A\times B\times S} = 4.27 \div 10 = \mathbf{.43}$

$MS_S = SS_S \div df_S = 63.22 \div 5 = \mathbf{12.64}$

Step 5: ANOVA Summary Table

Source	SS	df	MS	F	F_{crit}
A	13.44	1	13.44	1.26	6.61
A × S	53.23	5	10.65		
B	202.39	2	101.20	74.41 $p < .05$	4.10
B × S	13.61	10	1.36		
A × B	27.06	2	27.06	62.93 $p < .05$	4.10
A × B × S	4.27	10	.43		
S	63.22	5	12.64	**cannot be evaluated**	

2. Simple Effects Analysis: This illustration will work through a simple effect for **Factor A at level 1 of Factor B** (which you may note simply requires a single-factor, within-subject ANOVA). You may test yourself by completing the simple effects analysis, calculating a second F ratio for Factor A at level 2 of Factor B. For A at b_1 we will only be concerned with the b_1 scores in the first column (a_1b_1) and in the fourth column (a_2b_1) of the data set for 12.1 Model Problem 1.

Data for the b_1 Groups from Model Problem 1

	a_1b_1	a_2b_1	$\Sigma\Sigma A_jS_k$
s_1	9	15	**24**
s_2	11	15	**26**
s_3	11	14	**25**

s_4	12	17	**29**
s_5	12	11	**23**
s_6	11	16	**27**
Σ	**66**	**88**	**154**

To answer the simple-effects question (Is A significant at b_1?), we must calculate SS_A and $SS_{A\times S}$ for these data. With n = 6, the appropriate step-wise calculations follow:

Step 1: Basic Ratios

$[A] = (66^2 + 88^2) \div 6 = 12{,}100 \div 6 = \mathbf{2{,}016.67}$

$[S] = (24^2 + 26^2 + \ldots + 27^2) \div 2 = 3{,}976 \div 2 = \mathbf{1{,}988.00}$

$[T] = 154^2 \div 12 = 23{,}716 \div 12 = \mathbf{1{,}976.33}$

$[Y] = 9^2 + 11^2 + \ldots + 16^2 = \mathbf{2{,}044.00}$

Step 2: Sums of Squares

$SS_A = [A] - [T] = 2{,}016.67.00 - 1{,}976.33 = \mathbf{40.34}$

$SS_{A\times S} = [Y] - [A] - [S] + [T] = 2{,}044.00 - 2{,}016.67 - 1{,}988.00 + 1{,}976.33 = \mathbf{15.66}$

Step 3: Degrees of Freedom

$df_A = a - 1 = 2 - 1 = \mathbf{1}$; and $df_{A\times S} = (a - 1)(n - 1) = (1)(5) = \mathbf{5}$

Step 4: Mean Squares

$MS_A = SS_A \div df_A = 40.34 \div 1 = \mathbf{40.34}$

$MS_{A\times S} = SS_{A\times S} \div df_{A\times S} = 15.66 \div 5 = \mathbf{3.13}$

Step 5: ANOVA Summary Table

Source	SS	df	MS	F	F_{crit}
A	40.34	1	40.34	12.89 $p \leq .05$	6.61
A × S	15.66	5	3.13		

3. $A_{comp} \times B_{comp}$ Interaction Contrast:

An interaction contrast for a mixed two-factor design was illustrated using data limited to the a_1b_1, a_2b_1, a_1b_2, and a_2b_2 (columns 1, 2, 4, and 5) from the table for Model Problem 1. We will now repeat that analysis treating the data as if they came from a complete within-subject two factor design. For the mixed two-factor ANOVA, we found F_{AxB} (1, 10) = 21.19. Treating the data as if each of the six subjects provided four scores (a score for each a_ib_j group), we will need to calculate a new error term ($F_{A \times B \times S}$) with $df_{A \times B \times S}$ = 5. Also, whereas F_{crit} (1, 10) = 4.96 for the mixed two-factor design, F_{crit} (1, 5) = 6.61 for the within-subject two-factor design.

Step 1: Basic Ratios

[A] = **2,944.42** (same as for Model Problem 1.3)

[B] = **3,003.08** (same as for Model Problem 1.3)

$[S] = (40^2 + 44^2 + \ldots + 48^2) \div (a)(b) = 11{,}829 \div (2)(2) = 11{,}829 \div 4 = \mathbf{2{,}957.25}$

[T] = **2,926.04** (same as for Model Problem 1.3)

[AB] = **3,043.50** (same as for Model Problem 1.3)

[AS] = **3,015.50** (same as for Model Problem 1.3)

$[BS] = (24^2 + 26^2 + \ldots + 21^2) \div a = 6{,}083 \div 2 = \mathbf{3{,}041.50}$

[Y] = **3,125.00** (same as for Model Problem 1.3)

Step 2: Sums of Squares (All that is needed for $A_{comp} \times B_{comp}$ interaction is $SS_{A \times B}$ and $SS_{A \times B \times S}$)

$SS_{A \times B}$ = [AB] – [A] – [B] + [T] = **22.04** (same as for Model Problem 1.3)

$SS_{A \times B \times S} = [Y] - [AB] - [AS] - [BS] + [A] - [B] + [S] - [T]$

$SS_{A \times B \times S} = 3{,}125.00 - 3{,}043.50 - 3{,}015.50 - 3{,}041.50 + 2{,}944.42 + 3{,}003.08 + 2{,}957.25 - 2{,}926.04 = \mathbf{3.21}$

Step 3: Degrees of Freedom

$df_{A \times B} = (a - 1)(b - 1) = 1$ and $df_{A \times B \times S} = (a - 1)(b - 1)(n - 1) = (1)(1)(5) = \mathbf{5}$

Step 4: Mean Squares

$MS_{A \times B} = SS_{A \times B} \div df_{A \times B} = 22.04 \div 1 = \mathbf{22.04}$ (same as for Model Problem 1.3)

$MS_{A \times B \times S} = SS_{A \times B \times S} \div df_{A \times B \times S} = 3.21 \div 5 = \mathbf{.64}$

Step 5: $F = MSA \times B \div MSA \times B \times S = 22.04 \div .64 = 34.44$

ANOVA Summary Table for this $A_{comp} \times B_{comp}$ Interaction

Source	SS	df	MS	F	Fcrit
A × B	22.04	1	22.04	34.44 $p < .05$	6.61
A × B x S	3.21	5	.64		

12.3: Practice Problem 1. An experimenter was interested in determining if retention was better if rest intervals separated successive study periods. The task involved a series of study trials on a 30-word list with each word shown for 3 seconds. After the last study word was presented, subjects were allowed 2 minutes for a written recall test. The alternate study trial-test trial procedure continued until subjects recalled at least 25 of the 30 words. Twenty-four hours later students returned for a retention test. A total of 100 students participated in the experiment, with 20 randomly assigned to each of five groups. The groups differed in terms of length of rest after each test trial: $a_1 = 3$ sec, $a_2 = 30$ sec, $a_3 = 60$ sec, $a_4 = 120$ sec and $a_5 = 240$ sec. The 30-word list subjects had to learn consisted of a random order of 15 uncommon (b_1) and 15 common words (b_2). This is a 5 × 2 factorial design with n = 20. As it has been described, is this a two-factor mixed design or a two-factor within-subject design?

12.4: Practice Problem 2. An experimenter measured reaction times (RTs) in milliseconds for naming "long" words. Each subject (n = 8) saw 60 words presented successively, and the task involved saying a presented word aloud as quickly as possible. The experimenter was interested in whether certain characteristics of the **initial four letters** of each word influenced how quickly words could be named. One independent variable was whether the first four letters of each long word also formed a word: either a high-frequency word (HF), a low-frequency word (LF), or a non-word (NW). In addition, the experimenter was interested in "neighborhood density" of the four-letter units. Neighborhood density was defined as the number of different four-letter words that could be formed by just changing one letter in the initial four-letters of each long word (e.g., *must* is a high-frequency word with several neighbors, such as *bust*, *dust*, *mast*, and *muse*; *plan* is a high-frequency word with relatively few neighbors, such as *clan* and *play*). There were six conditions in the experiment defined by whether the first four letters of a long stimulus word formed an HF-large neighborhood (LN) word (e.g., *mustard*), an HF-small neighborhood (SN) word (e.g., *planet*), an LF-LN word (e.g., *gullet*), an LF-SN word (e.g., *piercing*), an NW-LN word (e.g., *villain*), or an NW-SN word (e.g., *crimson*). The experimenter measured RTs for naming the various long words. There were 10 exemplars for each category and eight subjects participated in the experiment. The 60 words each subject had to name were presented in a random order. Average RTs are reported in the following table. Were RTs influenced by Factor A (HF vs. LF vs. NW), by Factor B (LN vs. SN), and by the interaction? Check and report the results by repeating the analysis using SPSS (i.e., did the computer get it right?).

Mean RTs in Seconds for Naming Words

Subjects	a_1b_1 HF-LN	a_1b_2 HF-SN	a_2b_1 LF-LN	a_2b_2 LF-SN	a_3b_1 NW-LN	a_3b_2 NW-SN	ΣS_k
1	.61	.66	.60	.71	.62	.68	**3.88**
2	.55	.61	.59	.64	.58	.65	**3.62**
3	.66	.75	.70	.71	.74	.80	**4.36**
4	.60	.59	.62	.65	.63	.67	**3.76**
5	.52	.52	.50	.60	.59	.65	**3.38**
6	.55	.54	.60	.61	.59	.66	**3.55**
7	.61	.68	.67	.67	.64	.73	**4.00**
8	.50	.60	.52	.56	.51	.66	**3.35**
$\Sigma AB_{i,j}$	**4.60**	**4.95**	**4.80**	**5.15**	**4.90**	**5.50**	

12.5: Practice Problem 3. There are 24 scores in the data matrix that follows. This is a two-factor mixed design (a = 3, b = 2, and n = 4), as each subject serves in both B conditions, but in only one A condition. After you complete the ANOVA, test the simple effects of Factor B at each level of Factor A.

The Data Matrix

	a_1b_1	a_1b_2	$\Sigma\Sigma A_1S_k$		a_2b_1	a_2b_2	$\Sigma\Sigma A_2S_k$		a_3b_1	a_3b_2	$\Sigma\Sigma A_3S_k$
s_1	8	5	**13**	s_5	3	3	**6**	s_9	2	7	**9**
s_2	7	4	**11**	s_6	1	5	**6**	s_{10}	1	5	**6**
s_3	6	4	**10**	s_7	3	4	**7**	s_{11}	2	6	**8**
s_4	9	5	**14**	s_8	4	3	**7**	s_{12}	2	7	**9**
$\Sigma\Sigma A_iB_j$	**30**	**18**			**11**	**15**			**7**	**25**	

Do the appropriate ANOVA indicating what effects, if any, are significant at $p \leq .05$. Confirm your results by repeating the analysis using SPSS. After you complete the ANOVA, test the simple effects of Factor B at each level of Factor A.

12.6. Practice Problem 4. Practice problem 3 involved a two-factor mixed design [A × (B × S)]. Do an ANOVA for this problem again; however, this time treat the data as if they were from a two-factor, within-subject design. That is, assume each row in the data matrix represents scores for one subject, and that there are only four different subjects who participated in this experiment. After you complete the two-factor, within-subject analysis of variance, check your results by repeating the ANOVA using SPSS. Finally, determine whether the interaction contrast involving only levels 1 and 2 of Factor A and 1 and 2 of Factor B is significant at $p < .05$.

VIII. Practice Test 3: Chapters 9–12

(answers given in Appendix B)

I. Multiple choice: Circle the number for the best answer to each question

1. Two independent variables (Factor A and Factor B, with a = 2 and b = 2) are said to "interact" if

   a. $a_1b_1 - a_1b_2 + a_2b_1 - a_2b_2 \neq 0$
   b. $a_1b_1 + a_1b_2 - a_2b_1 - a_2b_2 \neq 0$
   c. $a_1b_1 - a_1b_2 - a_2b_1 + a_2b_2 \neq 0$
   d. $a_1b_1 + a_1b_2 + a_2b_1 + a_2b_2 \neq 0$

2. Following the analysis of variance for a two-factor, between-subjects design (A x B), analytical comparisons on Factor A (main comparisons) should be considered if

   a. $df_A > 2$
   b. F_A was significant
   c. $F_{A\times B}$ was not significant
   d. All of the above

3. A statistical test of the difference between a_1 and a_4 after a two-factor ANOVA in which the A x B interaction was not significant is an example of

   a. a simple comparison
   b. a main comparison
   c. a simple effect
   d. an interaction contrast

4. Given that $MS_A = 10$, $MS_S = 20$, and $MS_{A\times S} = 2$ from an A × S ANOVA, it follows that

   a. $F_A = .5$
   b. $F_S = 10$
   c. $F_{A\times S} = .1$
   d. none of the above

5. Which one of the following problems, if relevant, would pose the most serious threat to the validity of results from a within-subject design?

   a. practice effects
   b. carryover effects
   c. individual differences
   d. power

6. Four chimpanzees trained in American Sign Language are given signing vocabulary tests on four different occasions administered by their usual human companion researcher and with no one else present on one occasion, with a 2nd person who was a "known" researcher present on another occasion, with a 2nd person who was a stranger present on another occasion, or with another chimpanzee as the 2nd individual present on another occasion. The experimenter compared number of vocabulary items correctly named during the four test sessions. The type of experimental design described in this paragraph is

   a. an A × B independent-groups design
   b. an A × S single-factor, within-subject design
   c. an A × (B × S) mixed two-factor design
   d. an (A × B × S) two-factor, within-subject design

7. Ten students participated in an experiment during which they read 2 short stories (a sad [a_1] and a happy story [a_2]). Immediately after reading the 2 stories they were given 15 anagrams to solve (an anagram is a string of letters that can be rearranged to spell a word). The 15 anagrams were presented successively. Correct solution words for 5 anagrams had appeared earlier in the sad story (b_1); for another 5 anagrams, correct solution words had appeared in the happy story (b_2); and for another 5 anagrams, correct solution words had not appeared in either story (b_3). The response measure was the time taken to solve each anagram. This paragraph describes a

   a. one-factor, within-subject design
   b. two-factor, between-subject design
   c. two-factor, mixed design
   d. two-factor, within-subject design

8. Computing sums of squares for a within-subject A × S design involves operations similar to computing sums of squares for a two-factor between-subject design, given the simple substitution of the coding letter S for the coding letter B. However, the similarity breaks down when calculating $SS_{S/AS}$ and $SS_{S/AB}$. The reason for this is that

   a. [Y] – [AS] = 0
   b. $SS_{S/AB} = 1.0$
   c. [A] – [AS] – [AB] + [T] = 0
   d. [Y] – [T] = 1.0

9. For a two-factor mixed design [A × (B × S)], the appropriate error term for a simple effect of Factor B at level 1 of Factor A (B at a_1) is

   a. $MS_{S/A}$
   b. $MS_{B \times S/A}$
   c. $MS_{Bcomp\ x\ S\ at\ a1}$
   d. $MS_{B \times S\ at\ a1}$

10. For an experiment with a = 3, b = 2, and n = 5, which one of the analytical comparisons below identifies a follow-up "simple comparison"?

   a. A_1 vs. A_2
   b. $A_1 + A_2$ vs. A_3
   c. A at b_1
   d. A_1 vs. A_2 at b_1

II. Assume the data below are from a **two-factor between-subject** design.

DATA MATRIX (these are the individual scores, and as you can see, a = 2, b = 3, and n = 4). Some of the Basic Ratio computations necessary for completing an analysis of variance have been done and are given below the Data Matrix, so it is not necessary for you to repeat these calculations.

	b_1		b_2		b_3	
	a_1	a_2	a_1	a_2	a_1	a_2
	1	3	4	3	8	5
	0	2	5	2	6	3
	2	5	7	4	5	5
	4	3	4	5	7	2
ΣY_i	7	13	20	14	26	15
ΣY_i^2	21	47	106	54	174	63

[Y] = 465 [A] = 381.08 [B] = 404.62 [T] = 376.04

1. Do an Analysis of Variance, presenting your results in an ANOVA summary table.

2. Is the simple effect of Factor A at level 3 of Factor B significant (i.e., **is A at b_3 significant** at $p \leq .05$)?

3. Now for this problem assume the data in the matrix above came from a **two-factor mixed design**, with repeated measures on Factor A (each subject had an a_1 score, and a_2 score that appear within a given row of the Data Matrix), and different students were randomly assigned to each of the three b conditions (b_1, b_2, and b_3). With this design, it would be meaningful to determine a total score for each subject by summing his or her two a_i scores (e.g., $a_1b_1s_1 + a_2b_1s_1 = 1 + 3 = 4$). For this problem you **DO NOT** need to do a complete analysis of variance. The only thing you are being asked to calculate is the BS basic ratio. That is, **what does [BS] equal?**

III. The data below are from an A × S within-subject design with a = 4 and n = 5.

	a_1	a_2	a_3	a_4	ΣS_j
s_1	31	31	41	30	**133**
s_2	33	26	29	36	**124**
s_3	25	30	32	36	**123**
s_4	28	30	27	27	**112**
s_5	24	28	35	33	**120**
ΣY_i	**141**	**145**	**164**	**162**	
ΣY_i^2	**4035**	**4221**	**5500**	**5310**	

A. Do an appropriate analysis of variance to determine whether or not differences among the 4 groups are significant at the $p \leq .05$ level.

B. Regardless of whether or not Factor A is significant, calculate MS_{Acomp}, $MS_{Acomp \times S}$, and F_{Acomp} for an analytical comparison to determine if scores in the a_1 condition differ significantly ($p \leq .05$) from scores in the a_4 condition.

Chapter 13

Chi Square (χ^2)

Not everything that can be counted counts; and not everything that counts can be counted.
—Einstein

I. What Are Categorical Data?

Up to this point we have considered analyses of data that are observed behaviors reduced to numerical representations. Example data sets, both real and fictitious, involved a range of scores from subjects, such as estimates of the speed of a car seen in a film clip, number of false recognitions in a memory test, number correct on a vocabulary test, and length of time volunteers are willing to spend in a sensory-deprivation chamber. These are commonly referred to as 'parametric' tests: tests for which statistics provide estimates of population parameters. Data are presumed to be measured on an interval or ratio scale and should provide approximations to a normally distributed population. Not all data can be measured along a continuum; at times the best that can be done is to group or categorize performance by counting successes, agreements, preferences, and in general, behaviors that merely classify individuals into one or another category. The focus is not on converting behaviors of interest to individual scores along a scale of measurement with many possible outcomes, but on merely counting and classifying individuals. The chi square test provides one such example.

A published experiment referred to in Chapter 1 (Loftus and Palmer 1974) actually had examples of both types of response measures in the same study. Subjects were asked to estimate the speed of a car involved in a filmed accident (a more-or-less continuous scale of measurement) and later they responded to a question about whether broken glass was present at the scene of the accident. The response measure for this question was categorical, and respondents in each experimental condition were simply grouped into one of two categories, with one category for participants who responded "yes" to the question about presence of broken glass and the other category for individuals who responded "no" to the question.

Comparing categories of responses may be relevant for a variety of research questions. For example, public opinion polls represent a common venue for conveying information about approval or disapproval of programs and politicians, and information from such polling activities is based on a simple count of the number of individuals in different categories. Preferences for products or for political candidates are gauged by counting number of individuals preferring "A," "B," "C," etc. As categories of interest are defined and filled by research participants, statistical questions arise regarding whether category frequencies differ from **what would be expected** if random variation was the only factor responsible for observed differences. Statistical tests we have considered involved reference distributions that were based on underlying assumptions that are not met by dichotomous categorical data. The chi-square (χ^2) test does not require that data satisfy the same underlying assumptions as the F and t tests considered in earlier chapters, and χ^2 is an appropriate statistic for analyzing categorical data. The test statistic introduced in this chapter is commonly referred to as chi-square, symbolized with the Greek letter Chi, (χ^2), and we shall conform to this designation. However, technically, the chi-square distribution values are only approximated by this χ^2 test statistic. As a result, some writers (e.g., Siegel and Castellan, Jr. 1988) prefer designating the test statistic as X^2 as the approximation of χ^2.

II. The Test Statistic: An Approximation to χ^2

If statistical analysis involves comparing category frequency data collected from a research project, then the first hurdle in performing a chi-square test is to determine what frequencies should be **expected** in each category. The null hypothesis tested is that observed and expected category frequencies are identical (in the long run), and the logic underlying H_0 testing for χ^2 is no different from the underlying logic for t tests and analysis of variance. Thus, the entire process is dependent upon development of a meaningful set of expected category frequencies! Once we have these, we can calculate a χ^2 statistic and compare it to an appropriate χ^2 reference distribution: if the calculated χ^2 is a poor fit (meaning it falls within the extreme 5% of the reference distribution values of χ^2), then differences between observed and expected category frequencies are said to be significant, and the null hypothesis is rejected. The formula for calculating χ^2 is as follows: $\chi^2 = \Sigma([f_O - f_E]^2 \div f_E)$, where f_O refers to <u>F</u>requency <u>O</u>bserved in a category, and f_E refers to <u>F</u>requency <u>E</u>xpected in a category. The χ^2 formula instructs us to subtract each expected category frequency (f_E) from the corresponding observed category frequency (f_O), square the resulting

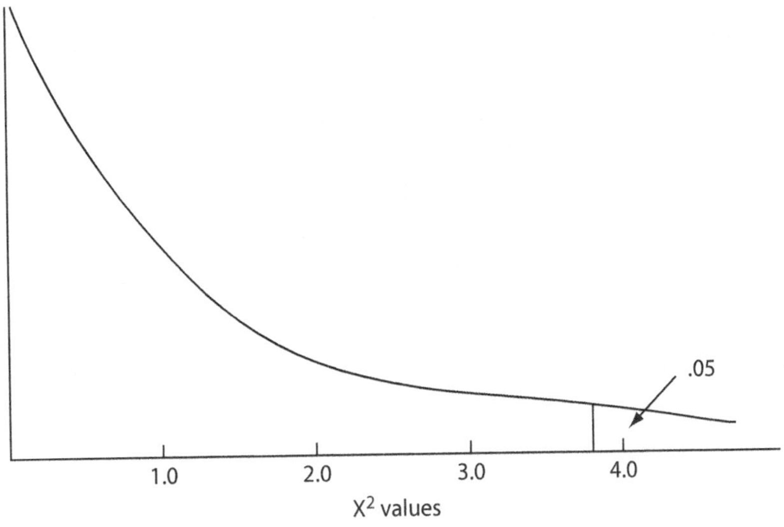

Figure 13.1: Sample χ^2 Reference Distribution (df = 1)

difference score, divide this squared difference score by the respective expected category frequency, and then sum all of the resulting computations. These operations must be performed for each category, and χ^2 is the sum of the resultant scores. When df = 1 (number of categories minus 1), the χ^2 distribution is a z^2 distribution. If you look in a z table (e.g., Aron and Aron 1997; or as illustrated in the excerpt from this table presented on page 66 you will see that $z_{.05} = \pm 1.96$ for a two-tailed test (.025 of the total distribution for $z \geq +1.96$ and .025 of the total distribution for $z \leq -1.96$), and $1.96^2 = 3.84$. Reference distributions for χ^2 are given in Table A-7 in Appendix A at the back of the book; you can see from this table that with df = 1, $\chi^2_{.05} = 3.84$. An approximate χ^2 reference distribution for df = 1 is sketched in Figure 13.1.

III. An Application with Non-Experimental Data

In non-experimental applications the researcher does not introduce an independent variable into the situation in which observations are recorded. For example, a consultant working for a Congress member may be engaged to find out whether the Congress member's constituents favor version A or version B of a universal healthcare plan (or neither). From a registered voter list, he or she randomly samples 300 individuals and conducts an opinion survey. Each individual is then classified as "Pro-A," "Pro-B," or "Neither." The number of individuals in each category represents the observed frequencies. However, in order to do a χ^2 test, we must also have expected frequencies for each category. Where do we find these? What would be expected will most likely be determined by a particular statistical or empirical model. For example, if the congress member wanted to know if these three categories were equally popular, then the expectation is that there would be the same number of individuals in each category. With a total of 300 individuals observed, this equal popularity assumption translates to an expectation of 100 individuals in each of the Pro A, Pro B, and Neither categories.

If we make up some data for this situation, we can illustrate the steps involved in calculating χ^2 to test if observed category frequencies differ significantly from expected category frequencies based on an H_0 that the three response categories are equally popular. The data in Table 13.1 are fictitious observed category frequencies, with corresponding expected category frequencies (based on the "equal popularity" assumption) in italics below the observed frequencies.

Table 13.1: Observed (f_O) and Expected (f_E) Frequencies for Survey of 300 Registered Voters

	For Plan A	For Plan B	For Neither
Observed Frequency	85	95	120
Expected Frequency	*100*	*100*	*100*

If you can follow the instructions given in the χ^2 formula, computing χ^2 when there are only three categories is not too complicated: $\chi^2 = \Sigma([f_O - f_E]^2 \div f_E)$: For Plan A, $f_O = 85$ and $f_E = 100$; for Plan B, $f_O = 95$ and $f_E = 100$; and for Neither Plan, $f_O = 120$ and $f_E = 100$. Substituting these values in the χ^2 formula and doing the necessary subtraction, squaring, division, and summing, we have:

$$\chi^2 = ([85 - 100]^2 \div 100) + ([95 - 100]^2 \div 100) + ([120 - 100]^2 \div 100)$$

$$\chi^2 = ([-15]^2 \div 100) + ([-5]^2 \div 100) + ([20]^2 \div 100) = (225 \div 100) + (25 \div 100) + (400 \div 100)$$

$$\chi^2 = 2.25 + .25 + 4.00 = \mathbf{6.50}$$

This χ^2 example has three categories, thus df = 3 – 1 = 2. From Table A-7 in Appendix A, for $p \leq .05$, $\chi^2_{crit}(2) = 5.99$. Given that the obtained $\chi^2 = 6.50$ exceeded this critical value of $\chi^2_{crit} = 5.99$ from the appropriate reference distribution, the null hypothesis that Plan A, Plan B, and Neither Plan are equally preferred alternatives is rejected.

For a second χ^2 illustration let's consider a situation similar to a two-factor ANOVA, *viz.*, a situation involving two classification variables. For the preceding problem we considered preferences for "universal health care" options (A, B, or Neither). Assume a researcher added a second classification variable, categorizing respondents based on self-reported annual income level. The categories distinguishing respondents were $40,000 or less; $40,001 to $80,000; $80,001 to – $120,000; and $120,001 or more. Fictitious data appear in the following table:

Table 13.2: Observed Frequencies for 300 Registered Voters across Four Income Levels

Income Level:	Plan A	Plan B	Neither Plan	Σ
\$40,000 or less	20	30	15	**65**
\$40,001 to \$80,000	30	35	25	**90**
\$80,001 to – \$120,000	20	20	35	**75**
\$120,001 or more	15	10	45	**70**
Σ	**85**	**95**	**120**	**300**

The first step in computing χ^2 is to determine an expected cell frequency for each of the 12 cells in Table 13.2. Each expected cell frequency may be calculated from the "marginal" sums. That is, referring to Table 13.2 we can see that a total of 85 of the 300 respondents favored Plan A. **If annual income level is unrelated to plan preference**, then proportion (85 ÷ 300 = .283) favoring Plan A should be similar across income levels. We should expect about 28.3% of the 65 respondents in the "less than \$40,000 income category to choose Plan A; 28.3% of the 90 respondents in the \$40,001–80,000 income category to choose Plan A, and the same for the remaining two income categories. Expectations for Plan B are based on the fact that 95 of 300 respondents (31.7%) favored Plan B. Thus, cell expectations for the four income levels favoring Plan B is 31.7% (e.g., an expected frequency favoring Plan B for the \$40,000 or less category of 65 × .317 = 20.6). One hundred and twenty respondents chose "Neither," hence 40.0% (120 ÷ 300 = .400) of respondents at each income level would be expected to select neither plan if income level is unrelated to plan preference. The 12 expected cell frequencies appear in Table 13.3.

Table 13.3: Expected Cell Frequencies for Table 13.2

Income Level:	For Plan A	For Plan B	For Neither Plan	Σ
\$40,000 or less	18.4	20.6	26.0	**65**
\$40,001 to \$80,000	25.5	28.5	36.0	**90**
\$80,001 to – \$120,000	21.2	23.8	30.0	**75**
\$120,001 or more	19.8	22.2	28.0	**70**
Σ	**85**	**95**	**120**	**300**

Solving for χ^2 involves a lengthy series of 12 appropriate substitutions in the formula for χ^2: $\chi^2 = \Sigma([f_O - f_E]^2 \div f_E)$.

$\chi^2 = (20 - 18.4)^2 \div 18.4 + (30 - 25.5)^2 \div 25.5 + (20 - 21.2)^2 \div 21.2 + (15 - 19.8)^2 \div 19.8 + (30 - 20.6)^2 \div 20.6 + (35 - 28.5)^2 \div 28.5 + (20 - 23.8)^2 \div 23.8 + (10 - 22.2)^2 \div 22.2 + (15 - 26.0)^2 \div 26.0 + (25 - 36.0)^2 \div 36.0 + (35 - 30.0)^2 \div 30.0 + (45 - 28.0)^2 \div 28.0$

$\chi^2 = .14 + .79 + .07 + 1.16 + 4.29 + 1.48 + .61 + 6.70 + 4.65 + 3.36 + .83 + 10.32 = \mathbf{31.40}$

Degrees of freedom for a two-factor χ^2 are determined by multiplying one minus the number of rows in the data matrix by one minus the number of columns: (r – 1)(c – 1), where r and c refer to rows and columns respectively. There are four rows and three columns in Table 13.2; thus, df = (4 – 1)(c – 1) = (3)(2) = 6. Critical values for χ^2 are given in Table A-7 in Appendix A; and with df = 6, $\chi^2_{crit} = 12.59$ for $p \leq .05$. Since χ^2 calculated from this data set exceeds χ^2_{crit}, H_0 is rejected, and the conclusion is that preferences for different plans are not the same across different income levels.

IV. An Application with Experimental Data

Researchers often collect categorical data in the context of a quasi-experiment (e.g., health plan preferences may be compared between self-selected groups of Democrats and Republicans) or in an experimental setting with research participants randomly assigned to different experimental groups. Individuals in the different groups are classified into different categories based on their behavior or responses to questions defined by the independent variable. As noted earlier, an experiment by Loftus and Palmer (1974) included categorical data: participants responded "yes" or "no" to a question about presence of broken glass at the scene of an accident. There were 50 subjects in each of three groups. All participants viewed a film clip of an automobile accident. On a post-viewing questionnaire, 50 subjects had a question asking for a speed estimate using the verb "smashed" to describe the collision, 50 subjects had the same question with "smashed" replaced by the verb "hit," and 50 subjects did not have a speed-estimate question. One week later all subjects were asked a yes–no question concerning the presence of broken glass; participants were entered into an appropriate category based on their responses. This three-group experiment will serve our purpose for illustrating χ^2 with experimental data. Loftus and Palmer's results for the broken-glass question are presented in Table 13.4.

Table 13.4: Observed Frequencies of Subjects Who Reported Seeing and Not Seeing Broken Glass (Data from Loftus and Palmer 1974).

	Verb in Speed-Estimate Question			
	Smashed	Hit	Control	Σ
Number Responding "Yes"	16	7	6	**29**
Number Responding "No"	34	43	44	**121**
Total	**50**	**50**	**50**	**150**

The essential key to calculating χ^2 is determining expected values. What is a reasonable expectation for the distribution of "yes" and "no" responding to the broken-glass question if H_0 is true? If the speed question and specific verbs have no influence on memory for whether broken glass was present at the scene of the accident, then a fair expectation is that the three groups will be roughly comparable in terms of number of "yes" and "no" respondents. However, just expecting them to be comparable does **not** mean that there should be 50% "yes" and 50% "no" respondents in each group. The problem is that we do not know what to expect with regard to proportion of subjects reporting broken glass after viewing this particular filmed accident scene, so we will determine expectations empirically from the data at hand. The reasoning for doing this is as follows: Based on the marginal totals (in bold), you can see that a total of 29 out of 150 subjects responded "yes" (**19.33%**), and 121 out of 150 subjects responded "no" (**80.67%**) to the broken-glass question. If H_0 were true, then responses from participants should have been the same for these three groups. Hence we should **expect** 19.33% of the subjects in each group to be in the "yes" category and 80.67% in each group to be in the "no" category. There were 50 subjects in each group, and 19.33% of 50 = 9.67, therefore the expected number of "yes" respondents in each group is 9.67. Similarly, 80.67% of 50 = 40.33, thus the expected number of "no" respondents in each group is 40.33. While obviously it is not possible to have a fraction of a subject in any category, it is normally okay to carry out expected cell frequencies to one or more decimal places. Also, an important check for expected cell frequencies is to make sure that the marginal totals (for rows and columns) for the table of expected cell frequencies are the same as the marginal totals for the observed cell frequencies (e.g., the observed "yes" and "no" responses in each condition equaled 50, and the same is true for the expected "yes" and "no" responses in each condition, because expected cell frequencies merely determine the approximate distribution of the 50 subjects in each of the three groups). Having determined expected category frequencies, we can now calculate χ^2 for these data:

$$\chi^2 = \Sigma([f_O - f_E]^2 \div f_E)$$

$$\chi^2 = ([16 - 9.67]^2 \div 9.67) + ([7 - 9.67]^2 \div 9.67) + ([6 - 9.67]^2 \div 9.67) + ([34 - 40.33]^2 \div 40.33) + ([43 - 40.33]^2 \div 40.33) + ([44 - 40.33]^2 \div 40.33)$$

$\chi^2 = 4.14 + .74 + 1.39 + .99 + .18 + .33 = \mathbf{7.77}$ (same as Loftus and Palmer 1974)

The significance of the χ^2 test is determined by comparing the value calculated from the data with the appropriate critical value in Table A-7 in Appendix A. We need to know degrees of freedom in order to enter the table at the appropriate location. For this example problem involving two rows and three columns of data, df = (2 – 1)(3 – 1) = 2, and $\chi^2_{crit}(2) = 5.99$. Since the χ^2 for this data set exceeded 5.99, the null hypothesis (that the three groups do not differ) was rejected.

We will complete this section with a slightly more tedious (computationally) example, also from a published study (Cialdini, Reno, and Kallgren 1990). Nothing really changes in the process for a χ^2 test with a larger data set with more categories, except that the string of observed minus expected scores can get rather long, so it is important to be very careful in transferring observed and expected frequencies to their respective places in the χ^2 formula.

One of the experiments reported in the Cialdini et al. article examined whether people were more likely to litter if the surrounding area was messy rather than neat. To test if this was the case, people were handed a piece of paper announcing a minor community event soon after parking at a large amusement park. Assume that as they turned the corner from the parking lot, the immediate area had 0, 1, 2, 4, 8, or 16 of these announcement flyers scattered on the ground. An observer counted the number of people who discarded their flyer in this area, and the number who did not. That is, individuals in these six experimental conditions, differentiated in terms of number of discarded flyers ("trash") already visible, were categorized as either "litterers" or "non-litterers" based on what they did with the flyers they were given. The appropriate test statistic is χ^2, and the results from the Cialdini et al. experiment are presented in Table 13.5.

Table 13.5: Number of "Litterers" Observed by Cialdini et al. (1990)

	Discarded Flyers in the Vicinity						
	0	1	2	4	8	16	Σ
Number of Litterers	11	6	12	14	25	24	**92**
Number of Non-Litterers	49	53	47	46	35	36	**266**
Σ	**60**	**59**	**59**	**60**	**60**	**60**	**358**

Since we do not know what to expect with regard to typical littering behavior, we will determine expected category frequencies empirically, testing, in effect, an H_0 that littering behavior is the same regardless of amount of litter in the vicinity.[1] The proportion of litterers is determined by total number of individuals seen littering (92) divided by total number of individuals observed in this study (358): 92 ÷ 358 = .257. Similarly, the proportion of non-litterers is determined by total number of individuals not littering (266) divided by total number of individuals observed (358): 266 ÷ 358 = .743. For χ^2, the following question was tested: Were these proportions the same (statistically) for all groups (regardless of amount of litter in the immediate vicinity)? Multiplying the proportions by number of subjects in each group gives the expected category frequencies as indicated in Table 13.6. Again, note that the marginal totals are the same for both observed and expected frequency tables. For this analysis, expected cell frequencies have been carried out to two-decimal place accuracy. Table 13.6 presents a theoretical H_0 expectation of how the 60 subjects in the "0" group should be distributed between litterers and non-litterers, how the 59 subjects in the "1" group should be distributed between litterers and non-litterers, etc.

Table 13.6: Expected Category Frequencies for Littering Study

	Discarded Flyers in the Vicinity						
	0	1	2	4	8	16	Σ
Number of Litterers	15.42	15.16	15.16	15.42	15.42	15.42	**92**
Number of Non-Litterers	44.58	43.84	43.84	44.58	44.58	44.58	**266**
Σ	**60**	**59**	**59**	**60**	**60**	**60**	**358**

There are 12 cells in the χ^2 data matrix from this study. Thus, calculating χ^2 will require a string of 12 terms: an $(f_O - f_E)^2 \div f_E$ for each of the corresponding 12 f_O and f_E cells. Substituting the observed

1 The expected category frequencies developed by Cialdini et al. were weighted in accordance with their theory, rather than derived as they were in this example. Consequently, if you check their 1990 publication you will see that there is a discrepancy between the χ^2 value they reported and the one computed for the purpose of this illustration.

and expected cell frequencies in the formula and solving for χ^2, we have: $\chi^2 = ([11 - 15.42]^2 \div 15.42) + ([6 - 15.16]^2 \div 15.16) + ([12 - 15.16]^2 \div 15.16) + ([14 - 15.42]^2 \div 15.42) + ([25 - 15.42]^2 \div 15.42) + ([24 - 15.42]^2 \div 15.42) + ([49 - 44.58]^2 \div 44.58) + [53 - 43.84]^2 \div 43.84) + ([47 - 43.84]^2 \div 43.84) + ([46 - 44.58]^2 \div 44.58) + ([35 - 44.58]^2 \div 44.58) + ([36 - 44.58]^2 \div 44.58)$

$$\chi^2 = 1.27 + 5.53 + .66 + .13 + 5.95 + 4.77 + .44 + 1.91 + .23 + .05 + 2.06 + 1.65$$

$$\mathbf{\chi^2 = 24.65}$$

df = (r – 1)(c – 1) = (2 – 1)(6 – 1) = **5**, and from Table A-7 in Appendix A, for the .05 significance level $\chi^2_{crit}(5) = 11.07$. A χ^2 larger than $\chi^2_{crit}(5) = 11.07$ is significant, hence the null hypothesis that littering behavior is not influenced by amount of litter in the surrounding area is rejected.

V. Combining and Isolating Categories for Follow-Up χ^2 Comparisons

Much like the analysis of variance, a significant χ^2 with more than two categories does not pinpoint differences. While Cialdini et al. (1990) could conclude that littering was influenced by the presence of litter in the surrounding area, the significant χ^2 did not indicate whether littering varied between specific groups or combinations of groups (i.e., analytical comparison questions). The analytical comparison approach with χ^2 may take two forms: (1) A specific subset of categories may be compared (e.g., is littering more or less likely to occur if there is no litter in the area compared to a relatively neat area with only one piece of litter?); or (2) new **meaningful** categories may be created from the original data for comparison (e.g., is littering more or less likely to occur if there is only one or two pieces of litter in the vicinity compared to 16 pieces—meaningful categories of "a little" vs. "a lot" of litter present). In either case, the χ^2 test procedure is straightforward: simply prepare an appropriate table of observed category frequencies, then either derive expected category frequencies from the newly created table or derive them by using the proportions from the original data set (.257 litterers and .743 non-litterers for this data set) and solve for χ^2. We will illustrate procedures for doing these two specific analytical comparisons using the first alternative of deriving new expected cell frequencies from the data involved for the analytical comparison of interest. The first analytical comparison ("0" vs. "1") requires isolating two of the original six litter conditions. The second analytical comparison ("1 + 2" vs. "16") requires creating a new category (combining "1" and "2") to contrast with another category ("16" in this example; however, depending on the question of interest, the comparison category could also be a new combination, such as a combination of "8" and "16"). We will see later that the χ^2 approximation is not recommended when the expected frequency for one or more categories is 5 or fewer, thus there may be situations for which a researcher may decide to combine categories in a meaningful way before even initiating a χ^2 analysis.

1. Isolating Categories: Does the "0" group differ from the "1" group?

Step 1: Prepare a table of observed category frequencies (f_O) restricted to the groups of interest.

	Discarded Flyers in the Vicinity		
	0	1	Σ
Number of Litterers	11	6	**17**
Number of Non-Litterers	49	53	**102**
Σ	**60**	**59**	**119**

Step 2: Prepare a table of expected category frequencies derived from the observed frequencies. The expected category frequencies in the table that follow were determined from the proportion of litterers (17 ÷ 119 = **.143**) and the proportion of non-litterers (102 ÷ 119 = **.857**) derived from this table of observed category frequencies.

	Discarded Flyers in the Vicinity		
	0	1	Σ
Number of Litterers	(.143)(60) = 8.57	(.143)(59) = 8.43	**17**
Number of Non-Litterers	(.857)(60) = 51.43	(.857)(59) = 50.57	**102**
Σ	**60**	**59**	**119**

Step 3: Using the four observed minus expected category frequency scores, solve for χ^2.

$$\chi^2 = ([11 - 8.57]^2 \div 8.57) + ([6 - 8.43]^2 \div 8.43) + ([49 - 51.43]^2 \div 51.43) + ([53 - 50.57]^2 \div 50.57)$$

$$\chi^2 = .69 + .70 + .11 + .12 = \mathbf{1.62};\ df = (r - 1)(c - 1) = (2 - 1)(2 - 1) = \mathbf{1}$$

From Table A-7 in Appendix A, $\chi^2_{crit}(1) = 3.84$; and since χ^2 calculated from the restricted data set for this analytical comparison was 1.62 [less than $\chi^2_{crit}(1)$], the difference in littering between these two categories is not significant.

2. Combining categories: Does presence of small amounts of litter (categories "1" and "2" combined) differ from presence of large amounts of litter (category "16")?

Step 1: Prepare a table of observed category frequencies relevant for the contrast of interest.

	Discarded Flyers in the Vicinity		
	1 & 2	16	Σ
Number of Litterers	18 (6 + 12)	24	**42**
Number of Non-Litterers	100 (53 + 47)	36	**136**
Σ	**118**	**60**	**178**

Step 2: Prepare a table of expected category frequencies derived from the observed frequencies. The proportion of litterers is 42 ÷ 178 = **.236**; and the proportion of non-litterers is 136 ÷ 178 = **.764**. The table of expected category frequencies follows:

	Discarded Flyers in the Vicinity 1 & 2	16	Σ
Number of Litterers	(.236)(118) = 27.84	(.236)(60) = 14.16	**42**
Number of Non-Litterers	(.764)(118) = 90.16	(.764)(60) = 45.84	**136**
Σ	**118**	**60**	**178**

Step 3: Using the four observed minus expected category frequency scores, solve for χ^2.

$$\chi^2 = ([18 - 27.84]^2 \div 27.84) + ([24 - 14.16]^2 \div 14.16) + ([100 - 90.16]^2 \div 90.163) + ([36 - 45.84]^2 \div 45.84)$$

$\chi^2 = 3.48 + 6.84 + 1.07 + 2.11 = \mathbf{13.50}$; df = (r – 1)(c – 1) = (2 – 1)(2 – 1) = **1**

From Table A-7 in Appendix A, $\chi^2_{crit}(1) = 3.84$: Since $\chi^2 = 13.50$ for this restricted data set exceeds χ^2_{crit}, the difference in littering when small vs. large amounts of litter are in the immediate area is significant.

VI. Using SPSS with Chi Square

Doing Chi-Square problems by hand is tedious and can lead to simple errors due to numerous copying, squaring, dividing, and adding operations. Using SPSS software can eliminate these problems given careful inputting of data. As we encountered with inputting data for the between-subject ANOVA, Chi-Square makes use of "dummy" codes. Categories are assigned a unique code for each level of each categorization variable (e.g., "1" for a_1, "2" for a_2, "3" for a_3, etc.; and "1" for b_1, "2" for b_2, etc.). Entering data in an SPSS spreadsheet for Chi-Square only requires entering these "dummy" codes appropriately. For the example problem that follows, the first column has Factor A codes (the first 20 entries are "1" corresponding to the classification of "Democrat"). As you can see, the second column has the coded subject classifications for candidate preferences (Factor B), with the first 13 entries of "1" (Democrats for Candidate 1) and the next seven entries of "2" (Democrats for Candidate 2).

Below is the observed frequency table for a poll.

f_o	Democrat	Republican	Independent	**Total**
Candidate 1	13	11	6	**30**
Candidate 2	7	9	14	**30**
Total	**20**	**20**	**20**	**60**

a_1 = Democrat
a_2 = Republican
a_3 = Independent

b_1 = Candidate 1
b_2 = Candidate 2

Table 13.7: Data View Screen for SPSS for a Chi-Square

	Category	Response		Category	Response
1	1	1	31	2	1
2	1	1	32	2	2
3	1	1	33	2	2
4	1	1	34	2	2
5	1	1	35	2	2
6	1	1	36	2	2
7	1	1	37	2	2
8	1	1	38	2	2
9	1	1	39	2	2
10	1	1	40	2	2
11	1	1	41	3	1
12	1	1	42	3	1
13	1	1	43	3	1
14	1	2	44	3	1
15	1	2	45	3	1
16	1	2	46	3	1
17	1	2	47	3	2
18	1	2	48	3	2
19	1	2	49	3	2
20	1	2	50	3	2
21	2	1	51	3	2
22	2	1	52	3	2
23	2	1	53	3	2
24	2	1	54	3	2
25	2	1	55	3	2
26	2	1	56	3	2
27	2	1	57	3	2
28	2	1	58	3	2
29	2	1	59	3	2
30	2	1	60	3	2

To perform a Chi-Square, you begin by going to "Analyze" on the tool bar at the top of the screen. Scroll down to "Descriptive Statistics," then click on "**Crosstabs.**" This action will open a dialog box similar to the one in Table 13.8).

Table 13.8: SPSS Dialog Box for Crosstabs

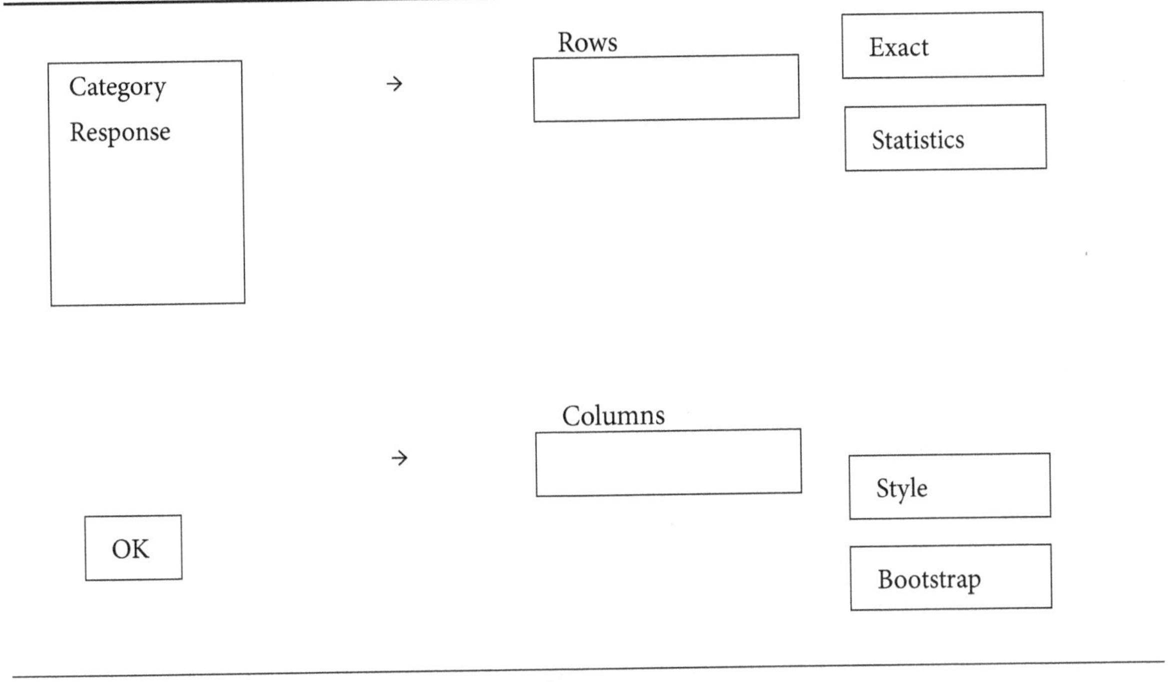

Since "Category" or Political Affiliation was Factor A, you need to highlight "Response" and click on the arrow to the left of the "Rows" box. Highlight "Category" and click on the arrow to the left of the "Columns" box. Click on the "Statistics" button to the right of the boxes and a new dialog box will open. Click on the "Chi-Square" box and then click on the "Continue" button. Move your cursor and click on "OK" to produce an observed frequency table and a Chi-Square test box.

Table 13.9: SPSS Output for Chi-Square

Response * Category Crosstabulation

Count

		Category			Total
		1	2	3	
Response	1	13	11	6	30
	2	7	9	14	30
Total		20	20	20	60

Chi-Square Tests

	Value	df	Asymp. Sig. (2-sided)
Pearson Chi-Square	5.200[a]	2	.074
Likelihood Ratio	5.320	2	.070
Linear-by-Linear Association	4.818	1	.028
N of Valid Cases	60		

a. 0 cells (0.0%) have expected count less than 5. The minimum expected count is 10.00.

The SPSS analysis show $\chi^2 = 5.200$, $p = .074$. Thus, the conclusion is that Democrats, Republicans, and Independents did not differ significantly (at $p \le .05$) in their preferences for these two candidates.

VII. Effect Size and Power

The metric commonly used for estimating effect size for χ^2 is ϕ for a 2 × 2 contingency table, and a 1946 modification by Cramér (ϕ_C) for contingency tables that are larger than 2 × 2 (see Sheskin 1997, pages 245–246). To provide general parameters for ϕ, Cohen (1988) offered the following framework: $\phi \ge .10$, but $< .30$ indicates a "small" effect of the independent variable transitioning to a medium effect; $\phi \ge .30$, but $< .50$ indicates a "medium" effect transitioning to a large effect; and $\phi > .50$ indicates a "large" effect. For experiments with two levels of the independent variable (forming a 2 × 2 contingency table, e.g., comparing litterers and non-litterers when either 0 or 1 piece of litter is visible in the surrounding area or when 1 or 2 pieces of litter are present compared to 16 pieces of litter present), $\phi = |\sqrt{(\chi^2 \div N)}|$; where **N is the total number of observations recorded summed across all categories** (not lower case "n," which would represent the number of individuals within a single category) and the two lines bracketing $\sqrt{(\chi^2 \div N)}$ indicate the absolute (positive) value of the square root. For contingency tables that involve more than two levels of the independent variable, ϕ_C ("C" for Cramér) $= |\sqrt{(\chi^2 \div [N][k - 1])}|$; where **k = number of levels of the independent variable.**

There were 6 levels of the independent variable in the Cialdini et al. experiment (0, 1, 2, 4, 8, or 16 pieces of paper in the litter-observation area). From the overall χ^2 analysis, $\chi^2 = 24.65$; $N = 358$; and $k = 6$. Thus, $\phi_C = |\sqrt{(24.65 \div [\{358\}\{5\}])}| = |\sqrt{(24.65 \div 1{,}790)}| = |\sqrt{.0138}| =$ **.12** (a small effect according to Cohen's guidelines). What was the size of the χ^2 effect reported in the follow-up analysis between small amounts of visible litter (the combination of the 1 and 2 pieces of litter groups) vs. large amounts of litter (16 pieces of litter in the area)? For that 2 × 2 contingency table, $N = 178$; $\chi^2 = 13.50$; and $\phi = |\sqrt{(13.50 \div 178)}| = |\sqrt{.0758}| =$ **.28** (close to a medium-sized effect). Procedures for assessing power and estimating sample size necessary to achieve desired levels of power for χ^2 are difficult computationally and conceptually (see Siegel and Castellan, Jr. 1988), and they are not detailed in this text. The interested reader may wish to review Keppel et al. (1992) for hand-calculation power analyses, or Erdfelder et al. (1996) for computer software alternatives.

VIII. Caveats

The distribution of a χ^2 statistic computed from a data set is an approximation to a theoretical χ^2 distribution, and the two distributions are identical for all intents and purposes when the sample data set is large. The χ^2 reference distribution is not appropriate with small data sets. Small is often interpreted to mean any data set for which one or more categories has an **expected frequency** of 5 or fewer cases. In addition, the χ^2 reference distribution is based on the assumption that category entries are independent of one another; consequently χ^2 is not appropriate for data sets that count the same individual in two or more categories. Over 50 years ago researchers were cautioned about misapplying the χ^2 test while emphasizing these points: (1) a reasonable size for expected cell frequencies, and (2) independence among category entries (Lewis and Burke 1949).

What alternatives should be considered when the requirements stated above are not met? Obviously if expected category frequencies are too low, it may be possible to solve this problem simply by collecting more data. If it appears that only 10% of people litter, then the research plan should include observing over 50 individuals in each group. An after-the-fact solution to increase group size (and expected cell frequencies) by combining groups could be considered when it can be done meaningfully. A correction procedure (Yates' correction for continuity) has been suggested for use with small samples. However, since Sheskin (1997) noted that some sources recommend that Yates' correction for continuity should always be employed, and other sources recommend that Yates' correction overcorrects and should never be used, this procedure will not be detailed here. Finally, an exact test of probabilities (the Fisher exact test) similar to the binomial procedures described in Chapter 4 may provide an acceptable alternative in some situations (see Sheskin 1997; Siegel and Castellan, Jr. 1988).

When the same subjects appear in two or more categories, a χ^2 modification may be considered (Cochran 1950). Since the independence problem represents a possible "misuse" of χ^2 often encountered in small-scale projects (e.g., student class projects), the Cochran Q test for χ^2 will be illustrated in Appendix XIII. A at the end of this chapter. The Cochran Q test requires additional information as it is necessary to have detailed tracking of how each individual is counted across categories.

IX. Summary

Not all psychological research involves behavioral measures that provide scores along a presumably continuous scale. For situations that involve assignment of individuals or events to categories, observed distributions of individuals across categories may be compared statistically to chance expectations. The chi-square statistic (χ^2) is appropriate in many cases for statistical tests of categorical data. A distinction was made between non-experimental and experimental data in terms of the inclusion of an independent variable in an experimental design. Chi-square is an appropriate test statistic for both non-experimental and experimental categorical data.

Illustrations of χ^2 computational procedures were provided using both published data and fictitious data. In addition, procedures for follow-up, detailed analyses (analytical comparisons) for χ^2 were illustrated. These procedures required an initial step of separating categories of interest or combining existing categories into new meaningful categories of interest. For both approaches, new categories and corresponding observed frequencies are used to determine appropriate expected category frequencies for calculating χ^2.

Effect size for χ^2 is estimated by ϕ, a metric based on dividing χ^2 by the total number of subjects (N) in the sample, and then taking the positive square root: $|\sqrt{(\chi^2 \div N)}|$. A modification of this estimate (Cramér's ϕ_C) was suggested for experimental data with more than two levels of the independent variable. Cohen provided a context for ϕ, identifying .10 as small, .30 as medium, and .50 as large effect sizes.

Researchers must recognize that there may be concerns about whether assumptions of the underlying model for χ^2 applications have been met. The two most notable among these underlying assumptions involve issues of sample size and independence of category entries. Design modifications should be considered when research projects are planned with small sample sizes that could likely result in one or more expected category frequencies having an estimate of five or fewer subjects. If sample size is a problem, one could increase number of subjects or combine categories in a meaningful way, or turn to alternative statistical procedures for testing H_0 (e.g., Fisher Exact Test). Also, if category frequencies are not independent (i.e., a given subject may be entered in two or more categories), χ^2 may be adjusted appropriately, as illustrated with the Cochran Q test.

X. Appendix 13.A: The Cochran Q Test

Assume an experimenter allows subjects one minute to solve 6 anagram problems. Also, assume there are 30 individuals, and each person is tested on three different occasions. Thus, each subject would attempt to solve a total of 18 anagrams during the course of the experiment (6 anagrams on each testing occasion). Each anagram has five letters, which when rearranged form a common English word (e.g., the letters D E U J G can be rearranged to spell the word JUDGE). We will assume that the experimenter is interested in whether relations between positions of letters in anagrams and their respective solution words affect problem solving. One response measure she used was a dichotomous score based on whether all six problems were solved within a one-minute period. On one occasion, the six anagrams had no common letter positions with their solution words (as for D E U J G—JUDGE). On another testing occasion the final two letter positions were common for respective anagrams and solution words (e.g., R A F M E—FRAME); and for the remaining testing condition, the initial two letter positions were common (e.g., D A E C N—DANCE). We will assume that test order and anagram problems were properly balanced across the experimental conditions.

The Cochran Q test is illustrated with the fictitious data presented in Table 13.10. The primary data in Table 13.10 appear in the columns headed by abbreviations of the common letter positions between corresponding anagrams and solution words: "0 Common," "2 Final," and "2 Initial." The data entered are either the number "1," a category code indicating that all six anagrams were solved within one-minute time limit, or "0," a category code indicating that fewer than six anagrams were solved within a one-minute time limit. Since each subject was tested in each condition, each subject could be counted in more than one category. There are three key computational components in the Cochran Q test: (1) the number of categories (k), which equals 3 in this example problem; (2) the column (C_i) sums; and (3) the row (R_j) sums. The data for this illustration are presented in Table 13.10.

Table 13.10: Fictitious Anagram Data

Subjects	0 Common	Common Anagram/Solution Word Letter Positions 2 Final	2 Initial	Row Sum(R)	R^2
1	1	0	1	2	4
2	0	1	1	2	4
3	0	1	0	1	1
4	0	1	1	2	4
5	1	1	1	3	9
6	0	0	0	0	0
7	1	0	1	2	4
8	0	1	1	2	4
9	0	1	0	1	1
10	0	0	1	1	1
11	1	1	0	2	4
12	0	1	1	2	4
13	1	1	1	3	9
14	0	0	1	1	1
15	0	1	1	2	4
16	1	1	1	3	9
17	0	0	1	1	1
18	0	1	1	2	4
19	0	0	1	1	1
20	0	1	0	1	1
21	1	1	1	3	9
22	1	0	1	2	4
23	0	1	1	2	4
24	0	0	1	1	1
25	1	1	1	3	9
26	0	0	0	0	0
27	1	1	1	3	9
28	0	1	1	2	4
29	0	1	1	2	4
30	0	1	1	2	4
C_i(columns)	**10**	**20**	**24**	**54**	**118**
C_i^2	**100**	**400**	**576**	**and $\Sigma C^2 = 1{,}076$**	

Cochran's Q is an adjusted chi-square value, and $Q = (k - 1)(k[\Sigma C^2] - [\Sigma R]^2) \div (k[\Sigma R] - [\Sigma R^2])$; where k = # of categories = 3; $\Sigma C^2 = 1{,}076$; $\Sigma R = 54$; and $\Sigma R^2 = 118$. For this problem:

$$Q = (3 - 1)(3[1{,}076] - 54^2) \div (3[54] - 118) = 2(3{,}228 - 2{,}916) \div (162 - 118) = 624 \div 44$$

Q = 14.18

Cochran's Q is evaluated by comparing Q with χ^2_{crit} from the χ^2 table (Table A-7 in Appendix A), and with df = 2, a value of 5.99 or larger is necessary for significance at $p \leq .05$. Therefore, it may be concluded that number of subjects receiving perfect scores differed significantly across the three experimental conditions.

As a final exercise, we will use this fictitious data set to illustrate what the scope of the problem is in this situation if a χ^2 test had been done with subjects counted in more than one category. That is, what does χ^2 equal for this specific anagram problem? Table 13.11 is a χ^2 table showing the number of perfect scores (coded as "1" in Table 13.10) and the number of non-perfect scores (coded as "0" in Table 13.10) for each anagram-solving condition. The marginal totals in Table 13.11 indicate that 54 out of a total of 90 observations were perfect scores (i.e., solved all six anagrams in the set in the allocated one-minute time period), and 36 out of 90 observations were failures to solve at least one of the anagrams from a given set. The expected category frequencies in the table were determined by multiplying the total number of observations in each condition (30) by .60 (the proportion of perfect scores: 54 ÷ 90 = .60) and by .40 (the proportion of failures to achieve perfect scores: 36 ÷ 90 = .40). Our interest in this exercise is the comparison of χ^2 with Q. Thus, in addition to another χ^2 calculation opportunity, comparing the difference between χ^2 and Q allows us for this one case to gauge the extent of the problem for H_0 testing that would have resulted from applying χ^2 with subjects counted in more than one category.

Table 13.11: Observed and Expected (in parentheses) Frequencies for Fictitious Anagram Data

	Common Letters in Respective Anagrams and Solution Words			
	Ø Common	Final 2	Initial 2	Σ
Number of Perfect Scores	10 (18)	20 (18)	24 (18)	**54**
Number of Failures	20 (12)	10 (12)	6 (12)	**36**
Σ	**30**	**30**	**30**	**90**

$$\chi^2 = \Sigma([f_O - f_E]^2 \div f_E)$$

$$\chi^2 = ([10 - 18]^2\ 18) + ([20 - 18]^2 \div 18) + ([24 - 18]^2 \div 18) + ([20 - 12]^2 \div 12) + ([10 - 12]^2 \div 12) + ([6 - 12]^2 \div 12)$$

$\chi^2 = 3.56 + .22 + 2.00 + 5.33 + .33 + 3.00 =$ **14.44** (in this case, Cochran Q and χ^2 values differed by .26!)

13.1: Model Problem. We will use fictitious data from a contrived experiment to illustrate analyses of categorical data. Assume subjects are asked to participate in small groups to discuss relevant campus topics. The discussion groups consist of four individuals: one is the subject in the experiment and the other three are confederates of the experimenter. One confederate (AR) argues with points raised by the subject, another (AG) agrees with the subject, and the third confederate maintains a neutral position (N). Before dismissing the subject at the conclusion of a 30-minute discussion session, the experimenter indicates to the subject that he or she may be called back to participate in a two-person discussion. The subject is then asked to choose which fellow discussant he or she would prefer to meet with if a second session were to take place. There are three different discussion topics; one dealing with student government, one with a proposed tuition surcharge to support intercollegiate athletics, and one with a campus-wide smoking ban. Assume the results for the number of subjects selecting each confederate for a follow-up discussion were as follows:

	Discussion Topics			
Confederates	Student Government	Tuition Surcharge	Smoking Ban	Σ
AR	14	11	5	**30**
AG	16	19	30	**65**
N	15	15	10	**40**
Σ	**45**	**45**	**45**	**135**

1. Do a χ^2 test for these data to evaluate whether discussion topic affects the choice of an arguing, agreeing, or neutral partner for a possible future discussion. The first step requires creating a table of expected frequencies paralleling the table of observed frequencies. Without *a priori* knowledge of expected frequencies, they will be derived empirically based on the assumption that if choices for a partner for future discussions are not related to discussion topics, then choice of an arguing, agreeing, or neutral partner ought to be the same. Since 30 out of 135 participants (22%) selected the "arguing" confederate, then the expected frequency for subjects from each discussion topic condition is $(30 \div 135) \times 45 = 10$. Similarly, the "agreeing" confederate was the choice 65 out of 135 times, hence if discussion topic is not relevant for choices, the expected frequency of choosing the agreeing confederate for each discussion topic is $(65 \div 135) \times 45 = 21.67$. Finally, 40 out of 135 = 13.33, which is the number expected for selections of the neutral confederate. The table of expected frequencies follows:

	Discussion Topics			
Confederates	Student Government	Tuition Surcharge	Smoking Ban	Σ
AR	10.00	10.00	10.00	**30**
AG	21.67	21.67	21.67	**65**
N	13.33	13.33	13.33	**40**
Σ	**45**	**45**	**45**	**135**

Solving for $\chi^2 = \Sigma(f_O - f_E)^2 \div f_E$ we have:

$\chi^2 = ([14 - 10]^2 \div 10) + ([11 - 10]^2 \div 10) + ([5 - 10]^2 \div 10) + ([16 - 21.67]^2 \div 21.67) + ([19 - 21.67]^2 \div 21.67) + ([30 - 21.67]^2 \div 21.67) + ([15 - 13.33]^2 \div 13.33) + ([15 - 13.33]^2 \div 13.33) + ([10 - 13.33]^2 \div 13.33)$

$\chi^2 = 1.60 + .10 + 2.50 + 1.48 + .33 + 3.20 + .21 + .21 + .83 = \mathbf{10.46}$; and df = (r – 1)(c – 1) = **4**

From Table A-7 in Appendix A, for df = 4, $\chi^2_{crit(.05)} = 9.49$; therefore the observed category frequencies are significantly different from the expected category frequencies as $\chi^2_{obs} > \chi^2_{crit}$.

2. Assume the experimenter recorded the discussions and noted that the surcharge and smoking groups generated more emotion than the student government groups. Do a follow-up analysis to contrast the so called "low-emotion" and "high-emotion" topics. To test this difference it will be necessary to create a new χ^2 contingency table by combining the surcharge and smoking groups ("high emotion").

Discussion Topics and Observed Cell Frequencies

Confederates	Low Emotion	High Emotion	Σ
AR	14	16	**30**
AG	16	49	**65**
N	15	25	**40**
Σ	**45**	**90**	**135**

The corresponding expected category frequencies are determined from the proportions in the table of observed category frequencies (e.g., the arguing confederate was the choice on 30 out of 135 occasions, and this proportion translates to 10 of the 45 individuals in the low-emotion topic group):

Discussion Topics and Expected Cell Frequencies

Confederates	Low Emotion	High Emotion	Σ
AR	(30)(45) ÷ 135 = 10.00	(30)(90) ÷ 135 = 20.00	**30**
AG	(65)(45) ÷ 135 = 21.67	(65)(90) ÷ 135 = 43.33	**65**
N	(40)(45) ÷ 135 = 13.33	(40)(90) ÷ 135 = 26.67	**40**
Σ	**45**	**90**	**135**

Now solving $\chi^2 = \Sigma(f_O - f_E)^2 \div f_E$, we have:

$\chi^2 = ([14 - 10]^2 \div 10) + ([16 - 20]^2 \div 20) + ([16 - 21.67]^2 \div 21.67) + ([49 - 43.33]^2 \div 43.33) + ([15 - 13.33]^2 \div 13.33) + ([25 - 26.67]^2 \div 26.67)$

$\chi^2 = 1.60 + .80 + 1.48 + .74 + .21 + .10 = \mathbf{4.93}$; and df = (r – 1)(c – 1) = **2**

From Table A-7 in Appendix A, for df = 2, $\chi^2_{crit(.05)}$ = 5.99; therefore the new observed category frequencies are not significantly different from the expected category frequencies as $\chi^2_{obs} < \chi^2_{crit}$.

3. Using the metric ϕ, estimate the size of the significant χ^2 reported in Part A Since there are more than two levels of the independent variable selected for the comparison in Part A we will use ϕ_C instead of ϕ to estimate effect size where $\phi_C = |\sqrt{(\chi^2 \div [N][k - 1])}|$.

$$\phi_C = |\sqrt{\chi^2 \div [N][k- 1]}\ |$$

For the overall experiment, N = 135, k = 3, thus $\phi_C = |\sqrt{(10.46 \div [135][3 - 1])}| = |\sqrt{.0387}| = \mathbf{.20}$

13.2: Practice Problem 1. Assume five different statistics instructors taught the beginning course in statistics at a large Midwestern university. For a class project a student compared responses to the question: "Would you recommend this class to a friend?" Fictitious data (perhaps revealing the authors' bias that statistics classes are more popular than may truly be the case) for the number of "yes" and "no" responses from students in the various classes are presented in the following table. Is there a significant difference among instructors?

Observed Category Frequencies for Instructors A–E

Responses	A	B	C	D	E
YES	44	52	38	40	56
NO	8	10	6	15	9

13.3: Practice Problem 2. A researcher was interested in "change blindness," a phenomenon indicating that many people may be unaware of relatively large changes taking place around them. Assume an experimenter had a student confederate stop individuals on the street to ask for directions to a nearby shopping mall. As directions were being given, two other confederates carrying a door passed between the two parties having this conversation, and the confederate who originally asked for directions changed places with one of the door-carrying confederates as the door passed by. There were four conditions in this experiment, and the dichotomous behavior recorded was whether or not the subject providing directions indicated awareness of the change in the person asking for directions. The new person asking directions was either a student resembling (in clothes and features) the original student, a student who looked much different from the original student, a 45 year-old adult, or an individual in a gorilla costume. Imaginary data appear below.

Observed Category Frequencies for Change-Blindness Conditions

Responses	Similar	Dissimilar	Adult	Gorilla
Noticed	3	10	26	40
Not Noticed	45	34	20	4

A. What does χ^2 equal, and is it significant?

B. Provide an estimate of effect size based on the value of ϕ_C. Do you conclude that the χ^2 effect in this experiment is a "small" effect, a "medium" effect, or a "large" effect?

C. Do a follow-up χ^2 test to contrast the Similar Student category with the Dissimilar Student category, and provide an estimate of the effect size (e.g., ϕ) for this comparison.

D. Do an analytical comparison to contrast the two student-confederate groups combined with the group that had the 45-year old confederate.

13.4: Practice Problem 3. Disregard the small number of cases in the data set below (the purpose for presenting only a small data set is to keep the hand-calculation chore under control), and for practice, just proceed to test the null hypothesis using the Cochran Q adjustment for χ^2. Assume the data are from a series of four sign-language (American Sign Language) vocabulary tests given to five chimpanzees in a sign-language research project. The tests involved presenting pictures of 20 objects in succession; the chimpanzees' task was to sign the name of the object pictured. The first test was given when the chimps were approximately 6 months old, with successive tests given at 3-month intervals. A passing grade for the chimps was a score of 75% correct, and for each chimp at each test, a "0" indicates "failed" and a "1" indicates "passed." The data are presented in the table below. Is the Cochran Q significant?

Passing (1) and Failing (0) Test Scores as a Function of Age

	6 months	9 months	12 months	15 months
Chimp #1	0	0	1	1
Chimp #2	0	0	1	1
Chimp #3	0	1	1	1
Chimp #4	1	1	1	1
Chimp #5	0	0	0	1

Chapter 14

Correlation: Describing a Relation Between Two Variables

Smoking is one of the leading causes of statistics.
—Knebel

I. Correlational Research

Thus far the emphasis in this text has been on experimental research. Measures of some aspect of behavior are examined under different conditions (e.g., different subjective states such as happy or sad mood, different environmental contexts such as the presence or absence of distracting background noises during performance, different task characteristics such as memorizing lists of similar vs. lists of dissimilar words). However, many research activities do not involve experiments with manipulated independent variables. Researchers instead may be interested in behavior in more naturalistic settings and/or with simply discovering and **describing** relationships between different aspects of behavior (i.e., the extent to which two characteristics or behavioral measures of performance are related). In this regard, the focus is on the *correlation* between pairs of scores. An important feature for an index that purports to measure such "togetherness" is that the size of the index must reflect the tendency for individuals to have similar scores on the two measures, that is, with one set of scores labeled X and the other set of scores labeled Y, individuals who have high X scores will tend to have high Y scores, and individuals who have low X scores will tend to have low Y scores.

The correlational method is limited in the sense that **the demonstration of a correlation does not justify a clear cause–effect conclusion.** Just knowing that two aspects of behavior are related does not imply a causal relation, as the degree of successful performance in the area measured by X could be the cause for the degree of successful performance in the area measured by Y, or vice-versa. If this were the only problem, antecedent–consequent logic could be used to favor one of these two possibilities because the cause of a phenomenon would be expected to precede the appearance of the phenomenon (the effect). Using grades as a measure of learning, given a correlation between high school grades and college grades, it is rather difficult to argue that the later-occurring measure (reflecting what was learned in college) caused the earlier-occurring measure (reflecting what was learned in high school). However, the more significant problem in trying to infer causation from correlation is that some other unknown factor or factors may be the cause of variation in both the X and Y performance measures. Nonetheless, a correlation index is useful for describing a relationship between meaningful pairs of scores, and correlation information is especially relevant in the real world of predicting future behavior based on past behavior.

It is not uncommon in introductory statistics textbooks to encounter treatments of *correlation* paired with the related topic of *regression* in the same chapter (e.g., Anderson and Sclove 1986; Keppel and Saufley 1980; Keppel et al. 1992; Lockhart 1998; McClave and Sincich 2000). Assessing the relationship between two response measures is central to both topics. With regard to measuring correlation and regression, the reference is to **linear** correlation and regression (measuring the degree to which a straight line describes the relationship between two variables). For correlation, describing a relation is the primary interest; for regression, using the relation among two or more variables for prediction is the primary interest (e.g., having determined in general the degree to which X and Y are related, information on an X score for an individual can be used to predict his or her future Y score). A common measure for describing a correlation is the *Pearson Product Moment Correlation Coefficient* (commonly shortened to Pearson r or just r), and the common measure in regression analyses used for assessing the variability in Y that is accounted for by X is the squared correlation coefficient (the squared correlation coefficient, r^2, which commonly appears in upper case as R^2).

II. Developing a Descriptive Index of a "Co-Relation"

In the analysis of experiments, measurements of behavior (dependent variables) were examined as a function of independent variables (e.g., designed variations in characteristics of individuals, of tasks individuals were given to perform, and/or of the environment or context in which observations were recorded). Although correlations between an independent variable and a dependent variable can certainly be calculated, correlation indices are most commonly assessed between two performance measures (dependent variables). Up to this point, we have used Y_i as a code or label for the dependent variable (scores) in an experiment. The second dependent variable in correlational research is commonly coded by X_i; with the same subscripts for X and Y (e.g., X_1 and Y_1) representing two scores or measures obtained from the same individual (although scores from identical twins, fathers and daughters, or other meaningful pairings of individuals may be represented by corresponding X and Y scores). In calculating an index of correlation, it is essential to keep X and Y scores paired appropriately.

At this point we will try what is hoped will be viewed as a common-sense approach to developing a numerical index that is sensitive to the degree of relatedness between paired scores. "Relatedness" simply means that if an individual has a high score on one measure (X), he or she will also have a high score on the other measure (Y), and similarly, low scores on X will be associated with low scores on Y. As we will see later, relatedness in an inverse sense (low scores on one measure associated with high scores on the other measure) is equally informative for predictive purposes. However, for now the focus will be on the logic that an **index** sensitive to relatedness must vary systematically as a function of different degrees of relatedness, having a high value when paired X and Y scores are similar in order of magnitude within their respective distributions and a low value when paired X and Y scores are mismatched.

A key component for an index that is sensitive to degree of relatedness involves *cross products* of corresponding X and Y scores. The largest sum of cross products that can be obtained for any set of paired scores occurs when scores on both measures are ordered in terms of magnitude. This relationship may be illustrated by calculating a sum of cross products for all combinations of just three pairs of scores. There are six combinations to consider. We will use X scores of 1, 2, 6; and Y scores of 2, 7, 9 for this illustration. The six pairing combinations appear in the following table, with Set I matching magnitude of X and Y scores directly and Set VI showing inverse matched pairings.

Set I		Set II		Set III		Set IV		Set V		Set VI	
X	Y	X	Y	X	Y	X	Y	X	Y	X	Y
1	2	1	2	1	7	1	7	1	9	1	9
2	7	2	9	2	2	2	9	2	2	2	7
6	9	6	7	6	9	6	2	6	7	6	2

The sums of the cross products for each set are as follows, and although we cannot say what these values will be without doing the calculations, we can assert that $\Sigma X_i Y_i$ will be largest for Set I and smallest for Set VI. Thus, $\Sigma X_i Y_i$ is a key component for determining the degree to which there is a direct relation in the ordering of scores within their respective distributions. The sums of the cross products for each set are as follows:

Set I: (1)(2) + (2)(7) + (6)(9) = 2 + 14 + 54 = **<u>70</u>**

Set II: (1)(2) + (2)(9) + (6)(7) = 2 + 18 + 42 = **62**

Set III: (1)(7) + (2)(2) + (6)(9) = 7 + 4 + 54 = **65**

Set IV: (1)(7) + (2)(9) + (6)(2) = 7 + 18 + 12 = **37**

Set V: (1)(9) + (2)(2) + (6)(7) = 9 + 4 + 42 = **55**

Set VI: (1)(9) + (2)(7) + (6)(2) = 9 + 14 + 12 = **35**

While sums of cross products accomplish the desired objective for measuring relatedness between paired scores, they only provide the starting point. We will consider three alterations to the scores in Set I to illustrate problems that need to be avoided as this index is developed further. The first alteration shown in Set IA has six paired scores. This set of six scores was created by simply repeating the original three pairs of scores in Set I. The alteration for Set IB results in a new set of three paired scores created by adding a constant k to each X and Y score in the original Set I, where k = 10. The alteration for Set IC also results in a new set of three paired scores; however, this set was created by multiplying each X and Y score in the original Set I by the constant k, again with k = 10. The sums of the cross products (ΣX_iY_i) for these three modified sets are quite different from the original set; however, for each modified set, the matching relationship of paired X and Y scores has not been violated.

Set IA		Set IB		Set IC	
X	Y	X	Y	X	Y
1	2	11	12	10	20
2	7	12	17	20	70
6	9	16	19	60	90
1	2				
2	7				
6	9				

Solving ΣX_iY_i for the three modified sets of paired scores, we have:

Set I: (1)(2) + (2)(7) + (6)(9) = 2 + 14 + 54 = **70**

Set IA: (1)(2) + (2)(7) + (6)(9) + (1)(2) + (2)(7) + (6)(9) = 2 + 14 + 54 + 2 + 14 + 54 = **140**

Set IB: (11)(12) + (12)(17) + (16)(19) = **640**

Set IC: (10)(20) + (20)(70) + (60)(90) = 200 + 1400 + 5400 = **7,000**

Although the matching relationship did not change for Sets IA, IB, and IC, you can see that an index based only on the sum of cross products of paired scores changed rather dramatically with an increase in number of scores (Set IA); when a constant was added to each score (Set IB), and when each score was transformed by multiplying by a constant (Set IC).

The first problem that increasing number of paired scores affects the sum of the cross products can be solved by adding one more step to the calculation, viz., taking the average of the sum of cross products.

Dividing the sum of cross products by the number of pairs (mean of the cross products) results in an index that is the same for Set I and Set IA (for Set I, n = 3, and $\Sigma X_iY_i \div n = 70 \div 3 =$ **23.33**; for Set IA, n = 6, and $\Sigma X_iY_i \div n = 140 \div 6 =$ **23.33**).

The solution for removing the discrepancy in cross products between Set I and Set IB also requires an additional step which happens to be identical to one taken earlier when an index for calculating variances was developed, *viz.*, we considered each score in relation to the mean of its respective distribution. A similar step for cross products results in a sum of the cross products of *mean deviation scores* rather than a direct sum of cross products. Combining both operations (averaging cross products of deviation scores) results in identical index values for Sets I, IA, and IB; however, because of a considerable increase in the variability in the Set IC scores, a measure based only on $(\Sigma[X_i - \bar{X}][Y_i - \bar{Y}]) \div n$ is still problematic. One more modification is necessary to eliminate this last problem. The following $\Sigma(X_i - \bar{X})(Y_i - \bar{Y}) \div n$ computations for all four example sets confirm the claim that Set I = Set IA = Set IB, while Set IC remains problematic: For Sets I and IA, $\bar{X} = 3$ and $\bar{Y} = 6$; For Set IB, $\bar{X} = 13$ and $\bar{Y} = 16$; and for Set IC, $\bar{X} = 30$ and $\bar{Y} = 60$. For each set, the "mean deviation scores" are as follows:

Set I: ([1 – 3][2 – 6] + [2 – 3][7 – 6] + [6 – 3][9 – 6]) ÷ 3 = ([-2][-4] + [-1][1] + [3][3]) ÷ 3 = 16 ÷ 3 = **5.33**

Set IA: ([1 – 3][2 – 6] + [2 – 3][7 – 6] + [6 – 3][9 – 6] + [1 – 3][2 – 6] + [2 – 3][7 – 6] + [6 – 3][9 – 6]) ÷ 6 = ([-2][-4] + [-1][+1] + [3][3] + [-2][-4] + [-1][+1] +[(3][3]) ÷ 6 = 32 ÷ 6 = **5.33**

Set IB: ([11 – 13][12 – 16] + [12 – 13][17 – 16] + [16 – 13][19 – 16]) ÷ 3 = ([-2][-4] + [-1][+1] + [3][3]) ÷ 3 = 16 ÷ 3 = **5.33**

Set IC: ([10 – 30][20 – 60] + [20 – 30][70 – 60] + [60 – 30][90 – 60]) ÷ 3 = ([-20][-40] + [-10][+10] + [30][30]) ÷ 3 = 1,600 ÷ 3 = **533.33**

You can see that Set IC presents a rather large problem for the index developed to this point. Even though the transformation did not change the relationship between pairs of X and Y scores, an index measuring correlation in terms of average cross products of mean deviations scores is 100 times larger for Set IC than it is for Sets I, IA, and IB. When each score in a distribution is multiplied by a constant k, the variance among the set of scores is increased by the constant squared (k^2). This last contaminating factor may be removed by an operation that will convert distributions of scores to new distributions with common variances. If each score in a distribution is divided by its standard deviation (an operation we encountered earlier with the z transformation), then s^2 and s = 1.0 for the new distribution, resulting in distributions with equal variances given this transformation.

This long-winded development of a measure of **co-variation** captures the essential ingredients for the index of correlation developed by the British statistician Karl Pearson (1857–1936), where $r = (1 \div [n - 1])(\Sigma[\{X_i - \bar{X}\} \div s_X][\{Y_i - \bar{Y}\} \div s_Y])$. An equivalent working formula for the Pearson r using the sum of squares approach familiar from the earlier ANOVA chapters is given by $r = SP \div \sqrt{([SS_X][SS_Y])}$, where SP is called the **sum of products**, and is equivalent to a sum of squares calculation with X_is multiplied by corresponding Y_is, replacing the squaring operations with SS calculations (X_is multiplied by X_is and Y_is multiplied by Y_is). The formulas below for SS_X and SS_Y show the squaring operation as $(X_i)(X_i)$ and $(Y_i)(Y_i)$ instead of the more conventional X_i^2 and Y_i^2 respectively, to emphasize the similarity between calculating sums of squares and sums of products (SP or SP_{XY}). One additional point that must be emphasized is that unlike sums of squares, which can only take on positive values, **SP can be either a positive number or a negative number.** A positive-signed SP indicates a positive correlation or direct relationship between corresponding X and Y scores, and a negative-signed SP indicates a negative correlation or inverse relationship between X and Y scores.

$$SS_X = \Sigma(X_iX_i) - ([\Sigma X_i][\Sigma X_i] \div n) = \mathbf{\Sigma X_i^2 - ([\Sigma X_i]^2 \div n)}$$

$$SS_Y = \Sigma(Y_iY_i) - ([\Sigma Y_i][\Sigma Y_i] \div n) = \mathbf{\Sigma Y_i^2 - ([\Sigma Y_i]^2 \div n)}$$

$$SP = \mathbf{\Sigma X_iY_i - ([\Sigma X_i][\Sigma Y_i] \div n)}$$

We will use the data in Sets I, IA, IB, and IC to illustrate the computational requirements for SS_X, SS_Y, and SP. The correlation coefficient (r) between respective X and Y scores may then be determined from these computations. While there is a certain amount of redundancy in this exercise, the point of calculating the index of correlation (r) for all four sets is to show that since the differences in the data sets did not affect the relationship between X and Y scores, this index of correlation (the value of r) between X and Y will be the same for the four data sets. (As you will see, **r = +.839**.)

Set I:

$SS_X = 1^2 + 2^2 + 6^2 - (9^2 \div 3) = 41 - 27 = \mathbf{14}$

$SS_Y = 2^2 + 7^2 + 9^2 - (18^2 \div 3) = 134 - 108 = \mathbf{26}$

$SP = (1)(2) + (2)(7) + (6)(9) - ([9][18] \div 3) = 70 - 54 = \mathbf{16}$

$\mathbf{r = SP \div \sqrt{([SS_X][SS_Y])} = 16 \div \sqrt{([14][26])} = 16 \div \sqrt{364} = 16 \div 19.08 = \underline{+.839}}$

Set IA:

$SS_X = 1^2 + 2^2 + 6^2 + 1^2 + 2^2 + 6^2 - (18^2 \div 6) = 82 - (324 \div 6) = 82 - 54 = \mathbf{28}$

$SS_Y = 2^2 + 7^2 + 9^2 + 2^2 + 7^2 + 9^2 - (36^2 \div 6) = 268 - (1296 \div 6) = 268 - 216 = \mathbf{52}$

$SP = (1)(2) + (2)(7) + (6)(9) + (1)(2) + (2)(7) + (6)(9) - [(18)(36) \div 6] = 140 - 108 = \mathbf{32}$

$\mathbf{r = 32 \div \sqrt{([28][52])} = 32 \div \sqrt{1456} = 32 \div 38.16 = \underline{+.839}}$

Set IB:

$SS_X = 11^2 + 12^2 + 16^2 - (39^2 \div 3) = 521 - 507 = \mathbf{14}$

$SS_Y = 12^2 + 17^2 + 19^2 - (48^2 \div 3) = 794 - 768 = \mathbf{26}$

$SP = (11)(12) + (12)(17) + (16)(19) - [(39)(48) \div 3] = 640 - 624 = \mathbf{16}$

$\mathbf{r = SP \div \sqrt{([SS_X][SS_Y])} = 16 \div \sqrt{([14][26])} = 16 \div \sqrt{364} = 16 \div 19.08 = \underline{+.839}}$

Set IC: $SS_X = 10^2 + 20^2 + 60^2 - (90^2 \div 3) = 4100 - 2700 = \mathbf{1400}$

$SS_Y = 20^2 + 70^2 + 90^2 - (180^2 \div 3) = 13400 - 10800 = \mathbf{2600}$

$SP = (10)(20) + (20)(70) + (60)(90) - [(90)(180) \div 3] = 7000 - 5400 = \mathbf{1600}$

$\mathbf{r = 1600 \div \sqrt{([1400][2600])} = 1600 \div \sqrt{364000} = 1600 \div 1907.88 = \underline{+.839}}$

III. Two Correlation Coefficients: Pearson r and Spearman ρ (rho)

It is fair to say that the correlation coefficient developed by Pearson is the most widely used measure of linear correlation for assessing the degree to which pairs of scores are related. Actual X and Y scores are used in calculating r. An alternative correlation coefficient was developed by Charles Spearman (1863–1945) for use when scores only provide rank-order information. That is, for three scores of 7, 12, and 24, if all that can really be stated is that 7 is the lowest, 24 the highest, and 12 is more than 7 but less than 24, then you are dealing with ordinal data that only provide **rank-order information**. The Spearman rank-order correlation coefficient *rho* (ρ) may be used when data are presented in ranks, or when it is only meaningful to regard actual X and Y scores as providing information about relative standing or rank order.

Formulas for calculating Pearson r and Spearman ρ correlation coefficients are given by: $r = SP \div \sqrt{([SS_X][SS_Y])}$ and $\rho = 1 - (6\Sigma D^2 \div n[n^2 - 1])$, where "6" is a constant, D is the difference between the **ranks** of respective X and Y pairs (e.g., rank for X_i minus rank for Y_i), and n corresponds to the number of pairs. Calculating r and ρ for the same data set normally yields similar, but not identical, values. The application of each procedure is illustrated with data displayed in Table 14.1. For this problem assume that an instructor in a statistics class asked each of his students to predict the score he or she would receive on the first exam in the class. The predicted scores are the X_is, and the actual scores earned on the first exam are the respective Y_i scores. The data in this set are fictitious (however, the first author actually did this exercise in a class several years ago, and the resulting data were similar to those reported in Table 14.1). If you are not familiar with assigning ranks to tied scores, procedures are described in detail in Model Problem 14.1 B at the end of this chapter.

We will begin by going through the computational steps for calculating the Pearson r. For calculating r, only the data in the first two columns in Table 14.1 are used. The ranks for the X and Y scores within their respective distributions appear in Columns 3 and 4, and the information in these two columns will be used for determining D and D^2 scores (reported in Columns 5 and 6, respectively) that are necessary for calculating ρ.

Table 14.1: Fictitious Data for Predicted Exam Scores (X) and Actual Exam Scores (Y)

Students	X (Predicted Score)	Y (Actual Score)	Rank X	Rank Y	D	D^2
1	100	92	1.5	4	-2.5	6.25
2	90	100	4.5	1	3.5	12.25
3	60	72	14	12	2	4
4	95	80	3	9	-6	36
5	70	75	12.5	11	1.5	2.25
6	50	50	15	16	-1	1
7	80	78	8.5	10	-1.5	2.25
8	85	66	6.5	13	-6.5	42.25
9	85	94	6.5	3	3.5	12.25
10	100	99	1.5	2	-.5	.25
11	90	90	4.5	5	-.5	.25
12	75	82	10.5	8	2.5	6.25
13	80	85	8.5	7	1.5	2.25
14	75	89	10.5	6	4.5	20.25
15	25	65	16	14	2	4
16	70	60	12.5	15	-2.5	6.25
Σ	**1,230**	**1,277**	**136**	**136**	**0**	**158**
Σ^2	**100,250**	**105,045**				

1. Pearson r: r = SP ÷ √($[SS_X][SS_Y]$); thus, SS_X (based on Column 1 data), SS_Y (based on Column 2 data), and SP (the cross products based on Columns 1 and 2) must be calculated.

$$SS_X = \Sigma X_i^2 - ([\Sigma X_i]^2 \div n) = (100^2 + 90^2 + \ldots + 70^2) - ([100 + 90 + \ldots + 70]^2 \div 16)$$

$$SS_X = 100{,}250 - (1230^2 \div 16) = \mathbf{5{,}693.75}$$

$$SS_Y = \Sigma Y_i^2 - ([\Sigma Y_i]^2 \div n) = (92^2 + 100^2 + \ldots + 60^2) - ([92 + 100 + \ldots + 60]^2 \div 16)$$

$$SS_Y = 105{,}045 - (1277^2 \div 16) = \mathbf{3{,}124.44}$$

$$SP = \Sigma(X_i)(Y_i) - ([\Sigma X_i][\Sigma Y_i] \div n) = ([100][92] + [90][100] + \ldots + [70][60]) - ([1230][1277] \div 16)$$

$$SP = 101{,}160 - 98{,}169.38 = \mathbf{2{,}990.62}$$

Knowing SS_X, SS_Y, and SP, we can now solve for $r = SP \div \sqrt{([SS_X][SS_Y])}$

$$r = 2{,}990.62 \div \sqrt{([5{,}693.75][3{,}124.44])} = 2{,}990.62 \div \sqrt{17{,}789{,}780.25}$$

$$r = 2{,}990.62 \div 4{,}217.79 = \underline{\mathbf{.709}}$$

2. Spearman ρ: $\rho = 1 - (6\Sigma D^2 \div n[n^2 - 1])$, and D^2 values are given in the last column of Table 14.1: $\Sigma D^2 = 158$, and $n = 16$. (**Remember the computational work inside the parentheses is subtracted from "1" in the formula for ρ!**)

$$\rho = 1 - (6[158] \div 16[16^2 - 1]) = 1 - (948 \div 16[255]) = 1 - (948 \div 4{,}080)$$

$$\rho = 1 - .232 = \underline{\mathbf{.768}}$$

IV. Calculating r for the Relation between Independent and Dependent Variables

We will consider one more illustration for calculating r: an example problem involving experimental data. For this illustration we will use the data from Model Problem 5.1 (page 84). For that problem, a = 5 and n = 6. There are 30 Y scores, and each will have to be paired with an appropriate X score. However, we do not have 30 X scores to pair with the 30 Y scores. There are only five variations or differences for the independent variable. Thus, we need to create a 'number code' to represent each of the five X conditions. All we have to do is have five different numbers—a different number for each group. Every subject in a given group must have the same X code number (a "dummy" code). For this illustration, we will use a simple approach and arbitrarily set X = 1 for subjects in the a_1 group, X = 2 for the a_2 group, X = 3 for the a_3 group, X = 4 for the a_4 group, and X = 5 for the a_5 group. The 30 Y scores appear in the following Table, with each Y score paired with its respective X code.

X	Y	X	Y	X	Y	X	Y	X	Y
1	4	2	6	3	2	4	3	5	0
1	4	2	9	3	5	4	4	5	4
1	7	2	7	3	4	4	5	5	2
1	5	2	6	3	6	4	5	5	3
1	3	2	5	3	5	4	7	5	6
1	6	2	9	3	3	4	4	5	3

To calculate r, we need SS_Y, SS_X, and SP: $SS_Y = [Y] - [T_Y]$ (the Y subscript for T indicates that the basic ratio for T is calculated using only the Y scores). Since [Y] and [T] were calculated in Model Problem 5.1, it will not be necessary to repeat those computational steps here: From Model Problem 5.1, we have

$SS_Y = 788.00 - 672.13 =$ **115.87**. However, we still need to calculate the sum of squares for the X scores (SS_X) and the sum of products (SP).

$SS_X = [X] - [T_X]$

$[X] = 1^2 + 1^2 + \ldots + 5^2 = 330$

$[T_X] = (1 + 1 + \ldots + 5)^2 \div 30 = 90^2 \div 30 = 270$

$SS_X = 330 - 270 =$ **60**

$SP = [XY] - [T_{XY}]$

$[XY] = (1)(4) + (1)(4) + \ldots + (5)(3) = 390$

$[T_{XY}] = (142)(90) \div 30 = 426$

$SP = 390 - 426 =$ **-36** (Since SP is a negative number, $r_{X,Y}$ will be negative.)

$r = SP \div \sqrt{([SS_X][SS_Y])} = -36 \div \sqrt{([60][115.87])} = -36 \div 83.38 =$ **-.43**

This negative correlation of r = -.43 indicates that Y scores tended to be smaller across the five increasing levels of X.

V. Doing Pearson r and Spearman ρ the SPSS Way

Letting SPSS do the r and ρ calculations can save considerable time and effort. The application is straightforward. When you open the SPSS spreadsheet, you only need to enter the X scores in one column and the Y scores in a second column. By clicking your mouse on "variable view" in the lower left-hand corner of the screen, you may label each column appropriately (e.g., X and Y or names identifying the variables). After you have labeled the two columns, return to the lower left-hand corner of the screen, click on "data view," and type in X scores in the first column and corresponding Y scores in the second column. You must be careful to: (1) make sure that X and Y pairs are not mixed up—that is, each row (an X and Y score) has to correspond to an appropriate X-Y pair from the data set (e.g., two scores from the same individual), and (2) **proof** what you have entered to make sure there are no typing errors. Entering wrong scores or mixing up pairs will result in incorrect r and ρ indices.

With the data entered and checked, you are now ready to command the computer to calculate r and ρ. You begin this process by going to "Analyze" on the tool bar at the top of the screen. Clicking the mouse on "Analyze" opens a menu of several options. About a third of the way down this menu you will see "Correlate." Clicking on this option opens a small menu to the right providing options of "**bivariate**," "partial," and "distance." Click on "bivariate" (two variables), which opens a dialog box similar to the following illustration. The two labels for your scores (X and Y) appear in an inset on the left side of this dialog box. A small box with an arrow appears to the right of this inset, and to the right of the arrow is a larger rectangle below the word "Variables." The "Variables" rectangle is blank, and you need to move X and Y from the left-hand box to the "Variables" box. This move is accomplished in two steps: First, highlight "X" and click on the arrow. The "X" will disappear from the left-hand box and appear in the "Variables" box. Second, do the same for "Y." Below these two boxes there is a row of three small circles and a second row of two small circles. In the first row, "Pearson" follows the first circle and "Spearman" follows the third circle. A check mark is present in the "Pearson" circle, indicating that r will be calculated (the default correlation computation). To add the ρ value, place a check mark in the "Spearman" circle by moving the cursor to the circle and clicking on it. A two-tailed test (same meaning as it has for a t test) is the default test, and the check mark should automatically appear in that circle. A small rectangle is near the bottom left corner of this dialog box with "ok" inside.

All you have to do to calculate correlation coefficients for r and ρ is click on "ok," and the correlations will appear as illustrated in the diagrams that follow the dialog box. The values for r and ρ that appear in the two correlation tables represent the SPSS analysis of the data from Table 14.1. The two SPSS correlation tables indicate that the Pearson **r = .709** (in bold), is significant at $p = .002$ and the Spearman **ρ = .767** (in bold), is significant at $p = .001$. These two tables also show that there were 16 pairs in the data set ($N = 16$). Just ignore the X-X and Y-Y "self-correlation" values of 1.000.

Dialog Box

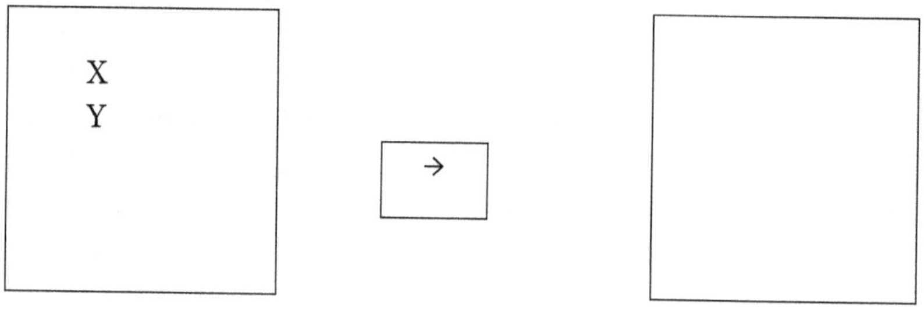

Correlation Coefficients:

O Pearson O Kendall's tau-b O Spearman

Tests of Significance

O two-tailed O one-tailed

OK

Correlation Table for Pearson r

Correlations:		X	Y
X	Pearson Correlation	1.000	**.709****
	Sig. (2-tailed)		.002
	N	16.000	16
Y	Pearson Correlation	**.709****	1.000
	Sig. (2-tailed)	.002	
	N	16	16.000

Correlation Table for Spearman ρ

Nonparametric Correlations:			X	Y
Spearman's ρ	X	Correlation Coefficient	1.000	**.767****
		Sig. (2-tailed)		.001
		N	16	16
	Y	Correlation Coefficient	**.767****	1.000
		Sig. (2-tailed)	.001	
		N	16	16

VI. Significance Tests and Effect Size

Measures of correlation primarily provide descriptive information about association or the degree to which paired X and Y scores agree or co-vary. A correlation coefficient of r = + .60 between height and weight for a sample of 20 25-year-old males describes an association ("taller is associated with heavier") between these two measures. However, it is an imperfect association, as it is not always the case that, for example, a 6'0" male will weigh more than a 5'9" male. Obviously, other factors may "tip the scales" one way or the other, and, as you may expect by now, "other factors" originate from both known and unknown sources. An estimate of the **size** of a correlation, in terms of the proportion of the total variance that is common to both X and Y measures, is given by the square of the correlation coefficient (r^2). Thus, an r = +.60 indicates that the magnitude of the correlation effect between height and weight is r^2 = .36, or, loosely speaking, there is a 36% overlap in height and weight variability.

The square of the correlation coefficient is an extremely important index for prediction purposes, as we will see in the next related chapter on *regression*. For regression analyses, one variable (normally the one that has temporal priority) is regarded as the "predictor" variable and designated with the letter code "X," with the second dependent variable in the temporal order and the one measuring the behavior of primary interest representing the "predicted" responses and designated with the letter code "Y." **The proportion of variability in Y accounted for by X is the square of the correlation coefficient**, and in regression analyses this index is often represented by using an upper case coding letter for the squared correlation coefficient, with subscripts identifying the related variables in question ($R^2_{Y.X}$).

Although the primary role for correlation assessment is the description of a relationship between two dependent variables, it is reasonable to address the *reliability* question for r and ρ as was done for differences among sample means based on results of t, F and χ^2 tests. For example, the correlation between two dependent variables may be r = .30, indicating that for the sample data the variables had r^2 = 9% of their variance in common; however, it may be important to know whether or not observing a correlation index as large as .30 between X and Y would be likely based only on "chance" or random error (an H_0 for correlation). Tables of appropriate reference distributions (based on the degrees of freedom, which for correlation is equal to the **number of paired scores minus two**) for both r and ρ are available. Critical cut-off values for $p \leq .05$ and $p \leq .01$ for the Pearson r are given in Table A-8 in Appendix A at the back of the book. For comparison purposes, a sample subset of critical values for r and ρ are presented in Table 14.2. Two points may be noted from this comparison: (1) the critical cut-off points for r and ρ converge as df increases, and (2) the Pearson r is a little more powerful than the Spearman ρ. That is, for a given df level, it takes a larger ρ value for "significance" compared to r. For example, with df = 10 and $p \leq .05$, $r_{crit} = .58$ and $\rho_{crit} = .65$.

Table 14.2: Examples of Critical Values of Pearson r and Spearman ρ for Significance at $p \leq .05$ and $p \leq .01$ (the selected values for ρ were taken from Kiess and Green 2010)

	Pearson r		Spearman ρ	
df_r	.05	.01	.05	.01
8	.63	.76	.74	.88
10	.58	.71	.65	.79
12	.53	.66	.59	.78
13	.51	.64	.57	.74
14	.50	.62	.54	.72
15	.48	.61	.52	.69
16	.47	.59	.51	.67
18	.44	.56	.48	.62
20	.42	.54	.45	.59
25	.38	.49	.40	.53
30	.35	.45	.36	.48

The earlier example problem involving the fictitious guessed exam scores and actual exam scores resulted in correlation coefficients rounded off to two decimal places of r = .71 and ρ = .77. There were 16 paired scores, thus df = 16 – 2 = 14. With 14 degrees of freedom, the critical value for r at $p \leq .05$ from Table 14.2 is .50. The critical value for ρ at $p \leq .05$ from Table 14.2 is .54. Since both r and ρ exceeded r_{crit} and ρ_{crit}, respectively, the conclusion with each of these two statistics was that there was a **significant** correlation at the .05 level between a score a student guessed he or she would receive and his or her actual exam score.

The concept of power for a correlation coefficient is a relevant consideration, just as it is for other statistical tests we have considered (t, F, χ^2). Power refers to the probability that the null hypothesis will be rejected (a decision that r or ρ is significant) when the two variables in question are linearly related. As with the other considerations of power, one may estimate the number of subjects necessary to detect a significant linear relation for a desired power level. Procedures for performing power analyses with a hand-held calculator are not detailed in this text; however, the interested reader may wish to consult Keppel et al. (1992).

VII. Summary

The focus of correlational research is on the assessment of the relationship between two or more characteristics of individuals (physical, personality, performance, etc.). The two descriptive indices of correlation discussed in this chapter, r and ρ, describe the linear relationship between dependent variables, and they may range in values from -1, indicating a perfect inverse relationship (e.g., large scores on one variable are associated with small scores on the other variable) to a +1, indicating a perfect direct relationship (e.g., large scores on one variable are associated with large scores on the other variable). The closer r values are to a -1 inverse relationship or a +1 direct relationship, the *stronger* the degree of association. That is, a correlation of $r = -.75$ indicates a relationship between X and Y that is just as strong and informative as a correlation of $r = +.75$ (i.e., for both correlations, $r^2 = .56$, indicating that X accounts for 56% of the variance in Y). It should be emphasized that no matter how strong the relationship between two variables, the mere demonstration of a correlation is not sufficient evidence for inferring that variation in one of the two measures **caused** the corresponding variation in the other measure, as it is quite reasonable that some third factor caused the two measured characteristics to co-vary. Although correlations most often provide an index of the degree of association between two dependent variables, it was also shown the correlation coefficients may be calculated between an independent variable and a dependent variable.

The Pearson Product Moment Correlation Coefficient (r) is the more frequently encountered index of correlation. An intuitive, common language approach was taken in describing the underlying logic of this index. Computing cross products (multiplying the two members of a pair) provides the core calculation for r, and this operation is sensitive to co-variation. For any set of X and Y numbers, if they are arranged so that the largest X is paired with the largest Y, the 2nd largest X paired with the 2nd largest Y, etc., then the sum of the cross products ($\Sigma[X][Y]$) will be larger than any other pairing arrangement. Problems for an index based only on a sum of cross products were identified. In order to "level the playing field," it was necessary (1) to consider each X and Y score relative to the mean of its respective distribution (deviations from the mean), (2) to use an average of the cross products of these deviation scores, and (3) to make an adjustment on X and Y scores so that all distributions have common variances of $s^2 = 1$.

Calculating the Pearson r for a data set was illustrated, and the same example data set was converted to ranks in order to illustrate computational procedures for the Spearman rank-order correlation coefficient (ρ). Normally, both r and ρ reflect similar degrees of association; however, it is obvious that r is based on more information (the actual X and Y scores are taken into consideration, not merely the ranking within the respective X and Y distributions that a particular score placed an individual), and as a result, the Pearson r is more powerful and preferred. Illustrations for computing r and ρ using a hand calculator were followed by a description for calculating these correlation coefficients using SPSS software.

Although correlation coefficients are calculated primarily for the descriptive purpose of reporting the degree to which two measures are associated, additional information is often provided to indicate whether an r or ρ value as large as or larger than the one reported could reasonably occur by random variation alone. Critical values for the Pearson r are presented in Table A-8 in Appendix A. In addition, Table 14.2 provided a limited sampling of critical ρ values appropriate for selected degrees of freedom (df = n – 2 for correlation reference distributions) to compare with critical r values. A calculated r that is equal to or larger than the appropriate critical r in the table would be expected to occur by chance alone infrequently (e.g., less than 5% of the time or less than 1% of the time), hence it would be regarded as significant in much the same way as significance is used in the context of t, F, and χ^2 tests. Finally, it was noted that the squared correlation coefficient (r^2 or R^2), rather than the actual value of r, represents the proportion of variance common to both X and Y scores, and as such, it provides an estimate of effect size for correlation (comparable to the use of ω^2 for estimating effect size with ANOVA and ϕ and ϕ_C for estimating effect size with the χ^2 test).

14.1: Model Problem. At the beginning of the 20th century, Karl Pearson reported correlations between the height of a parent and offspring (father–son; father–daughter; mother–son; and mother–daughter) (Pearson and Lee 1903). More than 4,000 individuals in more than 1,000 families provided data for the reported correlation coefficients. The reported r values were close to .50 and, as may be seen in Table A-8 in Appendix A, correlations of this magnitude would be significant with as few as 16 pairs of scores (with 16 pairs of scores, there would be n – 2 = 14 degrees of freedom for r). Even with just 14 degrees of freedom, r_{crit} (14) = .497 for the .05 significance level. To illustrate the steps for calculating r and ρ on a hand-held calculator we will use a small number of fictitious father–daughter pairs, rather than actual data from several hundred pairs. Assume the heights (in inches) in Table 14.3 below are from a sample of 10 father–daughter pairs.

Table 14.3: Height in Inches of Fictitious Father–Daughter Pairs

	Height of Father (X)	Height of Daughter (Y)
	67	67
	65	53
	75	65
	70	52
	70	66
	72	70
	71	64
	65	60
	62	59
	68	61
Σ	**685**	**617**
Σ²	**47,057**	**38,381**

A. To calculate r, three component calculations are necessary: $r = SP \div \sqrt{([SS_X][SS_Y])}$; thus we need to calculate SP, SS_X, and SS_Y.

Step 1: $SS_X = \Sigma X^2 - ([\Sigma X]^2 \div n)$

$\Sigma X^2 = 67^2 + 65^2 + \ldots + 68^2 = \mathbf{47{,}057}$

$(\Sigma X)^2 \div n = 685^2 \div 10 = 469{,}225 \div 10 = \mathbf{46{,}922.50}$

$SS_X = \Sigma X^2 - ([\Sigma X]^2 \div n) = 47{,}057 - 46{,}922.50 = \mathbf{134.50}$

Step 2: $SS_Y = \Sigma Y^2 - [(\Sigma Y)^2 \div n]$

$\Sigma Y^2 = 67^2 + 53^2 + \ldots + 61^2 = \mathbf{38{,}381}$

$(\Sigma Y)^2 \div n = 617^2 \div 10 = 380{,}689 \div 10 = \mathbf{38{,}068.90}$

$SS_Y = \Sigma Y^2 - ([\Sigma Y]^2 \div n) = 38{,}381 - 38{,}068.90 = \mathbf{312.10}$

Step 3: $SP = \Sigma XY - ([\Sigma X][\Sigma Y] \div n)$ (recall, to calculate a sum of products, the squaring operation used for SS_X and SS_Y is replaced by the multiplication of the two members of corresponding X and Y pairs)

$\Sigma XY = (67)(67) + (65)(53) + \ldots + (68)(61) = \mathbf{42{,}359}$

$(\Sigma X)(\Sigma Y) \div n = (685)(617) \div 10 = 422{,}645 \div 10 = \mathbf{42{,}264.50}$

$SP = \Sigma XY - [(\Sigma X)(\Sigma Y) \div n] = 42{,}359 - 42{,}264.50 = \mathbf{94.50}$

Step 4: Solve for r by substituting the values for SS_X, SS_Y, and SP in the formula for r:

$r = SP \div \sqrt{([SS_X][SS_Y])} = 94.50 \div \sqrt{([134.50][312.10])}$

$r = 94.50 \div \sqrt{42{,}289.55} = 94.50 \div 205.64 = \underline{\mathbf{.460}}$

B. To calculate ρ, the X and Y scores need to be converted to ranks, a difference in ranks (D) needs to be determined for each pair, and D scores need to be squared and summed, as $\rho = 1 - ([6][\Sigma D^2] \div [n][n^2 - 1])$.

Step 1 requires creation of a new Table (Table 14.4) to convert data in Table 14.3 to ranks. It is particularly important to pay attention to the procedure for converting tied scores to ranks. For example, the 4th highest X score is 70 inches, and two fathers were this tall. Technically, one of these individuals should be the 4th tallest and one should be the 5th tallest of the 10 fathers. Since we do not know which one should rank 4th and which one should rank 5th, they are both given the "average" of these two ranks or 4.5 ($[4 + 5] \div 2 = 4.5$). Since these two scores "use up" ranks 4 and 5, the next tallest father (68 inches in height) is given the rank of "6." One final note: if the 68-inch dad had also been 70 inches tall, then there would have been three fathers tied and three ranks (4, 5, and 6) that needed to be assigned. Using the averaging procedure, each of these three fathers would be given a rank of 5 ($[4 + 5 + 6] \div 3 = 5$); having thus "used up" ranks 4, 5, and 6, the next father in the order would be assigned the next available rank of 7.

Table 14.4: Rank Order of Heights of Father–Daughter Pairs

(X)	(Y)	Rank X	Rank Y		D	D^2
67	67	7	2		5	25
65	53	8.5	9		-.5	.25
75	65	1	4		-3	9
70	52	4.5	10		-5.5	30.25
70	66	4.5	3		1.5	2.25
72	70	2	1		1	1
71	64	3	5		-2	4
65	60	8.5	7		1.5	2.25
62	59	10	8		2	4
68	61	6	6		0	0
				Σ	0	**78**

Step 1. Determine values for D (difference score for each X and Y **rank**) and D^2 as calculated in the last two columns of Table 14.4.

Step 2. Substitute values for ΣD^2 and n in the formula: $\rho = 1 - ([6][\Sigma D^2] \div [n][n^2 - 1])$:

$$\rho = 1 - ([6][78] \div [10][100 - 1])$$

$$\rho = 1 - (468 \div 10[99]) = 1 - (468 \div 990)$$

$$\rho = 1 - .473 = \mathbf{.527}$$

(NOTE: a quick check of Table 14.2 indicates that for df = 8; [number of paired scores minus 2 (n – 2) = 10 – 2 = 8], an $r \geq .63$ and $\rho \geq .74$ are required for significance at $p \leq .05$. Thus for this small example set, the correlation between father and daughter heights is not significant. An r = .46 is similar to actual r values Pearson recorded; however, with hundreds of pairs of scores, a correlation this large would be well above the critical r required for significance at $p \leq .05$).

14.2: Practice Problem 1. What does r equal when calculated from the **ranks** of father–daughter heights rather than the actual heights in inches? (i.e., using columns 3 and 4 from Table 14.4, what does the Pearson r for the paired numbers in the Rank X and Rank Y columns equal?)

14.3: Practice Problem 2. Listed in Table 14.5 are two quiz scores obtained by 12 students: scores on a 15-point mathematics quiz (X) and scores on a 15-point geography quiz (Y).

Table 14.5: Math and Geography Quiz Scores

Math Score (X)	Geography Score (Y)
9	13
5	11
3	8
4	10
4	7
6	9
2	6
9	12
7	11
10	15
8	8
12	10

A. Compute the Pearson r for these data.

B. Is r significant at $p \leq .05$?

C. For r in Part A, what is the proportion of shared variability for the two quiz scores?

D. Confirm the r value calculated in Part A using SPSS.

14.4: Practice Problem 3. A high school basketball coach believes that his team scores more points when attendance at home games is relatively large. The number of points scored and attendance for 12 home games during one season are reported in Table 14.6. Determine the degree to which these two measures (points scored and attendance) are correlated by calculating ρ.

Table 14.6: Points Scored and Attendance Record for 12 Home Games

	Points Scored (X)	Attendance (Y)
	59	350
	80	320
	82	451
	75	250
	82	489
	53	450
	78	515
	67	375
	56	290
	79	480
	81	550
	65	470
Σ	**857**	**4,990**

14.5: Practice Problem 4. For this problem you are asked to calculate a correlation coefficient for experimental data. The independent variable represents different amounts of practice prior to playing a round of golf; the dependent variable is the score obtained on a subsequent round of 18 holes. Twenty professional golfers played a round of golf following different amounts of time "warming up" on the driving range and putting green. Assume that five golfers were randomly assigned to each of four conditions that differed in amount of warm-up time: Group a_1 warmed up for 15 minutes, Group a_2 warmed up for 30 minutes, Group a_3 warmed up for 45 minutes, and Group a_4 warmed up for 60 minutes. Although this description fits an experimental design, a correlation coefficient can be computed with X = minutes of warm-up time, and Y = the score for the round of golf. It will be necessary to use a dummy code for the X variable. For this problem we suggest simply using the actual number of minutes of practice as a dummy code for X; however, other dummy codes (e.g., "1" for Group a_1, "2" for Group a_2, "3" for Group a_3, and "4" for Group a_4) could be used following the same rules that were used for coding independent variables in SPSS. If you doubt this claim, simply substitute X scores of "1" for "15," "2" for "30," "3" for "45" and "4" for "60" in Table 14.7 to confirm that the same correlation coefficient indices (r and ρ) result with either set of dummy codes for X. From the fictitious data in Table 14.7, calculate r and ρ, and for r, estimate the proportion of variance in golf scores accounted for by amount of prior practice (15 to 60 minutes). After you calculate r, confirm your results using SPSS.

Table 14.7: Amount of Prior Practice and Golf Scores

Golfers	Minutes of Practice (X)	Golf Score (Y)
1	15	70
2	15	69
3	15	68
4	15	73
5	15	71
6	30	68
7	30	72
8	30	66
9	30	72
10	30	71
11	45	71
12	45	67
13	45	65
14	45	68
15	45	68
16	60	64
17	60	67
18	60	70
19	60	69
20	60	64
	Σ **750**	**1,373**

14.6: Practice Problem 5. For the 15 paired scores in this problem, determine if there is a significant correlation at the .05 level for r and for ρ.

Subjects	X	Y
1	10	11
2	17	14
3	15	16
4	12	13
5	20	19
6	16	18
7	15	20
8	15	17
9	21	15
10	13	10
11	11	14
12	14	12
13	18	13
14	16	22
15	25	21
Σ	**238**	**235**

Chapter 15

Linear Regression: Using Relations Between Variables for Prediction

... when I call for statistics about the rate of infant mortality, what I want is proof that fewer babies died when I was Prime Minister than when anyone else was Prime Minister. That is a political statistic.

—Churchill

I. Regression and Correlation
II. Determining a Line of "Best Fit"
III. Using the Regression Line to Minimize Prediction Error
IV. Multiple Correlation/Regression: Shouldn't Two Predictors Be Better than One?
  a. $R^2_{Y.X1.X2}$ by Hand Calculator
  b. $R^2_{Y.X1.X2}$ Using SPSS
V. Multiple Correlation/Regression Analysis and ANOVA
VI. Summary
VII. Practice Problems

I. Regression and Correlation

In Chapter 14, sets of paired numbers were used to demonstrate correlations between selected X and Y variables. Correlations measured were linear correlations (both r and ρ) assessing the extent to which the relationship between two variables could be represented by a straight line. The "straight-line" relationship is the only type of relationship that we will consider. For regression, the goal of prediction is to specify a best estimate of what *future* performance measures for individuals on the Y dependent variable will be. If we did not have any other information about individuals, the best predictions (in terms of minimizing error in the long run) that could be made about Y scores would be the mean Y score. However, with information about individuals' scores on an X measure plus knowledge about the degree to which X and Y scores are related (e.g., $r_{X.Y}$), a specific "best-fitting" straight line for a set of paired X and Y scores may be used to improve prediction accuracy. Once the best-fitting straight line

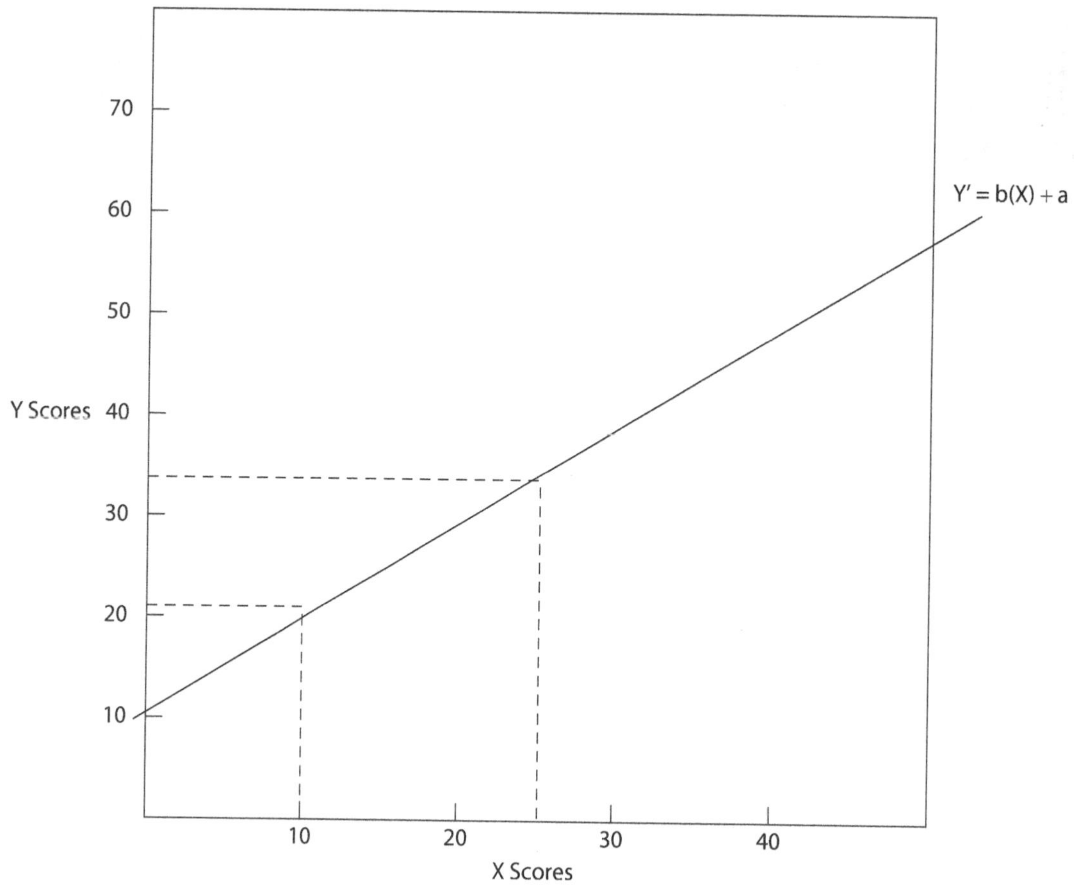

Figure 15.1: Graphical Illustration for Using a Regression Line to Predict Y Based on an X Score

is determined, the task of predicting a Y score (coded as Y') based on a given X score merely involves locating the Y point on the regression line that corresponds to the X point on this line (referred to as the X-Y coordinate).

Figure 15.1 is provided merely to serve as an illustration for how a regression line may be used for prediction. A graphical illustration such as this may be worth more than most of the preceding words for explaining the relation between regression lines and predicting future performance. Assume that the line [labeled Y' = b(X) + a] passing through the graph represents an appropriate regression line. If you know that an individual obtained a score of 10 on the X dependent variable, then based on this particular regression line, what Y score would you predict? To obtain a rough estimate (the formula for the regression line would have to be used to obtain a precise estimated Y score), simply extend a vertical line (shown as a dashed line in Figure 15.1) from the score of 10 on the X axis until it intersects the regression line. At this point of intersection, draw a horizontal line to the Y axis. The point at which the horizontal line intersects the Y axis is the predicted Y score, which for this set of paired scores is Y' ≈ 20 (predicted Y is approximately 20). Of course this prediction is just based on an eyeball test, hence an estimate of 20.5 might be just as reasonable as an estimate of 20. Similarly, you can see that for an individual who obtained a score of 25 on the X dependent variable, Y' ≈ 32.

The regression line in Figure 15.1 is simply provided for illustration: it is just an arbitrary line bearing no precise relation to any set of real or imaginary X and Y scores (i.e., we did not solve for the actual regression line).

II. Determining a Line of "Best Fit"

The regression line in Figure 15.1 was merely an arbitrary line drawn through the graph to illustrate how a predicted Y score is located based on information about a corresponding X score. In order to construct an actual regression line it is necessary to have a set of X and Y scores to work with. Thus, a regression analysis develops from correlations that have been established based on prior observations of paired X and Y scores. From these prior data, a regression line of best fit may be constructed, and the purpose of this section is to describe procedures for calculating the regression line. It should be noted that if there is little correlation between X and Y, a regression line will provide little improvement for predicting Y scores beyond what would be achieved by just using the mean ($\bar{Y}$) as the basis for predicting future scores.

Regression analyses are primarily based on linear (straight line) relationships between two dependent variables (this is not to say that it is impossible to do regression analyses based on curvilinear relations, with an independent variable for X rather than with two dependent variables, or with three or more variables—a situation that is quite common in applied research settings). Thus, deriving the appropriate straight line (regression line) for a set of paired X and Y scores requires our attention. You should know the formula for a straight line has the following form: **Y = b(X) + a**, and specific lines are defined by the *slope* of the line (the "b" constant) and the point that the line *intersects the Y axis* (the "a" constant). The slope provides information about how Y changes with respect to how X changes. The slope may be steep or shallow, and it may go in either an upward (positive) or downward (negative) direction. At some point, you can expect that a regression line will cross the Y axis, and knowing the location of this point is essential for identifying the one best regression line from among the set of parallel lines of the same slope. In this formula for the straight line, "b" is a constant indicating the slope of the line, and "a" is a constant indicating the point at which the line crosses the Y axis (Y intercept). You should note that "b" and "a" are simply coding letters representing the slope and Y intercept, respectively.

From a preliminary data set of paired X and Y scores, "b" and "a" constants are replaced by real numbers to define the regression line of interest. The specific regression constants simply identify the one line of best fit for the given data. In this context, "best-fitting" means that in terms of the extent to which actual data points fall above or below the line (an indication of prediction error), the regression line is the line that minimizes error in terms of the sum of the squared deviation scores (distances from the data points to the line). As was done when the concept of variance was introduced, we will again appeal to a "least-squares" solution so that the variability of actual X-Y points (coordinates) from the regression line will be smaller than the same measures taken from any other straight line. Thus, a specific regression line for a given data set is determined once the regression constants have been calculated by simply replacing "b" and "a" with their numerical values in the formula for a straight line. It is true that a second regression line can be calculated for predicting X scores from given Y scores; however, that is not of interest for our purposes.

If this description of a regression line for predicting Y scores from the relationship between X and Y scores seems somewhat obtuse, perhaps an illustration with a small number set will help make things a little clearer. In calculating two regression constants, always calculate the slope of the line (b) first, which is done using computational procedures you should be familiar with from doing analysis of variance calculations and calculating the Pearson r correlation coefficient, *viz.*, b = SP ÷ SS_X. Once b has been calculated, the Y intercept (a) is calculated from a formula that requires calculating means for the X and Y scores and some simple substitution and arithmetic: a = $\bar{Y}$ – (b)($\bar{X}$). Calculating means at this stage should not be a daunting task, so the only relatively tedious computations involve the b constant. The required computational steps will be illustrated by referring to the fictitious data set in Table 15.1.

Table 15.1: Determining the Regression Line for 10 Paired Scores

	X	Y	XY
	20	15	300
	8	8	64
	14	10	140
	9	12	108
	16	16	256
	11	7	77
	9	6	54
	5	8	40
	18	20	360
	9	11	99
Σ	**119**	**113**	**1,498**

Using the data in Table 15.1, we will begin with the least squares solution for determining the slope of the regression line: b = SP ÷ SS_X.

Step 1: Calculate $SS_X = \Sigma X^2 - ([\Sigma X]^2 \div n)$

$$\Sigma X^2 = 20^2 + 8^2 + \ldots + 9^2 = 1{,}629$$

$$(\Sigma X)^2 \div n = (119)^2 \div 10 = 1{,}416.10$$

$$SS_X = 1{,}629 - 1{,}416.10 = \mathbf{212.90}$$

Step 2: Calculate $SP = \Sigma XY - ([\Sigma X][\Sigma Y] \div n)$

$$\Sigma XY = (20)(15) + (8)(8) + \ldots + (9)(11) = 1{,}498$$

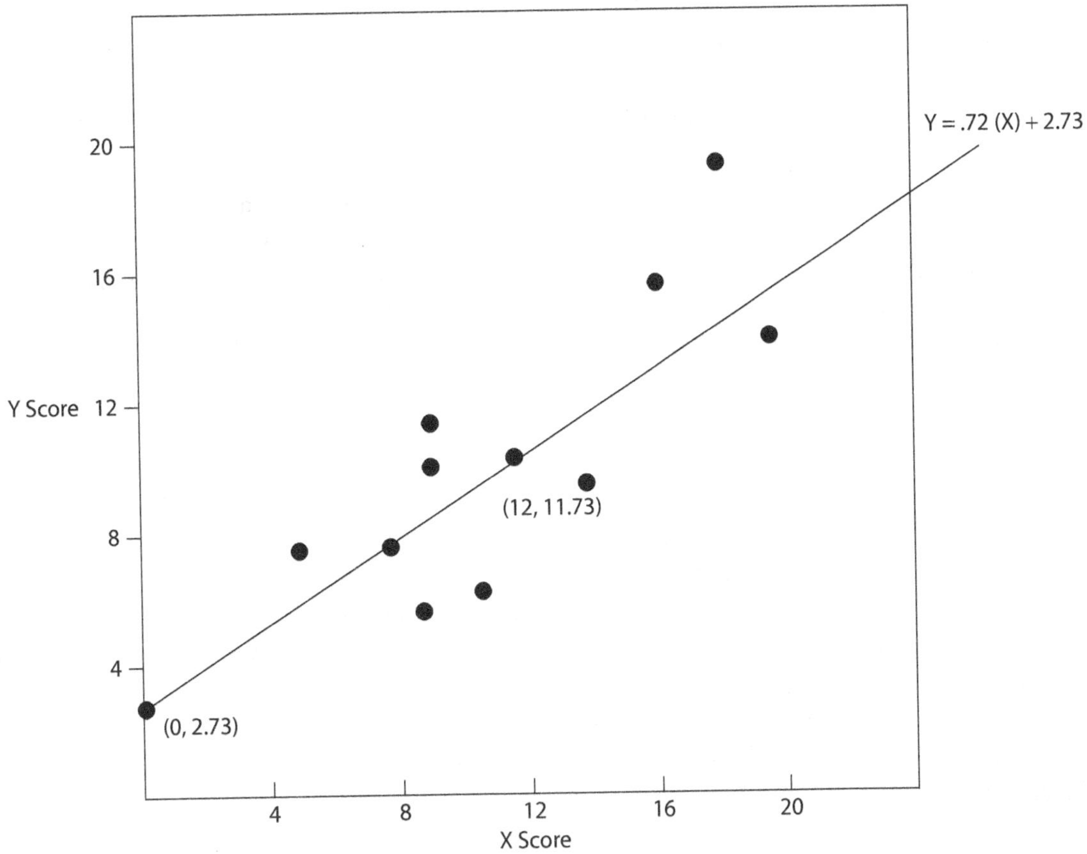

Figure 15.2: The 10 Paired Scores and Regression Line from Table 15.1

$(\Sigma X)(\Sigma Y) \div n = (119)(113) \div n = 13{,}447 \div 10 = 1{,}344.70$

$SP = 1{,}498 - 1{,}344.70 = \mathbf{153.30}$

Step 3: Determine b by dividing SP by SS_X

$b = 153.30 \div 212.90 = \mathbf{.72}$

Given that b = .72, we can now proceed to calculate the Y intercept: $a = \bar{Y} - (b)(\bar{X})$

Step 1: Calculate the mean for X and the mean for Y

$\bar{X} = 119 \div 10 = 11.90$ and $\bar{\mathbf{Y}} = 113 \div 10 = 11.30$

Step 2: Substitute b, $\bar{X}$, and $\bar{Y}$ in the formula for a: $a = \bar{Y} - (b)(\bar{X})$

$$a = 11.30 - (.72)(11.90) = 11.30 - 8.57 = 2.73$$

Substituting the slope (b = .72) and Y intercept (a = 2.73) in the formula for a best-fitting regression line (regressing Y on X) for this fictitious data set, we have: **Y' = (b)(X) + a = (.72)(X) + 2.73**. Plotting this specific regression line requires that you determine two points that fall on the line because you only need to know the location of two points on a line in order to draw the line (you simply connect the two points with a ruler or straight edge to draw the regression line). Finding two points on a line is relatively easy. Just pick two arbitrary values for X and solve the equation for Y. The good news is that we have already solved this equation for one point on the regression line (one X-Y coordinate): Do you know what this X-Y coordinate is? We just determined the two regression constants for this data set, where b (slope) = .72 and a (Y intercept) = 2.73. What does Y-intercept mean? It is the point where the regression line crosses the Y axis (hence the point on this line with an X coordinate = 0 and a Y coordinate = 2.73). That is, the regression line intersects the Y axis at 2.73; knowing that X = 0 at the Y axis, we have one point that is on the regression. One point is entered in Figure 15.2 at this coordinate of X = 0, Y = 2.73. Since two points are necessary to draw a straight line, one more point on the line is all we need, and any point will do. In Figure 15.2, an X point of 12 was arbitrarily selected, and for X = 12, Y' = (.72)(12) + 2.73 = 11.73. Thus, the second point used for drawing the regression line has X and Y coordinates of 12 and 11.73, respectively. The regression line and the data from Table 15.1 have been entered in Figure 15.2.

III. Using the Regression Line to Minimize Prediction Error

For any new individual who comes along, if information is available regarding how he or she performed on one behavioral measure (X), a more accurate prediction (compared to just guessing or using the average Y score as a general predictor) may be made regarding a future Y score. The prediction is accomplished by merely substituting the observed X score into the regression equation and solving for the predicted Y: $Y_i' = b(X_i) + a$. For the preceding illustration with b =.72 and a = 2.73, the predicted score for an individual with an X score of 20 is Y' = .72(20) + 2.73 = 17.13. One cautionary note that should be obvious is that trying to predict Y from X would not make much sense if there was not a reasonable correlation between the Y and X variables. Although a correlation coefficient was not reported for the data in Table 15.1, two of the three required calculations for r were completed. That is, $r = SP \div \sqrt{([SS_X][SS_Y])}$, and both SP and SS_X were calculated in determining the slope of the regression line (SP =153.30 and $SS_X = 212.90$). One more calculation (SS_Y) is the only major computation left that is necessary for computing r: $SS_Y = \Sigma Y^2 - ([\Sigma Y]^2 \div 10)$. Solving for ΣY^2 and $(\Sigma Y)^2$ we have:

$$\Sigma Y^2 = 15^2 + 8^2 + \ldots + 11^2 = 1{,}459$$

$$(\Sigma Y)^2 \div 10 = 113^2 \div 10 = 12{,}769 \div 10 = 1{,}276.90$$

$SS_Y = 1{,}459 - 1{,}276.90 = \mathbf{182.10}$

Combining this result with the earlier results for SS_X and SP, we can compute the correlation coefficient for these data:

$r = 153.30 \div \sqrt{([212.90][182.10])}$

$r = 153.30 \div \sqrt{38{,}769.09} = 153.30 \div 196.90$

$\mathbf{r = .779}$

In behavioral research a correlation of this magnitude would be reasonably impressive. Recall that the squared correlation coefficient provides a measure of effect size, and for this fictitious set of just 10 X and Y scores, $r^2 = .61$ indicating that approximately 61% of the variability in Y was accounted for or explained by variation in X. The squared correlation coefficient (r^2 or $\mathbf{R^2}_{Y.X}$) is the index used in regression analyses for evaluating prediction.

IV. Multiple Correlation/Regression: Shouldn't Two Predictors Be Better than One?

The question posed in this section (shouldn't two predictors be better than one?) would seem to have an obvious answer of "yes." If information about the relation between a score of interest (Y) and how individuals perform on one predictor variable (X_1) improves prediction, it certainly seems logical that prediction accuracy would be even further enhanced if information was also available on how individuals performed on a second predictor variable (X_2) related to Y scores. While there are a couple of "givens" that must be considered here, this expectation is the basis for *multiple-regression* applications.

Two or more predictor variables, each correlated with Y, will in most cases result in better prediction than either one alone. However, the improvement gained by including a second predictor variable may be far less than one might expect. If $r_{Y.X1} = .70$ and $r_{Y.X2} = .70$, both X_1 and X_2 account for 49% of the variability in Y (squaring the correlation coefficient provides an index of the proportion of variability in Y that is accounted for by variability in X, and for this example, $r^2_{Y.X1} = r^2_{Y.X2} = .49$). However, it is most likely the case that X_1 and X_2 in combination account for considerably less than 98% of the variability in Y [unfortunately, in most cases calculating the multiple-regression coefficient involves more complex operations than merely summing two squared correlation coefficients (e.g., it is likely that $R^2_{Y.X1.X2} \neq r^2_{Y.X1} + r^2_{Y.X2}$)]. The reason that the squared multiple-correlation coefficient is not simply the sum of the squared correlations between X_1 and Y and X_2 and Y is that some of the variability in Y that is accounted for by X_1 is redundant with variability in Y accounted for by X_2. That is, X_2 usually accounts for some of the variability in Y that has already been accounted for by X_1. The only condition for which $R^2_{Y.X1.X2} = r^2_{Y.X1} + r^2_{Y.X2}$ occurs for an X_1 and X_2 that are independent with no variation in common (i.e., $r_{X1.X2} = 0$). In the typical situation, variance in Y common to X_1 and X_2 would be included twice if $r^2_{Y.X1}$

and $r^2_{Y.X2}$ were simply summed, hence it is necessary to correct this problem, which is accomplished by the following formula:

$$R^2_{Y.X1.X2} = (r^2_{Y.X1} + r^2_{Y.X2} - [2][r_{Y.X1}][r_{Y.X2}][r_{X1.X2}]) \div (1 - r^2_{X1.X2}).$$

With a little algebra, you can see that substituting 0 for $r_{X1.X2}$ in the formula reduces the equation to $R^2_{Y.X1.X2} = ([r^2_{Y.X1} + r^2_{Y.X2}] \div 1)$.

a. $R^2_{Y.X1.X2}$ *by Hand Calculator.*

The formula for the multiple regression coefficient involves addition, subtraction, multiplication, and division; however, these operations are performed on various correlation coefficients and squared correlation coefficients requiring computational procedures covered in the last chapter. It should also be clear that both subscripts and exponents require careful attention to ensure that each correlation coefficient and squared correlation coefficient is properly entered into the equation (i.e., correlation coefficients are squared or not squared as called for). With only two predictor variables, each individual provides three scores: a score for each of the two predictor variables (X_1 and X_2) and a score for the performance measure of interest (Y). Calculating correlation coefficients with a hand calculator can be a tedious process, and for problems with two predictor variables, **three correlation coefficients** must be calculated. In this section we will work through the same example problem twice: first with a hand calculator, and second by using SPSS. Identical results will be produced (within rounding error). The computer procedure is recommended once you get past an introductory statistics course (and for an introductory statistics course with a computer laboratory), as it is the only realistic option for research involving fairly large sample sizes that have two or more X predictor variables (the data set for this illustration has 20 sets of scores, and probably exceeds what most students would consider as a "reasonable" hand-calculation exercise). The fictitious data for the present illustration appear in Table 15.2. For the purpose of this example problem, assume that a law firm was interested in predicting future "success" of 20 junior members in the firm (the Y variable in this example is a rating score given by senior members of the law firm on a 7-point rating scale). Two predictor variables they wish to examine are (1) the LSAT law test scores (X_1) and (2) the weekly billing hours (X_2) these 20 attorneys averaged during their first year of employment with the firm. To what extent do LSAT scores and billing hours in combination predict lawyer ratings?

Solving this problem for $R^2_{Y.X1.X2} = (r^2_{Y.X1} + r^2_{Y.X2} - [2][r_{Y.X1}][r_{Y.X2}][r_{X1.X2}]) \div (1 - r^2_{X1.X2})$ requires calculating three correlation coefficients: first, using the numbers in Columns 1 and 2 for $r_{Y.X1}$, second, using the numbers in Columns 1 and 3 for $r_{Y.X2}$; and third, using the numbers in Columns 2 and 3 for $r_{X1.X2}$.

1. $\mathbf{r_{Y.X1} = SP_{Y.X1} \div \sqrt{([SS_{X1}][SS_Y])}}$

   $SS_{X1} = 5{,}599{,}300.00 - (10{,}470^2 \div 20)$

   $SS_{X1} = 5{,}599{,}300.00 - 5{,}481{,}045.00 = \mathbf{118{,}255.00}$

Table 15.2: Lawyer Ratings Related to Examination Scores and Average Billing Hours

	Ratings (Y)	Test Scores (X_1)	Billing Hours (X_2)
	1	450	40
	6	700	60
	7	640	60
	4	580	55
	3	490	50
	5	550	55
	4	440	60
	3	470	35
	5	490	40
	6	500	55
	3	450	40
	6	560	60
	5	580	60
	4	590	50
	3	440	45
	2	430	45
	6	600	55
	5	590	55
	4	500	50
	2	420	55
Σ	**84**	**10,470**	**1,025**
Σ²	**402**	**5,599,300**	**53,725**

$SS_Y = 402 - (84^2 \div 20)$

$SS_Y = 402.00 - 352.80 = \mathbf{49.20}$

$SP_{Y.X1} = 45{,}830.00 - ([10{,}470][84] \div 20)$
$SP_{Y.X1} = 45{,}830.00 - 43{,}974.00 = \mathbf{1{,}856.00}$

$r_{Y.X1} = 1{,}856.00 \div \sqrt{([118{,}255.00][49.20])}$

$r_{Y.X1} = 1{,}856.00 \div 2{,}412.08 = \mathbf{.769}$

2. $\mathbf{r_{Y.X2} = SP_{Y.X2} \div \sqrt{([SS_{X2}][SS_Y])}}$

$SS_{X2} = 53725.00 - (1{,}025^2 \div 20)$

$SS_{X2} = 53{,}725.00 - 52531.25 = \mathbf{1{,}193.75}$

$SP_{Y.X2} = 4{,}460.00 - ([1{,}025][84] \div 20)$

$SP_{Y.X2} = 4{,}460.00 - 4305.00 =$ **155.00**; and from #1 above, $SS_Y = 49.20$; therefore,

$r_{Y.X2} = 155.00 \div \sqrt{([1{,}193.75][49.20])} = 155.00 \div 242.35 = \mathbf{.640}$

3. $\mathbf{r_{X1.X2} = SP_{X1.X2} \div \sqrt{([SS_{X1}][SS_{X2}])}}$

From Parts 1 and 2 we already know SS_{X1} = **118,255.00** and SS_{X2} = **1,193.75**, hence the only remaining calculation is the sum of the X_1 and X_2 cross products ($SP_{X1.X2}$):

$SP_{X1.X2} = 543{,}600.00 - ([10{,}470][1{,}025] \div 20)$

$SP_{X1.X2} = 543{,}600.00 - 536{,}587.50 = \mathbf{7{,}012.50}$

$r_{X1.X2} = 7{,}012.50 \div \sqrt{([118{,}255.00][1{,}193.75])} = 7{,}012.00 \div 11{,}881.37 = \mathbf{.590}$

Separately, these component calculations indicate that test scores (represented by X_1) account for approximately 59% ($r^2_{Y.X1} = .769^2 = .591$) of the variability in lawyer ratings (Y), and billing hours (represented by X_2) account for approximately 41% ($r^2_{Y.X2} = .640^2 = .410$) of the variability in lawyer ratings. From these calculations, it is also apparent that X_1 and X_2 are correlated, because if $r_{X1.X2} = 0$, then we would have the illogical conclusion that X_1 and X_2 combined to account for over 100% of the variability in Y (since $r^2_{Y.X1} + r^2_{Y.X2} = .591 + .410 = 1.001$). To determine the proportion of variability in Y accounted for by both predictor variables in combination, we must solve the multiple regression equation for $R^2_{Y.X1.X2}$. Solving this equation requires careful attention to the correlation coefficients ($r_{Y.X1}$, $r_{Y.X2}$, and $r_{X1.X2}$) and squared correlation coefficients ($r^2_{Y.X1}$, $r^2_{Y.X2}$, and $r^2_{X1.X2}$) that were calculated.

$$R^2_{Y.X1.X2} = (r^2_{Y.X1} + r^2_{Y.X2} - [2][r_{Y.X1}][r_{Y.X2}][r_{X1.X2}]) \div (1 - r^2_{X1.X2})$$

Substituting appropriate r and r^2 values from the preceding calculations into this formula, we have:

$$R^2_{Y.X1.X2} = (.769^2 + .640^2 - [2][.769][.640][.590]) \div (1 - .590^2)$$

$$R^2_{Y.X1.X2} = (.591 + .410 - [2][.769][.640][.590]) \div (1 - .348)$$

$$R^2_{Y.X1.X2} = (1.001 - .581) \div (.652) = .420 \div .652 = \mathbf{.644}$$

Given $R^2_{Y.X1.X2}$ = **.644**, the multiple correlation coefficient $r_{Y.X1.X2} = \sqrt{.644} =$ **.802**

As you can see, the law test scores accounted for about 59% of the variability in lawyer ratings, thus the added information about billing hours (which by itself accounted for approximately 41% of the variability in lawyer ratings) resulted in only an increment of a little over five percentage points, such that the combination of the X_1 and X_2 predictor variables accounted for 64.4% of the variability in Y. This procedure can be extended to problems involving more than two predictor variables, as an expanded regression equation can be derived following the form of $Y' = b_1(X_1) + b_2(X_2) + a$. However, hand calculations of the "b_1, b_2, … , b_j" and "a" constants are complicated and will not be attempted in this text (see Keppel and Zedeck 1989; McClendon 1994). Since the computer is essential in multiple correlation/regression analyses, the steps involved in solving $R^2_{Y.X1.X2}$ using SPSS software are described in the next section.

b. $R^2_{Y.X1.X2}$ Using SPSS

The most time-consuming operation when using SPSS to solve multiple correlation/regression problems involves entering the data into an SPSS spreadsheet. For this example problem the SPSS spreadsheet will require 20 rows (one for each subject) and three columns (for corresponding Y, X_1, and X_2 scores), thus the 20 sets of lawyer ratings, LSAT scores, and average weekly billing hours must be entered before we can instruct the computer to calculate the squared multiple-correlation coefficient. However, once the data have been entered, you simply proceed to the tool bar at the top of the screen and click on "analyze." The entry needed from the draw-down menu is "regression," and as you place your cursor on "regression," a box to the right will open. Move your cursor to the right and click on "linear." This action will open a dialog box similar to the one sketched here, given that your columns are labeled X1, X2, and Y. All that is left to do is to highlight "Y" and click on the arrow to the left of the "Dependent" box. The dependent variable "Y" will now appear in the "Dependent" variable box. Next, highlight "X1" and click on the arrow to the left of the "Independent" box to copy "X1" to that location, and then do the same for "X2." Clicking on "OK" reveals the value for R^2.

Dialog Box

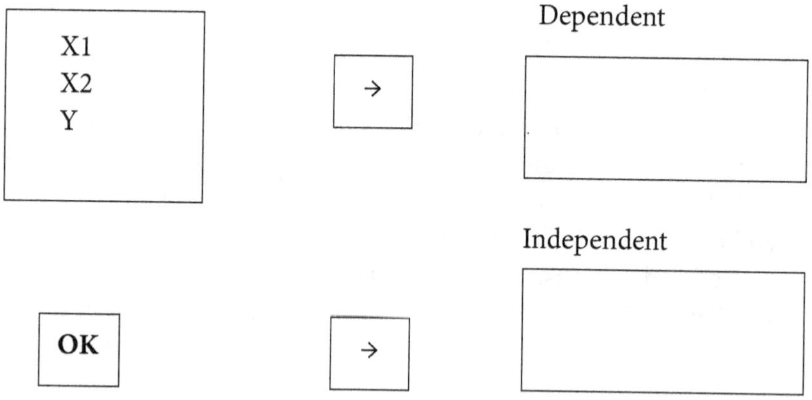

The correlation coefficients for r and ρ ($r_{Y.X1}$, $r_{Y.X2}$, and $r_{X1.X2}$) appear in the upper portion of Table 15.3 as they would on an SPSS printout. The regression portion of the SPSS printout (with only $r_{Y.X1.X2}$ and $R^2_{Y.X1.X2}$) appears below the table. As you can see, the computer values match the hand calculations:

$r_{Y.X1}$ = **.769**; $r_{Y.X2}$ = **.640**; $r_{X1.X2}$ = **.590**; $R_{Y.X1.X2}{}^2$ = **.644**; and $r_{Y.X1.X2}$ = **.802**

Table 15.3: Correlations and the Corresponding Multiple-Regression Coefficient from SPSS (indicates that the correlation is significant at the .01 level for a two-tailed test):**

		Y	X_1	X_2
Y	Pearson Correlation	1.000	.769**	.640**
	Sig. (2-tailed)		.000	.002
	N	20	20	20
X_1	Pearson Correlation	.769**	1.000	.590**
	Sig. (2-tailed)		.000	.006
	N	20	20	20
X_1	Pearson Correlation	.640**	.590**	1.000
	Sig. (2-tailed)	.002	.006	
	N	20	20	20

Regression Summary:

R	R^2
.803	**.645**

It was noted earlier that correlation and regression analyses may be performed with experimental data, where X values correspond to a "dummy" coding that represents the different levels of an independent variable. The analysis of variance is actually a special case of multiple regression analysis, and although computations on a hand-held calculator are more difficult and complex with the regression approach, analytical outcomes are the same (see Keppel and Zedeck 1989), as it can be shown that $\mathbf{F = (r^2 \div df_A) \div ([1 - r^2] \div df_{S/A})}$ (for a regression analysis, df_A and $df_{S/A}$ are more generally referred to as $df_{regression}$ and $df_{residual}$, respectively). For illustration purposes, consider the fictitious data from Table 6.3 (page 98) that were used for a single-factor between-subject ANOVA problem (n = 10, a = 2; thus $df_A = 1$ and $df_{S/A} = 18$). The data are presented again in the following table:

Group a_1	Group a_2
3	7
5	5
2	9
4	6
7	6
5	9
1	8
4	4
6	5
2	5

Remember correlation/regression analyses require two scores for each subject, so for this analysis one score will be a number reported in the table (e.g., "3" for the first subject, "5" for the second subject, "2" for the third subject, etc.) and the second score will be a number code representing each respective subject's group (we will use "1" for all subjects in Group a_1 and "2" for all subjects in Group a_2). Thus, the 10 scores in each group are presented in two sets of paired columns (corresponding X and Y scores) with 10 rows for the 10 subjects in Group a_1 and 10 rows for the 10 subjects in Group a_2. Of course, if you were using SPSS to solve this problem, these 20 X and Y pairings would be entered into two columns of 20 rows. Each score in Table 15.4 is paired with a number code representing the appropriate group designation. Once the data have been organized appropriately, all we need to do is calculate $r_{X.Y}$. From the ANOVA from Chapter 6, we know F = 9.16; since $F = (r^2 \div df_A) \div ([1 - r^2] \div df_{S/A})$, we can enter r^2 in this formula to confirm $(r^2 \div df_A) \div ([1 - r^2] \div df_{S/A}) = 9.16$ (within rounding error).

Table 15.4: Fictitious Experimental Data for Calculating r

Group a_1		Group a_2	
$\underline{Y}$	$\underline{X}$	$\underline{Y}$	$\underline{X}$
3	1	7	2
5	1	5	2
2	1	9	2
4	1	6	2
7	1	6	2
5	1	9	2
1	1	8	2
4	1	4	2
6	1	5	2
2	1	5	2

For all 20 Y scores, $\mathbf{\Sigma Y = 103}$, and for all 20 X scores, $\mathbf{\Sigma X = 30}$. Similarly, the sum of all 20 squared Y scores is $\mathbf{\Sigma Y^2 = 623}$, and the sum of all 20 squared X scores is $\mathbf{\Sigma X^2 = 50}$. Completing the computations necessary for calculating r, we have:

$$SS_Y = \Sigma Y^2 - ([\Sigma Y]^2 \div n) = 623 - (103^2 \div 20) = 623 - 530.45 = \mathbf{92.55}$$

$$SS_X = \Sigma X^2 - ([\Sigma X]^2 \div n) = 50 - (30^2 \div 20) = 50 - 45 = \mathbf{5.00}$$

$$SP = \Sigma XY - (\Sigma X \Sigma Y \div n) = 167 - ([103][30] \div 20) = 167 - 154.5 = \mathbf{12.50}$$

$$r = SP \div \sqrt{([SS_X][SS_Y])} = 12.50 \div \sqrt{([5][92.55])} = 12.50 \div 21.51 = \mathbf{.58}$$

$$F = (r^2 \div df_A) \div ([1 - r^2] \div df_{S/A}) = .336 \div (.664 \div 18) = .336 \div .037 = \mathbf{9.08}$$

If a third group is added to this data set (see Table 6.5, page 104), a multiple regression application may be illustrated. In rearranging these data for a correlation/regression analysis, it will be necessary to have three scores for each subject: a single column of the 30 Y scores, an X_1 number code for each score, and an X_2 number code for each score. To simplify computations, a special class of number codes will be used here. The number codes we will use for X_1 and X_2 in this problem are actually weighting coefficients, and they were selected to identify *orthogonal* (independent) comparisons. The reason for this is that for orthogonal comparisons, $r_{X1.X2} = 0$ (you can confirm this by doing the required calculations to show that $SP_{X1.X2} = 0$). We won't go into details for creating orthogonal coefficients; however, you can recognize that two sets of weighting coefficients are orthogonal if the sum of their cross products is equal to zero [i.e., $\Sigma(c_i)(c_j) = 0$]. With orthogonal coefficients, the formula for $R^2_{Y.X1.X2}$ is simplified because the final term in the numerator of the multiple regression equation $[(2)(r_{Y.X1})(r_{Y.X2})(r_{X1.X2})]$ drops out since $r_{X1.X2} = 0$, and the denominator is equal to 1.0 since $1 - r_{X1.X2} = 1.0$. From the analysis of variance on page 105, $F = 12.27$; thus, within rounding error, $(r^2 \div df_A) \div ([1 - r^2] \div df_{S/A})$ must equal 12.27. The rearranged scores and codes for X_1 and X_2 are presented below in the copy of Table 6.5.

Group a_1	Group a_2	Group a_3
3	7	6
5	5	7
2	9	11
4	6	10
7	6	8
5	9	6
1	8	5
4	4	9
6	5	9
2	5	10

We will rearrange the data for a correlation/regression analysis by using the following orthogonal coefficients for coding X_1 and X_2: for X_1, codes of +1 for Group a_1, -1 for Group a_2, and 0 for Group a_3; and for X_2, codes of +1 for the a_1 and a_2 groups, and -2 for Group a_3. As you can see, the sum of the X_1X_2 cross products is 0 (the last column in this table).

Y	X_1	X_2	X_1X_2
3	1	1	+1
5	1	1	+1
2	1	1	+1
4	1	1	+1
7	1	1	+1
5	1	1	+1
1	1	1	+1
4	1	1	+1
6	1	1	+1
2	1	1	+1
7	-1	1	-1
5	-1	1	-1
9	-1	1	-1
6	-1	1	-1
6	-1	1	-1
9	-1	1	-1
8	-1	1	-1
4	-1	1	-1
5	-1	1	-1

5	-1	1	-1
6	0	-2	0
7	0	-2	0
11	0	-2	0
10	0	-2	0
8	0	-2	0
6	0	-2	0
5	0	-2	0
9	0	-2	0
9	0	-2	0
10	0	-2	0
		$\Sigma \mathbf{X_1X_2}$	**0**

Since $r_{X1.X2} = 0$, $R^2_{Y.X1.X2} = r^2_{Y.X1} + r^2_{Y.X2}$, we only need to calculate $r_{Y.X1}$ and $r_{Y.X2}$.

For $r_{Y.X1}$:

$SS_Y = 1{,}316.00 - (184^2 \div 30) = 1{,}316.00 - 1{,}128.53 = \mathbf{187.47}$

$SS_{X1} = 20.00 - (0^2 \div 30) = \mathbf{20.00}$

$SP_{Y.X1} = -25.00 - ([184][0] \div 30) = \mathbf{-25.00}$

$r_{Y.X1} = SP_{Y.X1} \div \sqrt{([SS_{X1}][SS_Y])} = -25.00 \div \sqrt{([20][187.47])} = -25.00 \div 61.23 = \mathbf{-.408}$

For $r_{Y.X2}$, we know $SS_Y = \mathbf{187.47}$, and:

$SS_{X2} = 60.00 - (0^2 \div 30) = \mathbf{60.00}$

$SP_{Y.X2} = -59.00 - ([184][0] \div 30) = \mathbf{-59.00}$

$r_{Y.X2} = SP_{Y.X2} \div \sqrt{([SS_{X2}][SS_Y])} = -59.00 \div \sqrt{([60.00][187.47])} = -59.00 \div 106.06 = \mathbf{-.556}$

Since the dummy codes for X_1 and X_2 are orthogonal, $R^2_{Y.X1.X2} = R^2_{Y.X1} + R^2_{Y.X2}$; thus, $R^2_{Y.X1.X2} = (-.408)^2 + (-.556)^2 = .166 + .309 = \mathbf{.475}$; and calculating F from the regression analysis, we have: $F = (R^2 \div 2) \div ([1 - R^2] \div 27) = .238 \div (.525 \div 27) = .238 \div .019 = \mathbf{12.53}$ (Note: if you carry your calculations out to four decimal places, there will be a closer match between the F calculated using ANOVA procedures and this F calculated using the multiple correlation/regression procedures).

VI. Summary

This chapter began by differentiating the two related topics of correlation and regression. Both topics are concerned with a measurement of the relationship between two (or more) variables. A primary goal of a regression analysis is prediction. Specifically, two sets of scores (representing behaviors of interest) that normally occur in a chronological order are assessed to determine the degree to which they are linearly related. If the assessed relationship is reasonably high and reliable, then the measured relation coupled with information regarding position on one set of scores (the first set of scores in chronological order, referred to as the "predictor" variable and represented as the X set of scores) is used to estimate future performance on a second set of scores (the second set of scores in chronological order, represented as the Y set of scores and behavior of primary interest). An obvious first task then was to identify procedures employed in predicting performance on some criterion task based on information about performance on a relevant predictor task.

Prior information about the performance of individuals on relevant tasks is necessary to estimate the degree to which the two tasks are related. A large sample of pairs of scores (an X score and a Y score from each individual) may be plotted as points (corresponding X and Y coordinates for each point) on a two-dimensional graph. A line of best fit may be drawn through the set of points, with "best fitting" defined as the one straight line that results in the smallest sum of squared deviation scores (a deviation score refers to how far away a given X-Y coordinate point is from the best-fitting straight line derived from the data set). The resulting straight line is called the regression line and is used for predicting any future or unknown Y score for an individual based on his or her X score (performance on the predictor test). Given an X test score, the predicted Y score is simply the point on the regression line that corresponds to the X score that was obtained. Of course, prediction is not perfect; however, if there is a strong correlation between X and Y and a relatively small degree of variability (the data points from the original sample were tightly packed around the regression line), then prediction error should be small.

If using one performance measure to predict performance on a related second task makes sense, then it follows that basing predictions about performance on a criterion task would be even more precise if information were available about prior performance on two or more predictor tasks. While this is true, what is gained in predictive advantage with additional variables may be modest even when second, third, etc. predictor variables are highly correlated with the behavior of interest (Y). The reason that all of the correlation benefit from additional variables is not realized is that most likely there are relationships among sets of predictor variables (Xs). A sizeable proportion of the variability in Y may be accounted for by X_1, and a sizeable proportion of variability in Y may be accounted for by X_2. The problem is that variation common to X_1 and X_2 would be counted twice if proportions of variation in Y accounted for by X_1 and X_2 squared correlations were simply summed. Consequently, problems associated with determining the proportion of variability in Y that is accounted for by multiple predictor variables (multiple correlation/regression) are complex and become overwhelming for computations on a hand calculator with large data sets. This complexity may explain, in part, why the important topic of multiple predictor variables is usually not included in undergraduate introductory statistics textbooks in psychology. Realistically, when considering more than two predictor variables for regression analyses, the computer must come to the rescue.

Correlation and regression interests most often center on situations involving two or more dependent variables (response measures). However, levels of an independent variable (e.g., hours of sleep deprivation) may be coded numerically (e.g., "1" = 12 hours deprivation, "2" = 18 hours, "3" = 24 hours), and correlation and regression procedures may be used with experimental data. In an experimental context, each subject would contribute two or more scores. For example, a subject with a score of "10" on a test after 12 hours of sleep deprivation would have a Y score of "10" and an X score of "1," given that "1" was selected as the dummy code for the 12-hour group. A subject with a score of "8" on a test after 18 hours of sleep deprivation would have a Y score of "8" and an X score of "2." And to complete the example, a subject with a score of "6" on a test after 24 hours of sleep deprivation would have a Y score of "6" and an X score of "3." The chapter concluded with illustrations of two single-factor, between-subject designs, one with a = 2 and one with a = 3. The analysis of variance approach emphasized in earlier chapters is a special case of correlation/regression analysis adapted for experimental data, and these two illustrations concluded with a final step estimating F (the test statistic in the analysis of variance) from the squared correlation coefficients.

15.1: Model Problem 1. What is the formula for the best-fitting line (regression line) for the data in Practice Problem 5, Chapter 14 (page 315)? After you develop this formula, use it to predict a Y score (Y') for (1) X = 10; (2) X = 15; and (3) X = 20. The 15 paired scores from problem 14.6 are repeated below:

Subjects	X	Y
1	10	11
2	17	14
3	15	16
4	12	13
5	20	19
6	16	18
7	15	20
8	15	17
9	21	15
10	13	10
11	11	14
12	14	12
13	18	13
14	16	22
15	25	21
Σ	**238**	**235**
Σ^2	**3,996**	**3,875**

The regression line equation is $Y' = b(X) + a$; where $b = SP \div SS_X$ and $a = \bar{Y} - b(\mathbf{X})$.

Step 1: $SS_X = \Sigma X^2 - ([\Sigma X]^2 \div n) = 3{,}996 - (238^2 \div 15) = 3{,}996 - 3{,}776.27 = \mathbf{219.73}$

Step 2: $SP = \Sigma XY - ([\Sigma X][\Sigma Y] \div n) = 3{,}845 - ([238][235) \div 15) = 3{,}845 - 3{,}728.67 = \mathbf{116.33}$

Step 3: $b = 116.33 \div 219.73 = \mathbf{.53}$

Step 4: $a = \bar{Y} - (b)(\bar{X}) = 238 \div 15 - ([.53][235 \div 15]) = 15.67 - 8.41 = \mathbf{7.26}$

Step 5: Substitute numerical values for the regression constants into the straight line formula:

$$\mathbf{Y' = (.53)(X) + 7.26}$$

For the second part of this question, just replace "X" in the regression line formula:

(1) For X = 10, Y' = (.53)(10) + 7.26 = 5.30 + 7.26 = **12.56**

(2) For X = 15, Y' = (.53)(15) + 7.26 = 7.95 + 7.26 = **15.21**

(3) For X = 20, Y' = (.53)(20) + 7.26 = 10.60 + 7.26 = **17.86**

15.2: Model Problem 2. For the small data set below, calculate $\mathbf{R^2}_{Y.X1.X2}$ by hand and by using SPSS (Note: The SPSS printout that is copied at the end of this problem reports p values for a significance test. With n = 5, df = n – 2 = 3, you may expect that it will require a pretty substantial correlation for significance at the .05 level. Table A-8 in Appendix A reveals that $\mathbf{r_{crit}}$ = **.88** for a two-tailed test at $p \leq .05$ and df = 3.

X_1	X_2	Y	
7	7	25	$\Sigma X_1 = 38$; $\Sigma X_1^2 = 304$; $\Sigma X_2 = 25$; $\Sigma X_2^2 = 151$; $\Sigma Y = 155$; $\Sigma Y^2 = 4{,}975$
9	5	30	
10	8	40	$\Sigma X_1Y = 1{,}215$; $(\Sigma X_1)(\Sigma Y) = 5{,}890$; $\Sigma X_2Y = 790$; $(\Sigma X_2)(\Sigma Y) = 3{,}875$
5	3	25	
7	2	35	$\Sigma X_1X_2 = 203$; $(\Sigma X_1)(\Sigma X_2) = 950$

The basic computations for the various sums of squares and sums of products are shown to the right of this table. Given these results, you can proceed to compute three correlation coefficients: $r_{Y.X1}$, $r_{Y.X2}$, and $r_{X1.X2}$.

Step 1: Calculate $r_{Y.X1}$:

$SS_{X1} = 304 - (38^2 \div 5) = 304 - 288.80 = \mathbf{15.20}$

$SS_Y = 4{,}975 - (155^2 \div 5) = 4{,}975 - 4{,}805 = \mathbf{170.00}$

$SP_{Y.X1} = 1{,}215 - ([38][155] \div 5) = 1{,}215 - 1{,}178 = \mathbf{37.00}$

$r_{Y.X1} = 37.00 \div \sqrt{([15.20][170.00])} = 37.00 \div 50.83 = \mathbf{.728}$

Step 2: Calculate $r_{Y.X2}$:

$SS_{X2} = 151 - (25^2 \div 5) = 151 - 125 = \mathbf{26.00}$

$SS_Y = 170.00$ (from above)

$SP_{Y.X2} = 790 - ([25][155] \div 5) = 790 - 775 = \mathbf{15.00}$

$r_{Y.X2} = 15.00 \div \sqrt{([26.00][170.00])} = 15.00 \div 66.48 = \mathbf{.226}$

Step 3: Calculate $r_{X1.X2}$:

$SS_{X1} = 15.20$ and $SS_{X2} = 26.00$

$SP_{X1.X2} = 203 - ([38][25] \div 5) = 203 - 190 = \mathbf{13.00}$

$r_{X1.X2} = 13.00 \div \sqrt{([15.20][26.00])} = 13.00 \div 19.88 = \mathbf{.654}$

$R^2_{Y.X1.X2} = [.728^2 + .226^2 - (2)(.728)(.226)(.654)] \div (1 - .654^2) = .351 \div .572 =$ **.614** ($R^2 = .639$ in the accompanying SPSS table; a value that is matched if the preceding work is done to four-decimal-place accuracy)

SPSS Table Showing the Results of the Computer Calculated Correlations (with df = n – 2 = 5 – 2 = 3, none of the reported r values are significant at the .05 level):

		Y	X_1	X_2
Y	Pearson Correlation	1.000	.654	.728
	Sig. (2-tailed)		.231	.163
	N	5	5	5
X_1	Pearson Correlation	.654	1.000	.226
	Sig. (2-tailed)	.231		.715
	N	5	5	5
X_1	Pearson Correlation	.728	.226	1.000
	Sig. (2-tailed)	.163	.715	
	N	5	5	5

Regression Summary:

R	R^2
.800	**.639**

15.3: Practice Problem 1. Part of a college application process for high school seniors involves submitting scores on a standardized test such as the SAT or ACT. College admissions staff presumably uses test-score data in the admission decision process because it is assumed (or has been shown) that test scores are relevant for predicting college success. Assume the numbers in the tabled data for this problem represent high school ACT scores (X) from 16 students who just completed their freshman year at college. Along with the ACT scores are the freshman year GPAs (Y) for the same 16 students.

Data Matrix of ACT scores (X) and freshman GPAs on a 4-point scale

X (ACT)	Y (GPA)
18	4.0
15	2.8
20	3.5
17	2.2
20	3.8
19	4.0
18	3.0
15	4.0
21	2.0
16	3.1
21	3.6
18	1.8
19	3.2
16	2.8
17	2.4
20	3.2

A. Prepare a graph presenting a "scatterplot" of the 16 X-Y points (i.e., locate each of the 16 X scores on the horizontal axis with the 16 corresponding Y scores on the vertical axis). A point on a scatterplot corresponds to the location where an individual's X score intersects with his or her Y score, found by determining where a vertical line from the X score on the X axis intersects a horizontal line from the Y score on the Y axis. The point of intersection of these two lines is called the X-Y coordinate for that particular individual. Your graph will show one such point for each of the 16 students.

B. Write the formula for the regression line for these data. That is, for $Y' = b(X) + a$, replace the slope and Y-intercept regression constants ("b" and "a," respectively) with numbers.

C. Using the straight line formula you developed from the data in Part B, draw the best-fitting straight line through the points on your graph from Part A.

D. Based on these data, what freshman GPA would you predict for a future student who reported an ACT score of 15? What would you predict for a student who reported an ACT score of 20?

15.4: Practice Problem 2. It is fairly obvious that faculty salaries are related to years of service, and given the emphasis that many universities place on research, it is probably true that salaries are also related to publication records of faculty. Assume the data on the following page are from a university science department. Do the appropriate analyses to estimate (1) the extent to which years of post-doctoral experience (X_1) predict annual salaries (Y) ($r^2_{Y.X1}$ = ?), (2) the extent to which average number of publications per year (X_2) predict annual salaries ($r^2_{Y.X2}$ = ?), and (3) the extent to which years of post-doctoral experience and average number of yearly publications in combination predict annual salaries—i.e., estimate the proportion of the variability in faculty salaries (Y) that is accounted for by the combination of number of post-doctoral years (X_1) and average number of publications per year (X_2) (i.e., $R^2_{Y.X1.X2}$ = ?). This is a fairly large data set, and calculating $R^2_{Y.X1.X2}$ by hand may be a lot to ask. You should certainly solve this problem using SPSS or an alternative computer software package; however, completing this problem on a hand-held calculator will provide relevant practice for calculating and understanding the component correlation coefficients that determine R^2.

X_1 (Post-Doctoral Years)	X_2 (Average Publications)	Y (Salary in $1,000)
2	3.0	65
3	5.0	75
3	2.0	53
4	5.0	80
5	1.0	51
7	.5	49
8	3.0	75
9	1.5	66
12	1.0	65
14	.5	70
15	2.0	84
16	3.0	100
20	2.5	105
24	1.0	95
25	.2	87
30	1.2	91
35	2.0	110

Chapter 16

A Reprise and Practice Test for Chapters 13–16

All's Well That Ends Well.

— *Shakespeare*

I. A Reprise

Regardless of whether you have an impending comprehensive final exam or a final exam covering only the last few chapters (or even no final), it may be useful to review what was introduced in this textbook and acknowledge some traditional topics that were slighted. Decisions about topic selection, details covered, and emphases were driven by a goal of introducing statistical concepts and techniques relevant in today's research settings (with an emphasis on psychology experiments).

What this text did not do was focus on scales of measurement, various types of graphing techniques and curves, calculating percentiles, and various details involved in the organization of large bodies of data. Also, probability theory and concepts received scant attention, the t test statistic certainly had less text space devoted to it than is customary in introductory statistics textbooks, and analysis of variance was introduced without working up to it by considering the z distribution (unit-normal distribution) and the z test for comparing a sample mean with a population parameter (μ).

After briefly setting the research context by describing what an experiment entails and its underlying logic, two important uses for statistical applications were identified: (1) Descriptive statistics (effectively and efficiently summarizing large data sets), and (2) inferential statistics (ground rules agreed upon in advance for deciding whether or not research results should be regarded as reliable). Our consideration of descriptive statistics moved quickly to a review of two primary descriptive indices: one for identifying a center point for a data set, and one for characterizing the degree of dispersion or separation among scores in a data set. The most useful center point index for our purposes is the mean—a simple arithmetic average. A second center point index that is not encountered as much in experimental research is the median—a midpoint or a score numerically in the middle of a distribution of scores (a point for which there are an equal number of higher and lower scores). The median does not make use of as much information in a data set (actual numerical score values) as the mean; however, the median may have one advantage over the mean: e.g., a few unusually high scores will have little impact on a median index of central tendency; however, they may have a fairly substantial impact on a mean index of central tendency.

Measures of dispersion most frequently encountered in research reports are the related indices of variances, standard deviations (the square root of the variance), and standard errors. Means, variances, standard deviations, and standard errors are important single-number descriptors for a data set, and they are also centrally involved in inferential statistics (drawing inferences regarding the reliability of research results). While the descriptive function is arguably a necessary component of all data collection endeavors and arguably more central than the inferential function of statistics, the majority of chapters in the present textbook concerned statistical techniques in the service of inferential objectives.

Given time and space constraints set by a single course and a single textbook, only a sampling of the many inferential statistical tests was introduced. Considerable space was devoted to introducing the F test statistic in the analysis of variance, which constitutes a most widely used set of analytical procedures in published psychology experiments. A few years ago we did a count of the number of research reports published in a couple of the leading "experimental" psychology journals. That one-shot, arbitrary sampling revealed that the analysis of variance was used in 86% of the experiments reported.

There are a variety of experimental research paradigms varying in number of conditions and number of independent variables, coupled with whether a given subject in an experiment participates in only one condition or in more than one condition. For simple experiments involving only two conditions (e.g., an experimental group and a control group), a t test may be used instead of using the analysis of variance. Since the t test is occasionally encountered in the literature, it was briefly introduced. In addition, procedures were described for determining confidence intervals for the difference between two group means. Although the same conclusions about whether the difference between two group means is "significant" are arrived at with F, t, or confidence intervals, the latter is richer in information because it identifies a range of scores that likely include the true difference between two group means rather than just a yes–no decision about whether group means should be regarded as equivalent or different. The F test is more general and received the most attention as it is also applicable to experiments involving more than two groups or experimental conditions.

In complex experiments that involve two independent variables (or a single independent variable with more than two levels) it is important to extract from the data as much meaningful information as possible. With regard to statistical analysis, the analysis of variance often serves as just the first step in evaluating experimental results. Since an F test for experiments involving multiple groups provides only a general assessment regarding whether it is reasonable to attribute observed differences in performance to the independent variable (e.g., performance is not equivalent across all groups or experimental conditions), considerable attention was devoted to conducting both planned and unplanned analytical comparisons following an overall analysis of variance. Also, providing information about error variance and effect size (e.g., ω^2) is becoming more common in research reports, and these two topics were introduced. Importantly, the ability to estimate effect size enables researchers to attend to the important concept of power of an experiment for correctly detecting a significant effect when H_0 is false.

For some research problems in psychology, observations of behavior are in the form of frequency counts rather than numerical scores along some presumed continuum. For example, a researcher may be interested in comparing the extent to which individuals scatter litter in a high-rent district compared to a low-rent district, or whether cars come to a complete stop at an intersection stop sign, slow down without coming to a complete stop, or continue through the intersection with little change in speed. In these situations, data are frequency counts (e.g., number of individuals littering, number of cars stopping or slowing). Chi-square (χ^2) was introduced in Chapter 13 as the appropriate inferential test statistic for analyzing categorical data, and ϕ was introduced as a metric for estimating the magnitude of a χ^2 effect.

Chapters 14 and 15 introduced the related topics of correlation and regression. These are commonly used procedures for describing the relationship between two or more dependent variables (correlation) and applying that information to predict future performance on one of the measures (regression). The procedures may also be used with experimental data, in which case F can be calculated using the regression analysis procedures. Two correlation measures were introduced: the Pearson r for measuring the relationship between paired scores that are numerical representations along a presumed continuum, and the Spearman ρ for measuring relationships for data that are in the form of ranks (e.g., a "1" paired with a "5" for an individual with the highest score on one performance measure and the fifth highest score on the other performance measure). It was also noted that correlation coefficients (r and ρ) may be subjected to significance testing.

If relevant, a correlation between two variables enables one to use information gathered on one of the two measures to improve prediction about what scores on the other dependent variable will be. Of course, if improvement in prediction results from having information available on one predictor variable, then it makes sense that further improvement would result if information from two predictor variables was available to use for predicting scores on a third dependent variable measure. This multiple correlation/regression topic highlighted the final section of Chapter 15. Because of the computational complexity, multiple correlation/regression is a topic that is not often included in introductory statistics textbooks. Certainly any serious multiple regression analysis requires computer assistance with the calculations. Although using a hand-held calculator was stressed throughout the text presentations and for solving most practice problems, computer applications with SPSS were introduced with more complex applications. Doing statistical analyses with appropriate software is a necessary component in most research settings, and an essential one for analyzing large, complex data sets.

II. Practice Test for Chapters 13–16

(answers given in Appendix B)

I. Circle the letter of the best answer for each of the questions below:

1. A student did a project to determine if people who walked their dogs in a public park were more likely to 'pick-up' after them if the park had 6, 3, or 0 trash cans. He observed 60 dog walkers and simply recorded whether they picked up after their pets. The appropriate test statistic for this project is

   a. F
   b. χ^2
   c. r
   d. ρ

2. A sample of 45 men and a sample of 45 women (N = 90) differed significantly in the frequency with which they preferred 'reality' shows on TV compared to 'quiz' shows ($\chi^2 = 4.75$). Which value below represents the closest approximation for the size of this effect?

   a. .23
   b. .33
   c. .43
   d. .53

3. To use the Cochran Q adjustment for χ^2, you need

   a. an estimate of effect size
   b. independent cell frequencies
   c. information on how subjects were categorized each time they were observed
   d. to know all of the above

4. For 10 cancer patients who were smokers, their age at the time cancer was diagnosed was correlated with their age when they first started smoking. What must r equal or exceed to be significant at $p < .05$?

   a. .42
   b. .55
   c. .63
   d. .75

5. Squaring the correlation coefficient (r^2) provides a measure of

   a. association between F and r
   b. strength of relationship between X and Y
   c. similarity between regression and correlation
   d. the causal relationship

6. Given that a correlation between X and Y is r = .40, approximately what proportion of the variance in Y is accounted for by X?

   a. .80
   b. .63
   c. .20
   d. .16

7. If you have 14 subjects in your sample with each subject contributing an X score and a Y score, what is the critical value of r that the calculated correlation coefficient must equal or exceed to be significant at **$p \leq .01$**?

   a. .51
   b. .62
   c. .66
   d. .75

8. If the formula for a regression line is Y' = 1.5(X) – 25, the "Y" predicted score for a student who received an X score of 80 is:
   a. 80
   b. 85
   c. 90
   d. 95

9. If $r_{Y.X1} = .40$, $r_{Y.X2} = .30$, and $r_{X1.X2} = .20$, then it is probably true that $R^2_{Y.X1.X2}$
   a. is greater than .40
   b. is less than .40
   c. equals 0
   d. equals .90

10. The following pair-wise correlations between two predictor variables (X_1 and X_2) and between each predictor variable with Y are $r_{Y.X1} = .50$, $r_{Y.X2} = .30$, and $r_{X1.X2} = .40$. What is the proportion of variability in Y accounted for by these two predictor variables?
    a. .26
    b. .32
    c. .38

d. .44

II. After reading a brief description of a murder and a court case summary that ended in a guilty verdict, 200 individuals were asked whether they favored or opposed the death penalty for the convicted murderer. Fifty individuals were (1) not given any additional personal information about either the victim or the convicted murderer, (2) 50 others read a brief biography of the murderer prior to giving their response, (3) another 50 read a brief biography of the victim prior to giving their response, and (4) 50 others read both biographies prior to responding. The results in terms of the number of individuals favoring the death penalty and the number opposing the death penalty in this fictitious case are given in the following table.

	CATEGORIES			
	No Information	Murderer Biography	Victim Biography	Both Biographies
Favor Death Penalty	20	16	32	24
Oppose Death Penalty	30	34	18	26

A. Do a χ^2 test to determine if there is a significant difference ($p \leq .05$) among groups as a function of subjects' reading or not reading various biographies.

B. Do a follow-up comparison (an analytical comparison) to determine if the group that read only the convicted murderer's biography differed significantly (at $p \leq .05$) in terms of favoring or opposing the death penalty compared to the group who read only the victim's biography.

III. Below are 10 pairs of X and Y scores.

Subjects	SCORES X	SCORES Y
1	20	15
2	8	8
3	14	10
4	9	12
5	16	16
6	11	7
7	9	6
8	5	8
9	18	20
10	9	11

A. Is there a significant correlation as measured by r?

B. Is there a significant correlation as measured by ρ?

IV. Assume you have two predictor variables X_1 and X_2, with a correlation between Y and X_1 ($r_{Y.X1}$) equal to .56, and a correlation between Y and X_2 ($r_{Y.X2}$) equal to .70. If the correlation between the two predictor variables ($r_{X1.X2}$) is equal to .50, what does $R^2_{Y.X1.X2}$ equal?

Appendices

Appendix A: Statistical Tables

Directions: Critical values for F reference distributions that cut off the extreme 5% of the distribution ($p \leq .05$ in boldface type) and the extreme 1% of the distribution ($p \leq .01$) are located within the table. Rows correspond to the degrees of freedom for the error term (denominator in the F ratio) and columns correspond to the degrees of freedom for the effect (numerator in the F ratio).

df_{effect}

df_{error}	1	2	3	4	5	6	7	8	9	10	12	15	20
4	**7.71**	**6.94**	**6.59**	**6.39**	**6.26**	**6.16**	**6.09**	**6.04**	**6.00**	**5.96**	**5.91**	**5.86**	**5.80**
	21.2	18.0	16.7	16.0	15.5	15.2	15.0	14.8	14.7	14.6	14.4	14.2	14.0
5	**6.61**	**5.79**	**5.41**	**5.19**	**5.05**	**4.95**	**4.88**	**4.82**	**4.77**	**4.74**	**4.68**	**4.62**	**4.56**
	16.3	13.3	12.1	11.4	11.0	10.7	10.5	10.3	10.2	10.0	9.89	9.72	9.55
6	**5.99**	**5.14**	**4.76**	**4.53**	**4.39**	**4.28**	**4.21**	**4.15**	**4.10**	**4.06**	**4.00**	**3.94**	**3.87**
	13.8	10.9	9.78	9.15	8.75	8.47	8.26	8.10	7.98	7.87	7.72	7.56	7.40
7	**5.59**	**4.74**	**4.35**	**4.12**	**3.97**	**3.87**	**3.79**	**3.73**	**3.68**	**3.64**	**3.57**	**3.51**	**3.44**
	12.2	9.55	8.45	7.85	7.46	7.19	6.99	6.84	6.72	6.62	6.47	6.31	6.16
8	**5.32**	**4.46**	**4.07**	**3.84**	**3.69**	**3.58**	**3.50**	**3.44**	**3.39**	**3.35**	**3.28**	**3.22**	**3.15**
	11.3	8.65	7.59	7.01	6.63	6.37	6.18	6.03	5.91	5.81	5.67	5.52	5.36
9	**5.12**	**4.26**	**3.86**	**3.63**	**3.48**	**3.37**	**3.29**	**3.23**	**3.18**	**3.14**	**3.07**	**3.01**	**2.94**
	10.6	8.02	6.99	6.42	6.06	5.80	5.61	5.47	5.35	5.26	5.11	4.96	4.81
10	**4.96**	**4.10**	**3.71**	**3.48**	**3.33**	**3.22**	**3.14**	**3.07**	**3.02**	**2.98**	**2.91**	**2.85**	**2.77**
	10.0	7.56	6.55	5.99	5.64	5.39	5.20	5.06	4.94	4.85	4.71	4.56	4.41
11	**4.84**	**3.98**	**3.59**	**3.36**	**3.20**	**3.09**	**3.01**	**2.95**	**2.90**	**2.85**	**2.79**	**2.72**	**2.65**
	9.65	7.21	6.22	5.67	5.22	5.07	4.89	4.74	4.63	4.54	4.40	4.25	4.10

	df_{effect}												
df_{error}	1	2	3	4	5	6	7	8	9	10	12	15	20
12	**4.75**	**3.89**	**3.49**	**3.26**	**3.11**	**3.00**	**2.91**	**2.85**	**2.80**	**2.75**	**2.69**	**2.62**	**2.54**
	9.33	6.93	5.95	5.41	5.06	4.82	4.64	4.50	4.39	4.30	4.16	4.01	3.86
13	**4.67**	**3.81**	**3.41**	**3.18**	**3.03**	**2.92**	**2.83**	**2.77**	**2.71**	**2.67**	**2.60**	**2.53**	**2.46**
	9.07	6.70	5.74	5.21	4.86	4.62	4.44	4.30	4.19	4.10	3.96	3.82	3.66
14	**4.60**	**3.74**	**3.34**	**3.11**	**2.96**	**2.85**	**2.76**	**2.70**	**2.65**	**2.60**	**2.53**	**2.46**	**2.39**
	8.86	6.51	5.56	5.04	4.69	4.46	4.28	4.14	4.03	3.94	3.80	3.66	3.51
15	**4.54**	**3.68**	**3.29**	**3.06**	**2.90**	**2.79**	**2.71**	**2.64**	**2.59**	**2.54**	**2.48**	**2.40**	**2.33**
	8.68	6.36	5.42	4.89	4.56	4.32	4.14	4.00	3.89	3.80	3.67	3.52	3.37
16	**4.49**	**3.63**	**3.24**	**3.01**	**2.85**	**2.74**	**2.66**	**2.59**	**2.54**	**2.49**	**2.42**	**2.35**	**2.28**
	8.53	6.23	5.29	4.77	4.44	4.20	4.03	3.89	3.78	3.69	3.55	3.41	3.26
17	**4.45**	**3.59**	**3.20**	**2.96**	**2.81**	**2.70**	**2.61**	**2.55**	**2.49**	**2.45**	**2.38**	**2.31**	**2.23**
	8.40	6.11	5.18	4.67	4.34	4.10	3.93	3.79	3.68	3.59	3.46	3.31	3.16
18	**4.41**	**3.55**	**3.16**	**2.93**	**2.77**	**2.66**	**2.58**	**2.51**	**2.46**	**2.41**	**2.34**	**2.27**	**2.19**
	8.29	6.01	5.09	4.58	4.25	4.01	3.84	3.71	3.60	3.51	3.37	3.23	3.08
19	**4.38**	**3.52**	**3.13**	**2.90**	**2.74**	**2.63**	**2.54**	**2.48**	**2.42**	**2.38**	**2.31**	**2.23**	**2.16**
	8.18	5.93	5.01	4.50	4.17	3.94	3.77	3.63	3.52	3.43	3.30	3.15	3.00
20	**4.35**	**3.49**	**3.10**	**2.87**	**2.71**	**2.60**	**2.51**	**2.45**	**2.39**	**2.35**	**2.28**	**2.20**	**2.12**
	8.10	5.85	4.94	4.43	4.10	3.87	3.70	3.56	3.46	3.37	3.23	3.09	2.94
21	**4.32**	**3.47**	**3.07**	**2.84**	**2.68**	**2.57**	**2.49**	**2.42**	**2.37**	**2.32**	**2.25**	**2.18**	**2.10**
	8.02	5.78	4.87	4.37	4.04	3.81	3.64	3.51	3.40	3.31	3.17	3.03	2.88
22	**4.30**	**3.44**	**3.05**	**2.82**	**2.66**	**2.55**	**2.46**	**2.40**	**2.34**	**2.30**	**2.23**	**2.15**	**2.07**
	7.95	5.72	4.82	4.31	3.99	3.76	3.59	3.45	3.35	3.26	3.12	2.98	2.83

	df_{effect}												
df_{error}	1	2	3	4	5	6	7	8	9	10	12	15	20
23	**4.28**	**3.42**	**3.03**	**2.80**	**2.64**	**2.53**	**2.44**	**2.37**	**2.32**	**2.27**	**2.20**	**2.13**	**2.05**
	7.88	5.66	4.76	4.26	3.94	3.71	3.54	3.41	3.30	3.21	3.07	2.93	2.78
24	**4.26**	**3.40**	**3.01**	**2.78**	**2.62**	**2.51**	**2.42**	**2.36**	**2.30**	**2.25**	**2.18**	**2.11**	**2.03**
	7.82	5.61	4.72	4.22	3.90	3.67	3.50	3.36	3.26	3.17	3.03	2.89	2.74
25	**4.24**	**3.39**	**2.99**	**2.76**	**2.60**	**2.49**	**2.40**	**2.34**	**2.28**	**2.24**	**2.16**	**2.09**	**2.01**
	7.77	5.57	4.68	4.18	3.85	3.63	3.46	3.32	3.22	3.13	2.99	2.85	2.70
30	**4.17**	**3.32**	**2.92**	**2.69**	**2.53**	**2.42**	**2.33**	**2.27**	**2.21**	**2.16**	**2.09**	**2.01**	**1.93**
	7.56	5.39	4.51	4.02	3.70	3.47	3.30	3.17	3.07	2.98	2.84	2.70	2.55
40	**4.08**	**3.23**	**2.84**	**2.61**	**2.45**	**2.34**	**2.25**	**2.18**	**2.12**	**2.08**	**2.00**	**1.92**	**1.84**
	7.31	5.18	4.31	3.83	3.51	3.29	3.12	2.99	2.89	2.80	2.66	2.52	2.37
60	**4.00**	**3.15**	**2.76**	**2.53**	**2.37**	**2.25**	**2.17**	**2.10**	**2.04**	**1.99**	**1.92**	**1.84**	**1.75**
	7.08	4.98	4.13	3.65	3.34	3.12	2.95	2.82	2.72	2.63	2.50	2.35	2.20
120	**3.92**	**3.07**	**2.68**	**2.45**	**2.29**	**2.17**	**2.09**	**2.02**	**1.96**	**1.91**	**1.83**	**1.75**	**1.66**
	6.85	4.79	3.95	3.48	3.17	2.96	2.79	2.66	2.56	2.47	2.34	2.19	2.03
∞	**3.84**	**3.00**	**2.61**	**2.37**	**2.21**	**2.10**	**2.01**	**1.94**	**1.88**	**1.83**	**1.75**	**1.67**	**1.57**
	6.63	4.41	3.78	3.32	3.02	2.80	2.64	2.51	2.41	2.32	2.18	2.04	1.88

Table A-2: Critical Values of t Reference Distributions for "Two-Tailed" Tests

Directions: Critical values for different t reference distributions that cut off the extreme 5% of the distribution ($p \leq .05$ in boldface type) and the extreme 1% of the distribution ($p \leq .01$) are located within the table. The row numbers correspond to the degrees of freedom for the error term ($n_1 - 1 + n_2 - 1$).

df	p ≤ .05	p ≤ .01
2	**4.30**	9.92
3	**3.18**	5.84
4	**2.78**	4.60
5	**2.57**	4.03
6	**2.45**	3.71
7	**2.36**	3.50
8	**2.31**	3.36
9	**2.26**	3.25
10	**2.23**	3.17
11	**2.20**	3.11
12	**2.18**	3.06
13	**2.16**	3.01
14	**2.14**	2.98
15	**2.13**	2.95
16	**2.12**	2.92
17	**2.11**	2.90
18	**2.10**	2.88
19	**2.09**	2.86
20	**2.09**	2.84
21	**2.08**	2.83
22	**2.07**	2.82
23	**2.07**	2.81
24	**2.06**	2.80
25	**2.06**	2.79
30	**2.04**	2.75
40	**2.02**	2.70
60	**2.00**	2.66
120	**1.98**	2.62
∞	**1.96**	2.58

"Table 12," *Biometrika Tables for Statisticians*, vol. 1, 3rd ed., E. S. Pearson and H. O. Hartley, eds. opyright © 1970 by Oxford University Press. Reprinted with permission.

Table A-3: Coefficients for Assessing Linear, Quadratic, and Cubic Trend with Evenly Spaced Intervals along an Underlying Quantitative Dimension

Directions: Rows are organized in terms of number of experimental conditions (a) and functional relations (linear, quadratic, and cubic trends); columns identify the ordered coefficients.

		Coefficients								
a	Trend	c_1	c_2	c_3	c_4	c_5	c_6	c_7	c_8	c_9
3	Linear	-1	0	1						
	Quadratic	1	-2	1						
4	Linear	-3	-1	1	3					
	Quadratic	1	-1	-1	1					
	Cubic	-1	3	-3	1					
5	Linear	-2	-1	0	1	2				
	Quadratic	2	-1	-2	-1	2				
	Cubic	-1	2	0	-2	1				
6	Linear	-5	-3	-1	1	3	5			
	Quadratic	5	-1	-4	-4	-1	5			
	Cubic	-5	7	4	-4	-7	5			
7	Linear	-3	-2	-1	0	1	2	3		
	Quadratic	5	0	-3	-4	-3	0	5		
	Cubic	-1	1	1	0	-1	-1	1		
8	Linear	-7	-5	-3	-1	1	3	5	7	
	Quadratic	7	1	-3	-5	-5	-3	1	7	
	Cubic	-7	5	7	3	-3	-7	-5	7	

Table A-4: Critical q_T Values for a Tukey Test

Directions: Critical q_T values are located within the table at the intersection of the row corresponding to the degrees of freedom for the error term (e.g., $MS_{S/A}$) and the column corresponding to the number of groups or conditions in the original data set. The q_T value in boldface corresponds to .05 family-wise error (**α_{FW} = .05**) and the q_T value in regular type corresponds to a compromise α_{FW} = **.10**.

Number of Groups or Conditions (a)

df_{error}	2	3	4	5	6	7	8	9	10	12	15
4 (α_{FW} = .05)	**3.93**	**5.04**	**5.76**	**6.29**	**6.71**	**7.05**	**7.35**	**7.60**	**7.83**	**8.21**	**8.67**
(α_{FW} = .10)	3.02	3.98	4.59	5.04	5.39	5.68	5.93	6.14	6.33	6.65	7.03
5	**3.64**	**4.61**	**5.22**	**5.68**	**6.04**	**6.33**	**6.58**	**6.60**	**7.00**	**7.33**	**7.72**
	2.85	3.72	4.26	4.27	4.98	5.24	5.46	5.65	5.82	6.10	6.44
6	**3.46**	**4.34**	**4.90**	**5.31**	**5.63**	**5.90**	**6.12**	**6.32**	**6.50**	**6.79**	**7.14**
	2.75	3.56	4.07	4.44	4.73	4.97	5.17	5.34	5.50	5.76	6.08
7	**3.35**	**4.17**	**4.68**	**5.06**	**5.36**	**5.61**	**5.82**	**6.00**	**6.16**	**6.43**	**6.76**
	2.68	3.45	3.93	4.28	4.56	4.78	4.97	5.14	5.28	5.53	5.83
8	**3.26**	**4.04**	**4.53**	**4.89**	**5.17**	**5.40**	**5.60**	**5.77**	**5.92**	**6.18**	**6.48**
	2.63	3.38	3.83	4.17	4.43	4.65	4.83	4.99	5.13	5.36	5.64
9	**3.20**	**3.95**	**4.42**	**4.76**	**5.03**	**5.25**	**5.43**	**5.60**	**5.74**	**5.98**	**6.28**
	2.59	3.32	3.76	4.08	4.34	4.55	4.72	4.87	5.01	5.24	5.51
10	**3.15**	**3.88**	**4.33**	**4.66**	**4.91**	**5.13**	**5.31**	**5.46**	**5.60**	**5.83**	**6.12**
	2.56	3.27	3.70	4.02	4.26	4.47	4.64	4.78	4.91	5.13	5.40
12	**3.08**	**3.77**	**4.20**	**4.51**	**4.75**	**4.95**	**5.12**	**5.27**	**5.40**	**5.62**	**5.88**
	2.52	3.20	3.62	3.92	4.16	4.35	4.51	4.65	4.78	4.99	5.24

	Number of Groups or Conditions (a)										
df_{error}	2	3	4	5	6	7	8	9	10	12	15
14	**3.04**	**3.70**	**4.11**	**4.41**	**4.64**	**4.83**	**4.99**	**5.13**	**5.25**	**5.46**	**5.72**
	2.49	3.16	3.56	3.85	4.08	4.27	4.42	4.56	4.68	4.88	5.12
16	**3.00**	**3.65**	**4.05**	**4.33**	**4.56**	**4.74**	**4.90**	**5.03**	**5.15**	**5.35**	**5.59**
	2.47	3.12	3.52	3.81	4.03	4.21	4.36	4.49	4.61	4.81	5.04
18	**2.97**	**3.61**	**4.00**	**4.28**	**4.50**	**4.67**	**4.83**	**4.96**	**5.07**	**5.27**	**5.50**
	2.45	3.10	3.49	3.77	3.98	4.16	4.31	4.44	4.55	4.75	4.98
20	**2.95**	**3.58**	**3.96**	**4.23**	**4.45**	**4.62**	**4.77**	**4.90**	**5.01**	**5.20**	**5.43**
	2.44	3.08	3.46	3.74	3.95	4.12	4.27	4.40	4.51	4.70	4.92
22	**2.93**	**3.55**	**3.93**	**4.20**	**4.41**	**4.58**	**4.72**	**4.85**	**4.96**	**5.15**	**5.37**
	2.43	3.06	3.44	3.71	3.92	4.10	4.24	4.36	4.47	4.66	4.88
24	**2.92**	**3.53**	**3.90**	**4.17**	**4.37**	**4.54**	**4.69**	**4.81**	**4.92**	**5.10**	**5.32**
	2.42	3.05	3.42	3.69	3.90	4.07	4.21	4.34	4.45	4.63	4.85
30	**2.89**	**3.49**	**3.85**	**4.10**	**4.30**	**4.47**	**4.60**	**4.72**	**4.83**	**5.00**	**5.21**
	2.40	3.02	3.39	3.65	3.85	4.02	4.16	4.28	4.38	4.56	4.77
40	**2.86**	**3.44**	**3.79**	**4.04**	**4.23**	**4.39**	**4.52**	**4.64**	**4.74**	**4.90**	**5.11**
	2.38	2.99	3.35	3.61	3.80	3.96	4.10	4.22	4.32	4.49	4.69
60	**2.83**	**3.40**	**3.74**	**3.98**	**4.16**	**4.31**	**4.44**	**4.55**	**4.65**	**4.81**	**5.00**
	2.36	2.96	3.31	3.56	3.76	3.91	4.04	4.16	4.25	4.42	4.62
120	**2.80**	**3.36**	**3.69**	**3.92**	**4.10**	**4.24**	**4.36**	**4.47**	**4.56**	**4.72**	**4.90**
	2.35	2.93	3.28	3.52	3.71	3.86	3.99	4.10	4.19	4.35	4.54
∞	**2.77**	**3.31**	**3.63**	**3.86**	**4.03**	**4.17**	**4.29**	**4.39**	**4.47**	**4.62**	**4.80**
	2.33	2.90	3.24	3.48	3.66	3.81	3.93	4.04	4.13	4.28	4.47

Table A-5: Approximate Sample Sizes Needed to Achieve Power Levels of .50, .60, .70, .80, and .90 for Testing H_0 at $\alpha \leq .05$, Selected Effect Sizes (ω^2) Ranging from .01 to .20, and df_A Ranging From 1 to 8

FOR POWER ≈ .50

	df_A (a – 1)						
ω^2\	1	2	3	4	5	6	8
.01	190	164	143	129	115	107	92
.02	84	73	64	57	52	48	42
.03	63	55	48	43	38	35	31
.04	48	42	36	33	30	27	24
.05	38	33	29	26	24	22	19
.06	31	27	24	21	19	18	16
.08	22	19	17	15	14	13	11
.10	19	17	15	13	12	11	10
.12	16	14	12	11	10	9	8
.15	13	12	10	9	8	8	7
.20	9	8	7	6	6	5	5

FOR POWER ≈ .60

	df_A (a – 1)						
ω^2\	1	2	3	4	5	6	8
.01	240	205	177	157	143	130	113
.02	124	108	93	83	75	68	59
.03	83	71	61	55	49	45	40
.04	60	52	45	40	36	33	29
.05	48	41	36	32	29	26	23
.06	39	34	29	26	24	22	19
.08	29	25	21	19	18	16	14
.10	23	20	17	15	14	13	11
.12	19	16	14	13	12	11	9
.15	15	13	11	10	9	8	7
.20	11	9	8	7	6	6	5

Adapted from Cohen (1988, pp. 273–362).

FOR POWER ≈ .70

ω^2\	df_A (a − 1) 1	2	3	4	5	6	8
.01	300	256	218	192	173	157	138
.02	166	157	133	115	101	90	84
.03	104	88	76	68	59	55	47
.04	76	64	56	49	44	40	35
.05	60	51	44	39	35	32	28
.06	49	42	36	32	29	26	23
.08	36	30	26	23	21	19	17
.10	28	24	21	18	17	15	13
.12	23	20	17	15	14	13	11
.15	19	16	14	12	11	10	9
.20	13	11	10	9	8	7	6

FOR POWER ≈ .80

ω^2\	df_A (a − 1) 1	2	3	4	5	6	8
.01	384	317	271	240	215	196	168
.02	202	167	141	126	113	102	88
.03	132	110	93	83	74	67	58
.04	96	80	68	60	54	50	42
.05	76	64	54	48	43	39	34
.06	63	52	44	39	35	32	28
.08	46	38	32	28	26	23	20
.10	36	30	26	22	20	19	16
.12	29	25	21	19	17	15	13
.15	23	20	17	15	13	12	10
.20	16	14	12	10	10	9	8

FOR POWER ≈ .90

ω^2\	df_A (a − 1) 1	2	3	4	5	6	8
.01	520	417	350	310	275	250	214
.02	260	210	180	158	140	129	113
.03	180	145	121	106	96	87	74
.04	130	107	88	78	70	63	54
.05	103	87	71	62	56	50	43
.06	84	68	58	50	45	41	35
.08	61	50	42	37	32	30	26
.10	48	39	33	29	26	23	20
.12	39	32	27	24	21	19	16
.15	31	25	21	19	17	15	13
.20	22	18	15	13	12	11	10

Table A-6: Critical q_D Values for a Dunnett Test

Directions: Critical q_D values are located within the table at the intersection of the row corresponding to the degrees of freedom for the error term (e.g., $MS_{S/A}$) and the column corresponding to the number of groups or conditions in the original data set. The q_D value in boldface corresponds to .05 family-wise error (α_{FW} = .05); and the q_D value in regular type corresponds to a conservative α_{FW} = .01.

	Number of Groups or Conditions (a)									
df Denominator\	3	4	5	6	7	8	9	10	11	12
5 (α_{FW} = .05)	**3.03**	**3.29**	**3.48**	**3.62**	**3.73**	**3.82**	**3.90**	**3.97**	**4.03**	**4.09**
(α_{FW} = .01)	4.63	4.98	5.22	5.41	5.56	5.69	5.80	5.89	5.98	6.05
6	**2.86**	**3.10**	**3.26**	**3.39**	**3.49**	**3.57**	**3.64**	**3.71**	**3.76**	**3.81**
	4.21	4.51	4.71	4.87	5.00	5.10	5.20	5.28	5.35	5.41
7	**2.75**	**2.97**	**3.12**	**3.24**	**3.33**	**3.41**	**3.47**	**3.53**	**3.58**	**3.63**
	3.95	4.21	4.39	4.53	4.64	4.74	4.82	4.89	4.95	5.01
8	**2.67**	**2.88**	**3.02**	**3.13**	**3.22**	**3.29**	**3.35**	**3.41**	**3.46**	**3.50**
	3.77	4.00	4.17	4.29	4.40	4.48	4.56	4.62	4.68	4.73
9	**2.61**	**2.81**	**2.95**	**3.05**	**3.14**	**3.20**	**3.26**	**3.32**	**3.36**	**3.40**
	3.63	3.85	4.01	4.12	4.22	4.30	4.37	4.43	4.48	4.53
10	**2.57**	**2.76**	**2.89**	**2.99**	**3.07**	**3.14**	**3.19**	**3.24**	**3.29**	**3.33**
	3.53	3.74	3.88	3.99	4.08	4.16	4.22	4.28	4.33	4.37
12	**2.50**	**2.68**	**2.81**	**2.90**	**2.98**	**3.04**	**3.09**	**3.14**	**3.19**	**3.23**
	3.39	3.58	3.71	3.81	3.89	3.96	4.02	4.07	4.12	4.16

Adapted from C. W. Dunnett, New tables for multiple comparisons with a control, *Biometrics*, 1964, 20, 482–491

	Number of Groups or Conditions (a)									
df Denominator\	3	4	5	6	7	8	9	10	11	12
14	**2.46**	**2.63**	**2.75**	**2.84**	**2.91**	**2.97**	**3.02**	**3.07**	**3.11**	**3.14**
	3.29	3.47	3.59	3.69	3.76	3.83	3.88	3.93	3.97	4.01
16	**2.42**	**2.59**	**2.71**	**2.80**	**2.87**	**2.92**	**2.97**	**3.02**	**3.06**	**3.09**
	3.22	3.39	3.51	3.60	3.67	3.73	3.78	3.83	3.87	3.91
18	**2.40**	**2.56**	**2.68**	**2.76**	**2.83**	**2.89**	**2.94**	**2.98**	**3.01**	**3.05**
	3.17	3.33	3.44	3.53	3.60	3.66	3.71	3.75	3.79	3.83
20	**2.38**	**2.54**	**2.65**	**2.73**	**2.80**	**2.86**	**2.90**	**2.95**	**2.98**	**3.02**
	3.13	3.29	3.40	3.48	3.55	3.60	3.65	3.69	3.73	3.77
24	**2.35**	**2.51**	**2.61**	**2.70**	**2.76**	**2.81**	**2.86**	**2.90**	**2.94**	**2.97**
	3.07	3.22	3.32	3.40	3.47	3.52	3.57	3.61	3.64	3.68
30	**2.32**	**2.47**	**2.58**	**2.66**	**2.72**	**2.77**	**2.82**	**2.86**	**2.89**	**2.92**
	3.01	3.15	3.25	3.33	3.39	3.44	3.49	3.52	3.56	3.59
40	**2.29**	**2.44**	**2.54**	**2.62**	**2.68**	**2.73**	**2.77**	**2.81**	**2.85**	**2.87**
	2.95	3.09	3.19	3.26	3.32	3.37	3.41	3.44	3.48	3.51
60	**2.27**	**2.41**	**2.51**	**2.58**	**2.64**	**2.69**	**2.73**	**2.77**	**2.80**	**2.83**
	2.90	3.03	3.12	3.19	3.25	3.29	3.33	3.37	3.40	3.42
120	**2.24**	**2.38**	**2.47**	**2.55**	**2.60**	**2.65**	**2.69**	**2.73**	**2.76**	**2.79**
	2.85	2.97	3.06	3.12	3.18	3.22	3.26	3.29	3.32	3.35
∞	**2.21**	**2.35**	**2.44**	**2.51**	**2.57**	**2.61**	**2.65**	**2.69**	**2.72**	**2.74**
	2.79	2.92	3.00	3.06	3.11	3.15	3.19	3.22	3.25	3.27

Table A-7: Critical Values for χ^2 Reference Distributions

Directions: Critical values for different χ^2 reference distributions that cut off the extreme 5% of the distribution ($p \leq .05$ in boldface type) and the extreme 1% of the distribution ($p \leq .01$) are located within the table. The row numbers correspond to the degrees of freedom (R – 1)(C – 1), where R equals the number of rows and C equals the number of columns in the χ^2 data matrix.

df	p ≤ .05	p ≤ .01
1	**3.84**	6.63
2	**5.99**	9.21
3	**7.81**	11.34
4	**9.49**	13.28
5	**11.07**	15.09
6	**12.59**	16.81
7	**14.07**	18.48
8	**15.51**	20.09
9	**16.92**	21.67
10	**18.31**	23.21
11	**19.68**	24.72
12	**21.03**	26.22
13	**22.36**	27.69
14	**23.68**	29.14
15	**25.00**	30.58
16	**26.30**	32.00
17	**27.59**	33.41
18	**28.87**	34.81
19	**30.14**	36.19
20	**31.41**	37.57
21	**32.67**	38.93
22	**33.92**	40.29
24	**36.42**	42.98
26	**38.89**	45.64
28	**41.34**	48.28
30	**43.77**	50.89

Table A-8: Critical Values for the Pearson Correlation Coefficient (r)

Directions: Critical values for different r reference distributions that cut off the extreme 5% of the distribution ($p \leq .05$ in boldface type) and the extreme 1% of the distribution ($p \leq .01$) are located within the table. The different r reference distributions are determined by degrees of freedom, and for r, df = n – 2, where n is the number of paired scores.

df	$p \leq .05$	$p \leq .01$
2	**.950**	.990
3	**.878**	.959
4	**.811**	.917
5	**.754**	.875
6	**.707**	.834
7	**.666**	.798
8	**.632**	.765
9	**.602**	.735
10	**.576**	.708
11	**.553**	.684
12	**.532**	.661
13	**.514**	.641
14	**.497**	.623
15	**.482**	.606
16	**.468**	.590
17	**.456**	.575
18	**.444**	.561
19	**.433**	.549
20	**.423**	.537
25	**.381**	.487
30	**.349**	.449
35	**.325**	.418
40	**.304**	.393
45	**.288**	.372
50	**.273**	.354
60	**.250**	.325
70	**.232**	.302
80	**.217**	.283
90	**.205**	.267
100	**.195**	.254

Appendix B: Answers to Practice Exams

Practice Test 1: Chapters 1–5

I. Multiple Choice:

1.A.	3
1.B.	4
2.	a
3.	c
4.	b
5.	b
6.	d
7.	d
8.	c
9.	a
10.	c

II.

Interval	Frequency
35–38	3
31–34	2
27–30	4
23–26	6
19–22	5
15–18	4
11–14	1

III. $\bar{Y} = 14.625$; Mdn = 13.50; SS = 411.75; $s_E = \sqrt{(s^2 \div n)} = \sqrt{(27.45 \div 16)} = 1.31$

IV. $F = 3.24$

V. [A] = 292.14, [T] = 283.50, [Y] = 331.00

$SS_A = 8.65$

$SS_{S/A} = 38.86$

$SS_{Total} = 47.50$

Practice Test 2: Chapters 6–8

I. Multiple Choice:

1. a
2 c
3. b
4. d
5. c
6. b
7. d
8. c
9. b
10. d

II. ANOVA Summary Table

Source	SS	df	MS	F	
A	155.50	5	31.10	10.26	p < .05
S/A	54.50	18	3.03		

III. $SS_{ALin} = (4)(50)^2 \div 70 = 142.86$

$MS_{ALin} = 142.86 \div 1 = 142.86$

$F_{ALin} = 142.86 \div 3.03 = 47.15, p < .05$

IV. A. $F_{Acomp}(1, 12) = 13.89, p < .05$

B. $F_S = 7.78$

V. This problem may be solved estimating s_{diff} using only s_1^2 and s_3^2, in which case df = 8 and t (8) = 2.31; or it may be solved estimating s_{diff} using $MS_{S/A}$, in which case df = 12 and t (12) = 2.18.

A. $(\bar{Y}_{A3} - \bar{Y}_{A1}) - (2.31)\sqrt{([s_1^2 + s_3^2] \div n)} \le \mu_3 - \mu_1 \le (\bar{Y}_{A3} - \bar{Y}_{A1}) + (2.31)\sqrt{([s_1^2 + s_3^2] \div n)}$

$2.00 \le \mu_3 - \mu_1 \le 8.00$

B. $(\bar{Y}_{A3} - \bar{Y}_{A1}) - (2.18)\sqrt{([2][MS_{S/A}] \div n)} \le \mu_3 - \mu_1 \le (\bar{Y}_{A3} - \bar{Y}_{A1}) + (2.18)\sqrt{([2][MS_{S/A}] \div n)}$

$2.08 \le \mu_3 - \mu_1 \le 7.92$

Practice Test 3: Chapters 9–12

I. Multiple Choice:

1. c
2 d
3. b
4. d
5. b
6. b
7. d
8. a
9. d
10. d

II. 1.

ANOVA Summary Table

Source	SS	df	MS	F	
A	5.04	1	5.04	2.51	
B	28.58	2	14.29	7.11	$p < .05$
A × B	19.09	2	9.54	4.75	$p < .05$
S/AB	36.25	18	2.01		

2. $F_{A at b3}(1, 18) = 7.52, p < .05$

3. [BS] = 424.50

III. A. ANOVA Summary Table

Source	SS	df	MS	F

A	87.00	3	27.33	1.64	not significant
A × S	199.50	12	16.62		

B. $MS_{Acomp} = 44.10$; $MS_{Acomp \times S} = 15.60$; $F_{Acomp}(1, 4) = 2.83$ (not significant)

Practice Test 4: Chapters 13–16

I. Multiple Choice:

1. b
2. a
3. c
4. c
5. b
6. d
7. c
8. d
9. a
10. a

II. A. $\chi^2 = 11.26$, and for df = 3, it exceeds $\chi^2_{crit} = 7.81$ and is significant at the .05 level

B. $\chi^2_{comp} = 10.26$, and for $df_{comp} = 1$, it exceeds $\chi^2_{crit} = 3.84$ and is significant at the .05 level

III. r = .78; ρ = .66

IV. $r^2_{Y.X1} = .31$; $r^2_{Y.X2} = .49$; and $R^2_{Y.X1.X2} = .55$

References

Anderson TW, and Sclove SL. *The Statistical Analysis of Data* (2nd ed.). Palo Alto, CA: Scientific Press, 1986.

Aron A and Aron EN. *Statistics for the Behavioral and Social Sciences: A Brief Course.* Upper Saddle River, NJ: Prentice-Hall, 1997.

Ballardini N, Yamashita JA, and Wallace WP. "Presentation Duration and False Recall for Semantic and Phonological Associates." *Consciousness and Cognition* 17, (2008) 64–71.

Benazzi L, Horner RH, and Good RH. "Effects of Behavior Support Team Composition on the Technical Adequacy and Contextual Fit of Behavior Support Plans." *The Journal of Special Education* 40, (2006) 160–170.

Bird KD. "Confidence Intervals for Effect Sizes in Analysis of Variance." *Educational and Psychological Measurement* 62, (2002) 197–226.

Bird KD. *Analysis of Variance via Confidence Intervals.* London: Sage, 2004.

Box GEP, Hunter WG, and Hunter JS. *Statistics for Experimenters: An Introduction to Design, Data Analysis, and Model Building.* NY: Wiley, 1978.

Cialdini RB, Reno RR, and Kallgren CA. "A Focus Theory of Normative Conduct: Recycling the Concept of Norms to Reduce Littering in Public Places." *Journal of Personality and Social Psychology* 58, (1990) 1015–1026.

Cochran WG. "The Comparison of Percentages in Matched Samples." *Biometrika* 37, (1950) 256–266.

Cohen J. *Statistical Power Analysis for the Behavioral Sciences* (2nd ed.). Hillsdale, NJ: Erlbaum, 1988.

Cook TD and Campbell DT. *Quasi-Experimentation: Design and Analysis Issues for Field Settings.* Chicago: Rand McNally, 1979.

Dunnett CW. "A Multiple Comparison Procedure for Comparing Several Treatments with a Control." *Journal of the American Statistical Association* 50, (1955) 1096–1121.

Erdfelder E, Faul F, and Buchner, A. "GPOWER: A General Power Analysis Program." *Behavior Research Methods, Instruments, and Computers* 28, (1996) 1–11.

Faith MS, Allison DB, and Gorman BS. *Meta-Analysis of Single-Case Research.* In RD Franklin, DB Allison, and BS Gorman (eds.), *Design and Analysis of Single-Case Research.* Mahwah, NJ: Erlbaum, 1997.

Friedman, H. "Simplified Determinations of Statistical Power, Magnitude of Effect and Research Sample Sizes." *Educational and Psychological Measurement* 42, (1982) 521–526.

Friedman TL. *From Beirut to Jerusalem.* NY: Anchor, 1995.

Halcomb SH, Taylor JP, DeSouza KD, and Wallace WP. "False Recognition Following Study of Semantically Related Lists Presented in Jumbled Word Form." *Memory* 16, (2008) 443–461.

Hays WL. *Statistics* (4th ed.). Fort Worth, TX: Holt, Rinehart, Winston, 1988.

Hoekstra R, Morey RD, and Rouder, JN. 'Robust Misinterpretation of Confidence Intervals.' *Psychonomic Bulletin Review* 21, (2014) 1157-1164.

Hopkins BL, Cole BL, and Mason TL. "A Critique of the Usefulness of Inferential Statistics in Applied Behavior Analysis." *The Behavior Analyst* 21, (1998) 125–137.

Keppel G and Saufley WH Jr. *Introduction to Design and Analysis: A Student's Handbook*. NY: Freeman, 1980.

Keppel G, Saufley WH Jr., and Tokunaga H. *Introduction to Design and Analysis: A Student's Handbook* (2nd ed.). NY: Freeman, 1992.

Keppel G and Wickens TD. *Design and Analysis: A Researcher's Handbook* (4th ed.). Upper Saddle River, NJ: Pearson Prentice Hall, 2004.

Keppel G and Zedeck S. *Data Analysis for Research Designs: Analysis of Variance and Multiple Regression/Correlation Approaches*. NY: Freeman, 1989.

Keuls M. "The Use of the Studentized Range in Connection with the Analysis of Variance." *Euphytica Psychometrika* 60, (1952) 395–418.

Kiess HO and Green BA. *Statistical Concepts for the Behavioral Sciences* (4th ed.). Boston: Allyn and Bacon, 2010.

Kline RB. *Beyond Significance Testing: Reforming Data Analysis Methods in Behavioral Research*. Washington, DC: American Psychological Association, 2004.

Kraemer HC and Thiemann S. *How Many Subjects? Statistical Power Analysis in Research*. Newbury Park, CA: Sage, 1987.

Lewis D and Burke CJ. "The Use and Misuse of the Chi-Square Test." *Psychological Bulletin* 46, (1949) 433–489.

Lockhart RS. *Introduction to Statistics and Data Analysis*. NY: Freeman, 1998.

Loftus EF and Palmer JC. "Reconstruction of Automobile Destruction: An Example of the Interaction between Language and Memory." *Journal of Verbal Learning and Verbal Behavior* 13, (1974) 585–589.

Marascuilo LA and McSweeney M. *Nonparametric and Distribution-Free Methods for the Social Sciences*. Monterey, CA: Brooks/Cole, 1977.

McClave JT and Sincich T. *A First Course in Statistics* (7th ed.). Upper Saddle River, NJ: Prentice-Hall, 2000.

McClendon MJ. *Multiple Regression and Causal Analysis*. Itasca, IL: Peacock, 1994.

Mechanic A. "The Responses Involved in the Rote Learning of Verbal Materials." *Journal of Verbal Learning and Verbal Behavior* 3, (1964) 30–36.

Mill JS. *A System of Logic, Ratiocinative and Inductive: Being a Connected View of the Principles of Evidence and the Methods of Scientific Investigation* (8th ed.). NY: Harper, 1888.

Miller J. "What Is the Probability of Replicating a Statistically Significant Effect?" *Psychonomic Bulletin and Review* 16, (2009) 617–640.

Murphy KR and Myors B. *Statistical Power Analysis: A Simple and General Model for Traditional and Modern Hypothesis Tests* (2nd ed.). Mahwah, NJ: Erlbaum, 2004.

Newman D. "The Distribution of Range Samples from a Normal Population Expressed in Terms of an Independent Estimate of Standard Deviation." *Biometrika* 31, (1939) 20–30.

Pearson K and Lee A. "On the Laws of Inheritance in Man: I. Inheritance of Physical Characters." *Biometrika* 2, (1903) 357–462.

Postman L. *Short-Term Memory and Incidental Learning*. In AW Melton (ed.), *Categories of Human Learning* (145–201). NY: Academic Press, 1964.

Rosenthal R, Rosnow RL, and Rubin DB. *Contrasts and Effect Sizes in Behavioral Research: A Correlational Approach*. Cambridge, UK: Cambridge University Press, 2000.

Scheffé, H. "A Method for Judging All Contrasts in the Analysis of Variance." *Biometrika* 40, (1953) 87–104.

Sheskin DJ. *Handbook of Parametric and Nonparametric Statistical Procedures*. Boca Raton, FL: CRC Press, 1997.

Siegel S and Castellan NJ Jr. *Nonparametric Statistics for the Behavioral Sciences* (2nd ed.). New York: McGraw-Hill, 1988.

Thomas L and Krebs CJ. "A Review of Statistical Power Analysis Software." *Bulletin of the Ecological Society of America* 78, (1997) 126–139.

Winer BJ, Brown DR, and Michels KM. *Statistical Principles in Experimental Design* (3rd ed.). New York: McGraw-Hill, 1991.

Name Index

A

B

C

D

E

F

G

H

K

L

M

N

P

R

S

T

Y

Z

Subject Index

W

Y

Z

CPSIA information can be obtained
at www.ICGtesting.com
Printed in the USA
FSHW010430310821
84340FS